Fungi from Different Environments

Series on
Progress in Mycological Research

Fungi from Different Environments

Fungi from Different Environments

Editors
J.K. MISRA
S.K. DESHMUKH

Science Publishers
Enfield (NH) Jersey Plymouth

Science Publishers www.scipub.net
234 May Street
Post Office Box 699
Enfield, New Hampshire 03748
United States of America

General enquiries : *info@scipub.net*
Editorial enquiries : *editor@scipub.net*
Sales enquiries : *sales@scipub.net*

Published by Science Publishers, Enfield, NH, USA
An imprint of Edenbridge Ltd., British Channel Islands
Printed in India

ISBN: 978-1-57808-578-1

Library of Congress Cataloging-in-Publication Data

Fungi from different environments/edited by J.K. Misra,
S.K. Deshmukh.--1st ed.
p.cm. -- (Progress in mycological research)
Includes bibliographical references and index.
ISBN 978-1-57808-578-1 (hardcover)
1. Fungi--Ecology. 2. Fungi--Ecophysiology. 3. Mycology.
I. Misra, J.K. II. Deshmukh, S.K. (Sunil K.) III. Series.

QK604.2.E26F85 2009
597.5'17--dc22

2008041307

Contents

List of Contributors

Bradner, R.

Department of Chemistry and Biomolecular Sciences, Macquarie University, Sydney, NSW 2109, Australia

Buchalo, A.S.

N.G. Kholodny Institute of Botany, National Academy of Sciences of Ukraine, 2 Tereshchenkivska St., 01601 Kiev, Ukraine

Gostinčar, C.

University of Ljubljana, Biotechnical Faculty, Department of Biology, Večna pot 111, SI-1000 Ljubljana, Slovenia

Gunde-Cimerman, N.

University of Ljubljana, Biotechnical Faculty, Department of Biology, Večna pot, SI-1000 Ljubljana, Slovenia

Guzmán, G.

Instituto de Ecologia, Km 2.5 carretera antigua a Coatepec No. 351, Congregación El Haya, Apartado postal 63, Xalapa, Veracruz 91070, Mexico

Hoshino, T.

National Institute of Advanced Industrial Science and Technology, 2-17-2-1 Tsukisamu-higasi, Toyohira-ku, Sapporo 062-8517, Japan

Matsumoto, N.

National Agricultural Research Institute for Hokkaido Region, 1 Hitsujigaoka, Toyohira-ku, Sapporo 062-8555, Japan

Matsuura, K.

Laboratory of Insect Ecology, Graduate School of Environmental Science, Okayama University, Okayama 700-8530, Japan

Misra, J.K.

Mycological Research Unit, Department of Botany, Sri Jai Narain Postgraduate College, Lucknow 226001, India

Namikoshi, M.

Tohoku Pharmaceutical University, Komatsushima, Aoba-ku, Sendai 981-8558, Japan

Nevalainen, H.

Department of Chemistry and Biomolecular Sciences, Macquarie University, Sydney, NSW 2109, Australia

Nevo, E.

Institute of Evolution, University of Haifa, Mt. Carmel, Haifa 31905, Israel

Pant, A.

Mycological Research Unit, Department of Botany, Sri Jai Narain Post-graduate College, Lucknow 226001, India

Papini, R.

Dipartimento di Patologia Animale, Profilassi e Igiene degli Alimenti, Facoltà di Medicina Veterinaria, Viale delle Piagge 2, 56124 Pisa, Italy

Satyanarayana, T.

Department of Microbiology, University of Delhi, South Campus, Benito Juarez Road, New Delhi 110021, India

Singh, B.

Department of Microbiology, University of Delhi, South Campus, Benito Juarez Road, New Delhi 110021, India

Suzuki, A.

Department of Biology, Faculty of Education, Chiba University, 1-33 Yayoi-cho, Inage-ku, Chiba 263-8522, Japan

Te'o, J.

Department of Chemistry and Biomolecular Sciences, Macquarie University, Sydney, NSW 2109, Australia

Tilak, S.T.

Y.M. College, Erandwane, Bharati Vidyapeeth (Deemed University), Pune 411037, India

Tripathi, S.K.M.

Birbal Sahni Institute of Palaeobotany, 53 University Road, Lucknow 226007, India

Turk, M.

University of Ljubljana, Biotechnical Faculty, Department of Biology, Večna pot 111, SI-1000 Ljubljana, Slovenia

Ulfig, K.

West Pomeranian University of Technology, Polymer Institute, Dep. Biomaterials & Microbiological Technologies, al. Piastów 17, 70-310 Szczecin, Poland

Wasser, S.P.

Institute of Evolution and Department of Evolutionary and Environmental Biology, Faculty of Science and Science Education, University of Haifa, Mt. Carmel, Haifa 31905, Israel

Xu, Jin-Zhong

Pharmaceutical Informatics Institute, College of Pharmaceutical Sciences, Zhejiang University, Hangzhou, Zhejiang 310058, P.R. China

Yashiro, T.

Laboratory of Insect Ecology, Graduate School of Environmental Science, Okayama University, Okayama 700-8530, Japan

Preface

Fungi, the second largest group of organisms after insects, have been attracting scientists of various disciplines, besides mycologists, because of their fascinating nature and enormous capability to cope with and survive in many environments. Mycologists, according to one of several estimates, believe that some 1.5 million fungi exist in nature. The majority of these fungi, it is believed, may come from the tropical environment which is still under-explored.

A significant shift in the study of fungi (used here in the broad sense that includes not only those that have a monophyletic origin, but also some other fungus-like organisms such as Oomycota) has been taking place during the last few years with the advent of modern tools for study and advances of knowledge by using these tools. Now it is possible to look beyond morphological features and study fungi at the DNA level for a better understanding of their characteristics. Fungi are also now attracting the attention of scientists in various other disciplines such as biomedicine and biotechnology, engaged in the search for useful fungi in diverse environments to serve as sources for therapeutic agents and industrial enzymes. Fungi are known to produce low molecular weight compounds and several cholesterol-lowering ones like the statins. Compounds such as the cytochalasins, peptaibols, grisan and scirpene derivatives are found only in fungi. Not only this, but in recent years, the field of nanotechnology has opened many new areas of research among materials scientists who are attracted to exploring all possibilities of using microorganisms in the biosynthesis of nano-materials, including fungi. Scientists have found that fungi such as *Aspergillus fumigatus* and *Fusarium oxysporum* can be used as bionanofactories for the synthesis of silver nanoparticles.

All such areas of investigation and shifts in the priorities of fungal research have added many new and useful dimensions and information. In order to bring these advancements together, that are currently scattered in many journals and publications, a series of books is planned. This first volume, in a series of four in *Progress in Mycological Research*, aims to bring together what we know about the fungi from different environments. The present volume is comprised of 14 chapters written by experts in their chosen area of specialization and covers fungi from various environments such as air, water (freshwater and marine), palaeo-environment, and their influence on the environment and their management.

The editors are grateful to the contributors for providing the chapters and are also thankful to Dr. (Mrs.) Swati A. Piramal, Vice Chairperson, Dr. Somesh Sharma, Managing Director and Dr. H. Sivaramakrishnan, President, Piramal Life Sciences Limited, Mumbai for their help in various ways.

1

Fungi from Palaeoenvironments: Their Role in Environmental Interpretations

S.K.M. Tripathi
Birbal Sahni Institute of Palaeobotany,
53 University Road, Lucknow 226007, India
E-mail: skmtripathi@yahoo.com

Abstract

Fragments of fossil fungal remains are commonly seen in macerated residues prepared for palynological studies. These are less frequent in samples from Palaeozoic strata but are better represented in preparations of Lower Mesozoic sediments. A great spurt in fungal diversity is witnessed in the Tertiary Period. Except for some distinctive Tertiary forms, fossil fungal remains can not be generally ascribed to modern taxa, hence, their classification with living fungi is not possible. Fossil fungal remains are, therefore, described as Form Genera under the Artificial System of classification which is based on morphological characters only. However, wherever possible their affinities with extant forms are provided.

Innumerable variety of fossil fungal remains are described as spores, filaments, fruiting bodies and mycorrhiza which have been reported from Cretaceous to Tertiary sediments. Ascomycetous fungal remains got well-established during the Cretaceous time and became conspicuously abundant by the Tertiary Period. Enormous varieties of fossil fungal spores are reported from these sediments. Fossil spores are described under "Dispersed Spores" which include detached spores, microscopic sporangia and fragmented mycelia. Based on characters associated with size, symmetry, pores and septa, the spores are described under different morphologic groups. Fruiting bodies of epiphyllous fungi are commonly found in Tertiary sediments. These can be compared with extant forms with greater accuracy than the spores. The fruiting bodies or the ascocarps are variously shaped ostiolate or non-ostiolate bodies made up of radiating rows of mycelia giving an appearance of tissues arranged in a radiating fashion. The ascocarps contain asci.

Fossil fungi provide significant information about the past habitats and the hosts. Fossil epiphyllous fungi are more reliable and advantageous for palaeoclimatic interpretations. Occurrence of these fossils reflects moist and humid climate of tropical to subtropical belts. Studies particularly focusing the host fungus relationship are of great significance in attempting the palaeoenvironmental interpretations. This chapter encompasses the aforesaid aspects of fungi from palaeoenvironment.

INTRODUCTION

Palaeomycology, the study of fossil fungi, is still in its infancy. The main reason for this is the lack of adequate knowledge amongst the palaeontologists about fungal morphology. Furthermore, the general feeling that fungal remains are less useful for stratigraphical interpretations has been one of the many other factors for their neglect. Fossilized fungi are represented by their hyphae, fructifications and the dispersed spores. These have been recorded sporadically since long (Williamson, 1878, 1880; Kidston and Lang, 1921; Edwards, 1922). Most of the fossil fungal remains reported earlier are poorly described and badly illustrated. However, since the 1950s their study received more attention with the development of palynology. During the last couple of decades fossilized remains of fungi have been investigated with greater interest involving phylogenetic, biostratigraphic and palaeoenvironmental implications (Ramanujam, 1982; Taylor and White, 1989). Fungal remains are very resistant to chemical and biological degradation and are easily recoverable from rocks. These are commonly seen in palynological preparations. Fossil fungal remains are less common in Palaeozoic strata but become more frequent in Lower Mesozoic. A great sprut in the fungal diversity is witnessed in Tertiary Period. Records show that proliferation of fungi is linked with diversification of angiosperms. This chapter embodies the classification of fossil fungi, their geological records, the stratigraphical considerations and palaeoclimatic interpretations with suggestions for future research.

CLASSIFICATION OF FOSSIL FUNGI

Fossil fungi are fragmentary in nature and lack of characteristic features that are diagnostic of extant taxa. Except for some distinctive Tertiary forms, most of the fossil fungal remains can seldom be matched with modern taxa hampering their classification under the Natural System. Problems concerning the nomenclature and classification of fossil fungi have been discussed by various workers from time to time (Elsik, 1968, 1976; Pirozynski and Weresub, 1979). It has been argued by these workers that assigning most of the fossil forms a modern name will lead to taxonomic confusion and it will, therefore, be more convenient and logical to describe the fragmentary fossils as form genera on the basis of their morphological characters only. Under this scheme fossils are assigned artificial generic and specific names and wherever possible their affinities with extant taxa are ascribed. While dealing with the pertinent groups, characters taken into account for classification of different types of fungal remains have been elaborated. Fossil fungal remains have been dealt with here in two categories—the Fungal Spores, and the Epiphyllous Fungi. In addition to these, some specific and special fossil fungi described from Cretaceous and Tertiary sediments of India have been discussed.

A brief comment on some important fossil fungal remains highlighting their morphological features and the palaeoenvironmental significance has been mentioned in the present contribution.

As evidenced by the fossil records, Ascomycetes, the largest and most diversified group of modern fungi, got well-established during the Cretaceous time and became conspicuously abundant by the Tertiary Period (Jain, 1974; Jain and Kar, 1979; Jansonius, 1976; Ramanujam, 1982; Kalgutkar and Jansonius, 2000). Fungi of this group produced ascospores and coniodia which helped them to thrive in a variety of habitats as saprophytes, parasites, epiphytes and mycorrhiza. The Early Tertiary palynological assemblages are markedly characterized by increased number of epiphyllous fungi which were hosted by broad angiospermous leaves.

I. Fungal spores

Enormous varieties of fossil fungal spores are found in Late Cretaceous to Cenozoic sediments. These are described under 'dispersed spores' which include detached spores found in rocks, microscopic sporangia and hyphae or fragmented mycelia. In a classification system proposed by van der Hammen (1956) fossil fungal spores were grouped under various morphologic categories with the suffix 'Sporites'. Clarke (1965) proposed the suffix 'Sporonites' for naming the fossil fungal spores. Considering the characters such as; shape, size and symmetry of spores, absence/ presence and number of apertures, septa characters and the wall features, Elsik (1976) attempted to prepare a comprehensive and applicable taxonomy for the fossil spores. He proposed artificial supra-generic categories for classification of fossil fungal spores. These categories were primarily based on the cell number and presence or absence of apertures. Under these categories artificial genera and species could be conveniently described. Pirozynski and Weresub (1979) suggested a system called 'Saccardoan System' for classifying the fungal spore types. Under this scheme, based on number of cells and shape, the fungal spores may be classified as Amerospores (monocellate), Didymospores (dicellate), Phragmospores (tri- to multicellate), Dictyospores (muriform), Scolecospores (filiform), Helicospores (spirally coiled) or Staurospores (star-like). A brief description of different taxa classified under various groups of fungal spores is given in Table 1. The majority of fungal spores found in palynological preparations belong to Ascomycetes. Only a few questionable spores of Basidiomycetes have been described in some fossil assemblages. Palynological assemblages are often rich in different varieties of Conidia. These are produced by Fungi Imferfecti and the holomorphic ascomycetes. They may be one-celled to multi-celled and are of varied shapes.

Spores of some fungi, especially conidia and ascospores possess distinctive features leading to their identification and categorization with the extant forms. Fossil spores can be generally assigned to a natural class system of Phycomycetes, Ascomycetes or Basidiomycetes if the diagnostic morphological features are observable. Some fossil materials are assigned to the class Fungi-Imperfecti where spores or isolated structures (conidia, pycnidia or other sporangia, or isolated mycelia) are of exclusive morphology.

Table 1. Classification and diagnostic characters of fossil fungal spores

Spore type	*Taxa*	*Characteristic features*
Amerospores (monocellate)	*Basidiosporites* Elsik	Spores with single offset pore, unicellate, elongate, wall psilate, shape variable
	Diporisporites van der Hammen	Shape generally elongate, diporate, pores on opposite ends
	Exesisporites Elsik	Unicellate, lenticular, monoporate, pore small, pore surrounded by thickening
	Hypoxylonites Elsik	Oval to elongate, bilateral, psilate, provided with elongate scar, slit or furrow

Contd.

Table 1 continued

Spore type	*Taxa*	*Characteristic features*
Amerospores (monocellate)	*Inapertisporites* van der Hammen	Inaperturate, shape and size variable, psilate to variously ornamented
	Lacrimasporonites Clarke	Spatulate to elliptical in shape, wall psilate, monoporate, pore apical
	Monoporisporites van der Hammen	Spherical to sub-spherical, monoporate, psilate to finely punctuate
	Palaeoamphisphaerella Ramanujam and Srisailam	Shape elliptical, oblong or rhomboidal with rounded ends, provided with equatorial pore
Didymospores (dicellate)	*Dicellaesporites* Elsik	Two-celled, uniseptate, shape variable, inaperturate, psilate
	Didymoporisporonites Sheffy and Dilcher	Dicellate, uniseptate, apex of one cell provided with pore, psilate to punctuate
	Diploneurospora Jain and Gupta	Two-celled, unicellate, elliptical, upper cell prominent, thick-walled, sculptured with longitudinal ribs, lower cell small, hyaline with faint rib sculpture
	Dyadosporites van der Hammen ex Clarke	Diporate, with a single pore at each end, psilate to variously sculptured
	Fusiformisporites Rouse	Fusiform, inaperturate, the unit is split into equal halves by equatorial wall, bearing characteristic elongate striae, ribs, ridges or costae oriented parallel to longer axis

Contd.

Table 1 continued

Spore type	*Taxa*	*Characteristic features*
	Brachysporisporites Lang and Smith	Obovate, turbinate or pyriform, several-celled, cells broader than long, gradually diminishing in size towards the attachment cell which is the smallest, with very dark thick bands of septa similarly reducing in size.
	Cannanorosporonites Ramanujam and Rao	Tetracellate, barrel-shaped, basal and terminal cells smaller than central cells
	Diporicellaesporites Elsik	Elongate, diporate, one pore at each end of the spore
Phragmospores (three or more cellate)	*Foveoletisporonites* Ramanujam and Rao	Four or more celled, elongate, foveolate, foveolae irregularly aligned
	Multicellaesporites Elsik	Three or more celled, shape variable, inaperturate, psilate
	Ornasporonites Ramanujam and Rao	Fusiform, four-celled, diporate, basal and apical cells much small, one pore at each end
	Pluricellaesporites van der Hammen	Three or more celled, long, monoporate, psilate to scabrate
	Polycellaesporonites Chandra et al.	Elongate, multicellate, inaperturate, psilate, one end rounded the other end giving rise to a tube-like projection, cells arranged in clusters
Dictyospores (muriform)	*Spinosporonites* Saxena and Khare	Circular to sub-circular, inaperturate, multicellate, each cell giving rise to a robustly built spine
	Staphlosporonites Sheffy and Dilcher	Shape variable, four or more irregular cells arranged in clusters along more than one axis, inaperturate, psilate to punctate

Contd.

Table 1 continued

Spore type	*Taxa*	*Characteristic features*
Dictyospores (muriform)	*Tricellaesporonites* Sheffy and Dilcher	Shape variable, tri-cellate, inaperturate, cells along more than one axis, spore wall psilate to punctuate
Helicospores (coiled)	*Elsikisporonites* Kumar	Tubular and coiled in shape, monopore, pore at outer end, non-septate, spore wall smooth and hyaline
	Involutisporonites Clarke	Coiled, transversely septate, monoporate, psilate to variously ornamented
Staurospores (star-shaped)	*Frasnacritetrus* Taugourdeau	Main body rectangular, spherical or oval, psilate to variously ornamented, body provided with four unicellular processes
Miscellaneous	*Alleppeysporonites* Ramanujam and Rao	Spores branched, multicellate, septate, cells rectangular, basal and terminal cells provided with a conspicuous appendage
	Appendicisporonites Saxena and Khare	Subcircular, inaperturate, psilate, multicelate, each cell with a long process
	Rhizophagites Rosendahl	Nonseptate, thich-walled, hyphae with terminal sub-spherical vesicles of varying size

Fungal spore stratigraphy

Although most of the fungal spores are long ranging and do not bear any stratigraphical significance, but some are morphologically very distinct and have restricted range in geological time. Applicability of fungal spores has, therefore, increased with the record of such characteristic spores (Kalgutkar and Jansonius, 2000). Graham (1962) was amongst the pioneers to suggest the possibility of using fungal spores for supplementing age determinations in palynological studies. According to Elsik (1970) although variety of fungal spores are recorded from Mesozoic strata world over, their morphological complexity and frequency increases in Cenozoic. He noted that *Fusiformisporites* and similar longitudinally ribbed forms appear to be restricted to the Cenozoic. Elsik (*op. cit.*) further observed that fossil fungal spores described as *Exesisporites* which resemble with extant *Hypoxylon* type are more frequently recorded in Neogene sediments. Ramanujam (1982) opined that overall diversity in morphology of fungal spore was attained by late Cretaceous and Early Tertiary. While evaluating the stratigraphic potential of fungal remains in Indian sequences, he further observed that spores with relatively simpler morphology were recorded from early Mesozoic strata but in younger sediments ornamented spores with complex morphology were recorded. Kalgutkar and Jansonius (2000) while summarizing the stratigraphic significance of fossil fungal forms mentioned that as *Pesavis tagluensis* have a distinct morphology hence it is a stratigraphic marker. The dating potential of this fungal taxa in combination with other fungal forms has been established by several workers (Elsik and Jansonius, 1974; Jansonius, 1976; Staplin, 1976; Lang, 1978a,b; Norris, 1982; Young and McNeil, 1984; White, 1990).

II. Epiphyllous fungi

The epiphyllous fruiting bodies were amongst the first fungal groups that were unquestionably identified in the microfossil assemblages (Elsik, 1978a). The distinctive morphological features of these fossil fungal fructifications helped their comparison with extant counterpart with greater accuracy than the dispersed spores. Commonly occurring as parasites on the surface of leaves, stems and flowers of higher plants, these belong to the family Microthyriaceae (Ascomycetes). These have been extensively recorded from Neocomian to Quaternary of both the northern and southern hemispheres. Cookson (1947) pointed out the enormous diversity in fossils of Microthyriales and noticed their abundance in palynological preparations from mid Tertiary strata.

Microthyriaceous fungi have scutate fruit bodies called thyriothecia. In most of the cases thyriothecia possess radiating rows of mycelial cells giving an appearance of tissues arranged in radial fashion. These are the fruiting bodies or ascocarps and contain asci that are surrounded by or enclosed within a protective tissue. Ascocarps may be in the form of closed, globose structures, or flask-shaped bodies with an opening known as ostiole or the saucer shaped open structures. Modern Microthyriaceous fungi are classified on the basis of the mode of dehiscence of their fruiting bodies. Dehiscence is either through a regular or irregular cracking or by the formation of a central pore (ostiole). Other characters taken into consideration for distinguishing different taxa are the characters of mycelium, asci and ascospores.

Mycelia and spores are generally not found along with the fruiting bodies in palynologic residues. Absence of mycelial structures in most of the cases makes it extremely difficult to relate these with extant genera. However, several workers attempted to classify and formally describe the fossil thyrothecia (Edwards, 1922; Rosendahl, 1943; Cookson, 1947; Dilcher, 1965; Rao, 1959; Venkatachala and Kar, 1969, Jain and Gupta, 1970; Elsik, 1978b; Pirozynski, 1978). Fossil species of this fungal group are classified under the artificial system grouping them with Fungi

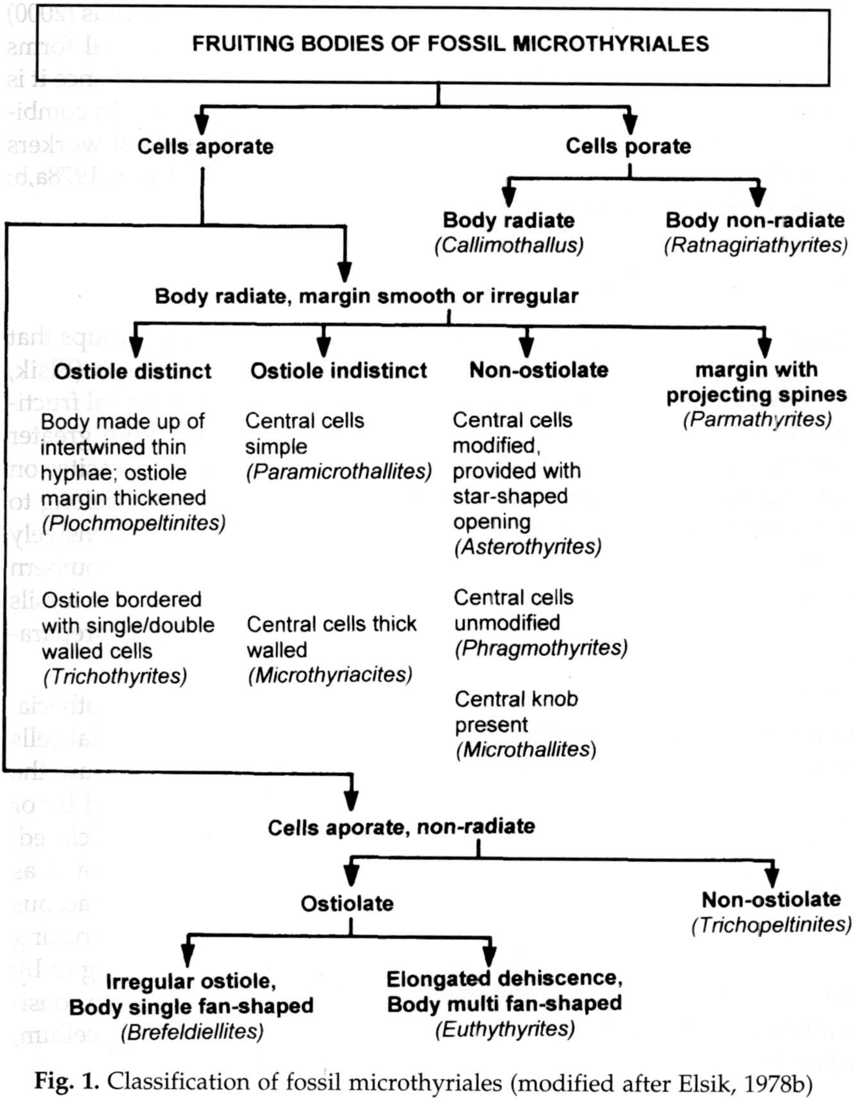

Fig. 1. Classification of fossil microthyriales (modified after Elsik, 1978b)

Imperfecti. The classification of dispersed fossil Microthyriales has been most comprehensively described by Elsik (1978b). The characteristic features considered for their classification are: shape and margin of the fruiting body, characters associated with the dehiscence mark, presence or absence of pores in individual cells and nature of the central part of the fruiting body. The classification of fossil fruiting bodies based on these characters has been summarized in Figure 1. The system is primarily based on porate or aporate individual cells of multicellular fruiting body. Forms with porate individual cells have been kept under the genus *Callimothallus*. The multicellular fruiting bodies without pores in individual cells are divided into Radiate and Non-radiate forms. The Non-radiate forms may be ostiolate or non-ostiolate. The Radiate forms are further divided into genera having smooth, fimbriate or spinose margins. The Radiate forms with smooth to fimbriate margins are further divided on the basis of the presence, absence or nature of ostiole. The size of fossil fruiting bodies generally ranges between 80 and 160 µm. Salient features of different genera are summarized in Table 2 and line diagrams of these genera have been provided in the plates.

Morphological diversity of epiphyllous fungi

As stated earlier, fruiting bodies of Microthyriaceae are most common in fossil assemblages. However, some other members of the epiphyllous fungi produce morphologically similar fructifications. The family Asterinaceae shows the presence of thyriothecium resembling those of Microthyriaceae. Fructifications of this family open by irregular crumbling, cracking or gelatization of the central area forming an irregular wide opening or stellate crack (Pirozynski, 1978). Fruiting bodies of Trichothyriaceae resemble those of Asterinaceae but are lenticular rather than scutelliform. The ostiole in these forms is often protruding and may be bordered by darkly pigmented cells which sometimes bear spine-like setae. This family is represented in fossil records by *Trichothyrites*. Thalli of the family Trichopeltinaceae, which are irregularly branched, membranous and are composed of regular cells arranged into orderly parallel or radiating patterns. The fructifications are in the form of circular ostiolate bulges in thallus. These common tropical epiphytes have fossil representatives assigned to the genera *Trichopeltinites* and *Brefeldiellites*.

Fructifications of the family Micropeltaceae are also shield-shaped and centrally ostiolate. Walls in these fruiting bodies are composed of haphazardly arranged indistinct hyphae forming a delicate hyphal reticulum at the margins. Members of this family are epiphytes growing on tropical evergreen plants. *Plochmopeltinites* are the fossil members of the family. Fruiting bodies of the epiphyllous ascomycetes of the family Parmu-

Table 2. Characteristic features of fossil fruiting bodies

Taxa	*Characteristic features*
Genus—***Asterothyrites*** Cookson Type species—*Asterothyrites minutus* Cookson (designated by Jansonius and Hills) **(Pl. 4, fig. 39)**	Ascomata round, flat, made up of radially arranged hyphae, cells isodiametric. Ascomata ostiolate, ostiole stellate in shape, probably formed by dissolution of central cells.
Genus—***Brefeldiellites*** Dilcher Type species—*Brefeldiellites fructiflabellus* Dilcher **(Pl. 5, fig. 42)**	Hyphae produce a large rounded membranous structure with marginal fertile areas or ascomata. Central ascoma cells break away as a dehiscence mechanism.
Genus—***Callimothallus*** Dilcher Type species—*Callimothallus pertusus* **(Pl. 4, fig. 40)**	Stroma round, radiate, no central dehiscence, individual cells may possess single pore.
Genus—***Euthythyrites*** Cookson Type species—*Euthythyrites oleinites* Lactotype selected by Jansonius and Hills **(Pl. 5, fig. 41)**	Ascomata linear, elliptical to oblong, ends rounded or flattened, lateral margins uneven, dehiscence by a longitudinal slit, cells radiating from mid-vertical line, hyphopodiate, hyphopodia small.
Genus—***Microthallites*** Dilcher Type species—*Microthallites lutosus* Dilcher **(Pl. 6, fig. 48)**	Stroma radiate, more or less round, ostiolate or non-ostiolate.
Genus—***Microthyriacites*** Cookson Type species—*Microthyriacites grandis* Cookson **(Pl. 4, fig. 37)**	Ascomata very large (1000-1200 µm), slightly convex. Central part constituted by thick isodiametric cells, peripheral cells elongated, radial.
Genus—***Paramicrothallites*** Jain and Gupta Type species—*Paramicrothallites spinulatus* (Dilcher) Jain and Gupta **(Pl. 5, fig. 45)**	Stroma radiate, more or less rounded, ostiolate, ostiole not surrounded by specialized cells.
Genus—***Parmathyrites*** Jain and Gupta Type species—*Parmathyrites indicus* Jain and Gupta **(Pl. 5, fig. 43)**	Ascomata flattened, non-ostiolate, more or less circular, hyphae radially arranged. Peripheral cells prominent with thickened radial walls, spines peripheral. Ostiole distinct.

Contd.

Taxa	*Characteristic features*
Genus—***Phragmothyrites*** Edwards Type species—*Phragmothyrites eocenica* Edwards **(Pl. 4, fig. 36)**	Ascomata sub-circular to circular with radially arranged hyphae, hyphal cells may be differentiated forming separate regions in the fruiting body. Central cells isodiametric.
Genus—***Plochmopeltinites*** Cookson Type species—*Plochmopeltinites* (Cookson) Jansonius and Hills **(Pl. 5, fig. 44)**	Ascomata of dimidiate form with ascomal membranes of sinuous plectenchyma.
Genus—***Ratnagiriathyrites*** Saxena and Misra Type species—*Ratnagiriathyrites hexagonalis* Saxena and Misra **(Pl. 6, fig. 47)**	Ascomata sub-circular or irregular in shape, margin thick, wavy, dark brown in colour, margin thick, wavy, non-ostiolate. Cells not arranged radially, porate. Pores generally distributed throughout stromata. Peripheral cells hexagonal, bigger, central cells small.
Genus—***Trichopeltinites*** Cookson Type species—*Trichopeltinites pulcher* Cookson **(Pl. 5, fig. 46)**	Ascomata developed as thickened areas of the thallus and dehiscing by an irregular ostiole as in *Trichopeltis* Theiss (Stevens).
Genus—***Trichothyrites*** Rosendahl Type species—*Trichothyrites pleistocaenica* Rosendahl **(Pl. 4, fig. 38)**	Thyriothecia disc- or saucer-shaped, made up of almost square radiating cells. Ostiolate, ostiole placed on an erect collar, made up of 2-6 tires of thick walled quadrilateral cells. Uppermost tire of cells may have short prolongations in some cases. Outline usually smooth but may appear lobate.

lariaceae superficially resemble those described earlier but are thicker and less distinctly cellular. Fossil representatives of these forms are *Callimothallus* and *Microthallites*.

STRATIGRAPHICAL CONSIDERATIONS

The earliest undisputed Microthyriaceous fungus, *Stomiopltis* is reported from Lower Cretaceous of Wealden, Isle of Wright (Alvin and Muir, 1970). A specimen referred to *Phragmothyrites* was described by Singh (1971) from Late Albian of Alberta. Stratigraphic record of fossil Microthyriaceous fungi shows that these occur in major parts of the Cenozoic, of these, *Callimo-*

thallus, Phragmothyrites, and *Trichothyrites* are most commonly found in palynological preparations. It is noticed that due to taxonomic confusion the stratigraphic application of different species of these genera is obscured. However, while making an assessment at generic level only an attempt has been made to summarize the stratigrapgic distribution of different fossil fruiting bodies recorded from Indian Tertiary sequences (Table 3). Taxa assigned to *Callimothallus* and *Cucurbitariaceites* are long ranging and are recorded from Palaeocene to Pliocene sediments. Different species of *Phragmothyrites* mark their presence in Palaeocene to Miocene, *Microthyriacites* in Eocene to Miocene and *Kutchiathyrites* in Oligocene to Miocene. Forms restricted to Miocene sequences only are: *Asterothyrites, Euthyrites Microthallites, Paramicrothallites, Parmathyrites, Plochmopeltinites, Ratnagiriathyrites, Trichopeltinites* and *Trichothyrites.* Forms assigned to *Siwalikiathyrites* are recorded from Miocene to Pliocene sediments.

Table 3. Stratigraphic distribution of fossil fruiting bodies in Indian tertiary sediments

Taxa	*Palaeocene*	*Eocene*	*Oligocene*	*Miocene*	*Pliocene*
Callimothallus Dilcher	▬	▬	▬	▬	▬
Cucurbitariaceites Kar et al.	▬	▬	▬	▬	▬
Phragmothyrites Edwards	▬	▬	▬	▬	
Microthyriacites Cookson		▬	▬	▬	
Kutchiathyrites Kar		▬	▬	▬	
Kalviwadithyrites Rao				▬	
Microthallites Dilcher				▬	
Paramicrothallites Jain and Gupta				▬	
Parmathyrites Jain and Gupta				▬	
Plochmopeltinites Cookson				▬	
Ratnagiriathyrites Saxena and Misra				▬	
Trichopeltinites Cookson				▬	
Tricothyrites Rosendahl				▬	
Asterothyrites Cookson				▬	
Euthythyrites Cookson				▬	
Siwalikiathyrites Saxena and Singh				▬	▬

III. Significant fungal remains from India

Diverse fungal remains have been described from Indian Cretaceous to Tertiary sequences (Potonie and Sah, 1960; Banerjee and Misra, 1968; Venkatachala and Kar, 1969; Jain and Gupta, 1970; Chitaley and Seikh, 1971; Chitaley and Patil, 1972; Kar et al., 1972; Kar and Saxena, 1976; Rao and Ramanujam, 1976; Chitaley, 1978; Chitaley and Yawale, 1978; Ramanujam and Rao, 1973, 1978; Patil and Ramanujam, 1980; Kumar, 1990; Rao, 1995; Tiwari and Tripathi, 1995; Tripathi, 2001; Kar et al., 2003, 2004a, b, 2005, 2006). A few of these are briefly discussed here.

Potonie and Sah (1960) described *Lirasporis intergranifer* (Pl. 3, fig. 30) from the Miocene Cannanore lignites of Kerala to accommodate the oval spores with notches at the ends and having parallel longitudinal ribs through the body. The size of the fossil ranges 69-103 µm × 116-134 µm. Jain and Kar (1979), amending the diagnosis of the taxa, described the form as a fungal body made up of long septate mycelia which run more or less parallel to each other from one end to other. The wall of the body generally laevigate but sometimes granulate. Kalgutkar and Jansonius (2000) commented that this form may have some stratigraphic significance.

Kar et al. (1972) described fossil fruiting body *Cucurbitariaceites* (Pl. 3, fig. 34) from early Tertiary sediments of Assam. The fruiting bodies are circular to sub-circular in shape, 40-120 µm in size, the outer region is dark in colour. The asci are up to 20 in number, cylindrical, generally developing from the inner region of the pseudoperithecia and mostly connect with each other from a broad polygonal area. In some cases the asci extend outwards crossing the external margin of the pseudoperithecia. A rupture is observed in some specimens in the central polygonal area bordered by basal parts of the asci. *Cucurbitariaceites* is distinguished from all other fossil genera of Microthyriales by its shape, darker outer layer, in the absence of true paraphyses and the presence of cylindrical asci. Kalgutkar and Jansonius (2000), while commenting on this genus stated that it shows affinity with the extant family Cucurbitariaceae belonging to the order Pseudosphaeriales. Most of the members of this order are confined to tropical areas though some are reported from temperate regions also.

Tiwari and Tripathi (1995) and Tripathi (2001) described a diversified fungal assemblage from Early Cretaceous Intertrappean beds of the Rajmahal Basin, Jharkhand. The assemblage shows the presence of many microthyriaceous fruiting bodies. Kar et al. (2003) reported a fruiting body assignable to Polyporaceae (Basidiomycetes) from the Lameta Formation exposed in Madhya Pradesh. This fossil, called *Lithopolyporales zeerabadensis* (Pl. 6, fig. 49), resembles the modern genus *Fomes* which are found as saprophytes on dead wood of various trees. Rao (2003) described a new fungal fruiting body *Kalviwadithyrites* from Sindhudurg Formation exposed at Kalviwadi, Sindhudurg District, Maharashtra. The Cleistotheicium (Pl. 6, fig. 50) is circular to subcircular in shape, dimidiate, non-ostiolate; the body made up of two sets of aporate cells, marginal cells rectangular to polygonal in shape, central cells isodiametric. A fossil fungus showing affinity to *Colletotrichum corda* belonging to the family Melanconiaceae (Deuteromycetes) was described from an Intertrappean bed located at Mohgaon-Kalan Village, Chindwara District, Madhya Pradesh by Kar et al. (2004a). The modern species of this genus causes red rot in the economically important plants. The fossil of this fungus shows the setae on the margins of the acervuli and was found to be preserved on a leaf cuticle. It was called *Protocolletotrichum deccanensis.* Kar et al. (2004b) described fossil parasitic fungi and epiphyllous fruiting bodies from the coprolite of dinosaurs. The coprolite yielding these fossils was collected from the Lameta Formation (Maastrichtian) of Central India. Occurrence of these fungi indicates that the leaves of plants infected by the recovered fungi were part of diet of the dinosaurs.

Mycorrhizal fungi constituted by fungal hyphae, auxillary cells, chlamydospores and a sporocarp belonging to the family Glomaceae were reported from Miocene sediments of Mizoram (Kar et al., 2005). Two types of fossil Ingoldian aquatic fungi were reported from Miocene sediments of Mizoram ((Kar et al., 2006). The first type of fossil (Pl. 6, fig. 51) is needle-shaped and belongs to the scolicospores. It is comparable to the extant genus *Tetrachaetum.* The other type of fossil, possessing globular to triangular body, belongs to staurospores (Pl. 6, fig. 52) and shows similarity with the extant genus *Ceratosporella.*

PALAEOCLIMATIC INTERPRETATIONS

Fungi are found in close association with specific plants and animals and if found in a fossil state are indicative of similar kind of situations during the geological past. Fossil fungi therefore, may provide useful information about the palaeoecology, past habitats and their hosts. In this regard fossil epiphyllous fungi can be more reliable and advantageous for palaeoclimatic interpretations. Occurrence of these fossils reflects moist and humid climate of tropical to subtropical belts (Prasad, 1986). The fossil peltate fungi are generally identified to the extant Microthyriaceae which are ectoparasites on leaves of higher plants of tropical to subtropical zones growing particularly in areas with high humidity. Edwards (1922) reported the occurrence of this group on conifer needles. Microthyriaceous fungi grow best in rain forests, rain forest margins and along creek banks (Ramanujam, 1982). Hence their presence is generally indicative of a wet tropical climate with heavy precipitation. The palaeohabitat interpretations based on fossil epiphyllous microthyriaceous fungi and their germlings is well-established through the studies on their modern equivalents growing on leaf litter from various Australian regions. These studies have shown the occurrence of microthyriaceous germlings in greater number on the plants growing in moist tropical habitats. Such studies have great potential in interpreting the palaeoclimate and should be undertaken for other geographical areas. However, the ecological interpretations based on epiphyllous fungi should be made with caution because some of these are reported to occur in wider latitudinal ranges (Dilcher, 1965; Selkirk, 1975). It is therefore, advisable to take into consideration the complete palynological assemblage for palaeoenvironmental interpretations. In most of the cases, coordinated studies of megafossils in association with palynological assemblages may provide more accurate information about the palaeoenvironmental conditions. Dilcher (1965) published an account of epiphyllous fungi thriving on leaves of different plants of Eocene age. Such studies bear great potential for determining the regional Palaeoclimate by comparing the fossils with extant taxa of known habitats. Environmental interpretations based on the presence of microthyriaceae may, however, sometimes be hampered due to the incorrect identification of the material. Their presence in dispersed fossil assemblage should, therefore, be ascertained before deciphering the past climate. The red alga *Caloglossa leprieurii*, generally found on grasses of brackish water marshes may be confused with *Trichopeltinites* due to morphological resemblance. Similarly, marine green alga *Ulvella lens* also resembles the fructifications of Microthyriaceae.

Studies particularly focusing on host fungus relationship are also of great significance in attempting the palaeoenvironmental interpretations. Chitaley (1978) and Chitaley and Yawale (1978) provided valuable palaeoecological information based on the presence of fossil fungal spores in petrified plant materials from the Deccan Intertrappean beds of India. Similar kinds of interpretations were published by Kar et al. (2004a, 2004b, 2005, 2006). These studies emphasize the importance of some fungal spores in evaluation of palaeoenvironment. Ramanujam and Srisailam (1980) noticed the prevalence of *Paleocirrenalia*, the hilicoid spore, in Neogene sediments of Kerala, South India and interpreted brackish to marine conditions by comparing them with modern fungi. Similarly, based on the presence of some other spores in the same strata a tropical climate has been interpreted by Ramanujam and Rao (1978) and Ramanujam and Srisailam (1980). A warm and humid environment has been interpreted by Kalgutkar and McIntyre (1991) in the Canadian Arctic due to the presence of helicosporous fungal types.

Studies of fossil fungal remains in coordination with micro- and megafossils of other groups have sometimes been used to infer the palaeoenvironment (Dilcher, 1973; Ramanujam, 1982). These assessments are based on the assumption that the palaeoclimatic sensitivity of fossil taxa was similar to that of the comparable modern counterparts. In this regard special stress was laid to explore the possibility of relating fossil fungal spores with those of modern fungi so as to realize their full potential in determining the ancient environment. However, only those types that could be related to the modern forms with certainty should be taken into account for this specific purpose.

GEOLOGICAL RECORDS OF FOSSIL FUNGI

Many coenocytic hyphae with fungal affinity are recorded from middle to late Precambrian rocks (Tyler and Barghoorn, 1954; Barghoorn and Tyler, 1965; Schopf, 1968; Schopf and Barghoorn, 1969). Other stray records of fossil fungi are reported from different parts of the Palaeozoic and Mesozoic eras (Harvey et al., 1969; Remy et al., 1994; Taylor and White, 1989). Kidston and Lang (1921) and Krassilov (1981) described different types of fossil fungi from Rhynie chert of the Devonian age. White and Taylor (1988, 1989) reported fungi from the Triassic rocks of Antarctica. Tiffney and Barghoorn (1974) and Pirozynski (1976) elucidated the antiquity of different fungal groups during the geological past. A vast variety of fossil fungal remains described as spores, filaments, fruiting bodies and mycorrhiza have been reported globally from Cretaceous to Tertiary sediments.

General remarks and futuristic approach

During the last four decades or so serious efforts were made towards the study of fossil fungi laying emphasis on their phylogenetic, stratigraphic and environmental considerations. Data generated on fossil fungi during this period is significant but it is only a good beginning. It will emerge as one of the exciting fields of research in the years to come. It has been observed that a number of fossil fungal forms recorded from Indian sediments need taxonomic revision, as these are either invalidly published or their diagnoses and status are not properly defined. Many species of different genera hence need to be recombined with some other suitable genera.

Acknowledgements: The author is thankful to Dr. Naresh Chandra Mehrotra, the Director, Birbal Sahni Institute of Palaeobotany, Lucknow for constant encouragement. Grateful acknowledgement is due to Dr. R.K. Saxena, Scientist F of this Institute for suggestions. Sincere thanks to Miss Divya Srivastava, SRF, at the Institute for help rendered.

References

Alvin, K.L. and Muir, M.D. 1970. An epiphyllous fungus from the Lower Cretaceous. Biological Journal of the Linnean Society, 2: 55-59.

Banerjee, D. and Misra, C.M. 1968. Cretaceous microflora from South India. Memoir of the Geological Society of India, 2: 99-104.

Barghoorn, E.S. and Tyler, S.A. 1965. Microorganisms from the Gunflint Chert. Science, 147: 563-577.

Chitaley, S.D. 1978. Fungal spores from the Deccan Intertrappean beds of Mohgaon Kalan, India. Proceedings of the IV International Palynological Conference, Lucknow (1976-77), 1: 305-311.

Chitaley, S.D. and Seikh, M.T. 1971. An infected grain from the Decccan Intertrappean cherts of Mohgaonkalan. The Journal of Indian Botanical Society, 50: 137-142.

Chitaley, S.D. and Patil, G.V. 1972. An ebenaceous fossil wood infected with deuteromycetous fungus from the Deccan Intertrappean beds of India. The Botanique, 3: 99-106.

Chitaley, S.D. and Yawale, N.R. 1978. Fungal remains from the Deccan Intertrappean beds of Mohgaonkalan, India. The Botanique, 7: 189-194.

Clarke, R.T. 1965. Fungal spores from Vermejo Formation coal beds (Upper Cretaceous) of central Colorado. Mountain Geologist, 2: 85-93.

Cookson, I.C. 1947. Fossil fungi from Tertiary deposits in the southern hemisphere. Part I. Proceedings of the Linnean Society of New South Wales, 72: 207-214.

Dilcher, D.L. 1965. Epiphyllous fungi from Eocene deposits in western Tennessee, U.S.A. Palaeontographica. Abt. B, 116: 1-54.

Dilcher, D.L. 1973. A revision of the Eocene flora of southeastern North America. Palaeobotanist, 20: 7-18.

Edwards, W.N. 1922. An Eocene Microthyriaceous fungus from Mull, Scotland. Transactions of the British Mycological Society, 8: 66-72.

Elsik, W.C. 1968. Palynology of a Paleocene Rockdale lignite, Milam County, Texas, I Morphology and taxonomy. Pollen et Spores, 10: 263-314.

Elsik, W.C. 1970. Palynology of a Paleocene Rockdale lignite, Milam County, Texas, III. Errata and taxonomic revisions. Pollen et Spores, 12: 99-101.

Elsik, W.C. and Jansonius, J. 1974. New genera of Paleocene fungal spores. Canadian Journal of Botany, 52: 953-958.

Elsik, W.C. 1976. Microscopic fungal remains and Cenozoic palynostratigraphy. Geoscience and Man, 15: 115-120.

Elsik, W.C. 1978a. Palynology of Paleocene Rockdale lignite, Milam County, Texas. I. Morphology and Taxonomy. Pollen et Spores, 10: 263-314.

Elsik, W.C. 1978b. Classification and geologic history of the Microthyriaceous fungi. Proceedings of the IV International Palynological Conference, Lucknow (1976-77), 1: 331-342.

Graham, A. 1962. The role of fungal spores in palynology. Journal of Paleontology, 36: 60-68.

Harvey, R., Lyson, A.G. and Lewis, P.N. 1969. A fossil fungus from Rhynie chert. Transactions of the British Mycological Society, 53: 155-157.

Jain, K.P. 1974. Fossil Fungi. pp. 38-46. In: Aspects and Appraisal of Indian Palaeobotany. K.R. Surange, R.N. Lakhanpal, and D.C. Bhardwaj, (eds.) Birbal Sahni Institute of Palaeobotany, Lucknow, India.

Jain, K.P. and Gupta, R.C. 1970. Some fungal remains from the Tertiaries of Kerala Coast. Palaeobotanist, 18: 177-182.

Jain, K.P. and Kar, R.K. 1979. Palynology of Neogene sediments around Quilon and Varkala, Kerala Coast, South India 1. Fungal remains. Palaeobotanist, 26: 105-118.

Jansonius, J. 1976. Palaeogene fungal spores and fruiting bodies of the Canadian Arctic. Geoscience and Man, 15: 129-132.

Kalgutkar, R.M. and McIntyre, D.J. 1991. Helicosporous fungi and early Eocene pollen, Eureka Sound Group, Axel Heiberg Island, Northwest Territories. Canadian Journal of Earth Sciences, 28: 364-371.

Kalgutkar, R.M. and Jansonius, J. 2000. Synopsis of Fungal spores, Mycelia and Fructifications. AASP Contribution Series 39: 1-423.

Kar, R.K. and Saxena, R.K. 1976. Algal and fungal microfossils from Matanomadh Formation (Palaeocene), Kutch, India. Palaeobotanist, 23: 1-15.

Kar, R.K., Singh, R.Y. and Sah, S.C.D. 1972. On some algal and fungal remains from Tura Formation of Garo Hills, Assam. Palaeobotanist, 19: 146-154.

Kar, R.K., Sharma, N., Agarwal, A. and Kar, R. 2003. Occurrence of fossil wood rotters (Polyporales) from Lameta Formation (Maastrichtian), India. Current Science, 85: 37-40.

Kar, R.K., Sharma, N., and Verma, U.K. 2004a. Plant pathogen *Protocolletotrichum* from a Deccan Intertrappean bed (Maastrichtian), India. Cretaceous Research, 25: 945-950.

Kar, R.K., Sharma, N., and Kar, R. 2004b. Occurrence of fossil fungi in Dinosaur Dung and its implication on food habit. Current Science, 87: 1053-1056.

Kar, R.K., Mandaokar, B.D. and Kar, R. 2005. Mycorrhizal fossil fungi from the Miocene sediments of Mizoram, Northeast India. Current Science, 89: 257-259.

Kar, R.K., Mandaokar, B.D. and Kar, R. 2006. Fossil aquatic fungi from the Miocene sediments of Mizoram, Northeast India. Current Science, 90: 291-292.

Kidston, R. and Lang, W.H. 1921. On Old Sandstone plants showing structure, from the Rhynie Chert Bed, Aberdeenshire. Part V. The thallophyta occurring in the peat-bed; the succession of the plants through a vertical section of the bed, and the conditions of accumulation and preservation of the deposit. Transactions of the Royal Society of Edinburgh, 52: 855-902.

Krassilov, V. 1981. *Orestovia* and the origin of vascular plants. Lethaia, 14: 235-250.

Kumar, P. 1990. Fungal remains from Miocene Quilon beds of Kerala State, South India. Review of Palaeobotany and Palynology, 63: 13-28.

Lang, R.T. 1978a. Southern Australian Tertiary epiphyllous fungi, modern equivalents in the Australian region, and habitat indicator value. Canadian Journal of Botany, 56: 532-541.

Lang, R.T. 1978b. Correlation of particular southern and northern hemisphere Palaeogene floras by the unusual fungal spores *Ctenosporites* and *Pesavis tagluensis.* Pollen et Spores, 20: 399-403.

Norris, G. 1982. Spore-pollen evidence for early Oligocene high latitude cool climatic episode in northern Canada. Nature, 297: 387-389.

Patil, R.S. and Ramanujam, C.G.K. 1980. Fungal flora of the carbonaceous clay from Tonakkal area, Kerala. Goelogical Survey of India, Special Publication 2(11): 261-270.

Pirozynski, K.A. 1976. Fungal spores in the fossil record. Biological Memoirs (In collaboration with International Society of Applied Biology), 1: 104-120.

Pirozynski, K.A. 1978. Fungal spores through the ages—a mycologist's view. Proceedings of the IV International Palynological Conference, Lucknow, (1976-77), 1: 327-330.

Pirozynski, K.A. and Weresub, L.K. 1979. The classification and nomenclature of fossil fungi. In: The whole fungus, the Sexual-Asexual Synthesis. B. Kendrick (ed.). Proceedings of the 2nd International Mycological Conference, University of Calgary, Kananaskis, Alberta, Canada, 2: 653-688.

Potonie, R. and Sah, S.C.D. 1960. Sporae dispersae of the lignites from Cannanore beach on the Malabar coast of India. Palaeobotanist, 7: 121-135.

Prasad, M.N.V. 1986. Fungal remains from Holocene peat deposits of Tripura state, northeastern India. Pollen et Spores, 28: 365-390.

Ramanujam, C.G.K. 1982. Recent advances in the study of fossil fungi.pp.287-301. In: Recent Advances in Cryptogamic Botany, Part II: Fossil Cryptogams. D.C. Bharadwaj, (ed.), The Palaeobotanical Society, Lucknow, India.

Ramanujam, C.G.K. and Rao, K.P. 1973. On some Microthyriaceous fungi from a Tertiary lignite of South India. Palaeobotanist, 20: 203-209.

Ramanujam, C.G.K. and Rao, K.P. 1978. Fungal spores from the Neogene strata of Kerala in South India. Proceedings of the IV International Palynological Conference, Lucknow, (1976-77), 1: 291-304.

Ramanujam, C.G.K. and Srisailam, K. 1980. Fossil fungal spores from the Neogene beds around Cannanore in Kerala state. The Botanique, 9: 119-133.

Rao, A.R. 1959. Fungal remains from some Tertiary deposits of India. Palaeobotanist, 7: 43-46.

Rao, K.P. and Ramanujam, C.G.K. 1976. A further record of Microthyriaceous fungi from the Nogene deposits of Kerala in South India. Geophytology, 6: 98-104.

Rao, M.R. 1995. Fungal remains from Tertiary sediments of Kerala Basin, India. Geophytology, 24: 233-236.

Rao, M.R. 2003. *Kalviwadithyrites*, A new fungal fruiting body from Sindhudurg Formation (Miocene) of Maharashtra, India. Palaeobotanist, 52: 117-119.

Remy, W., Taylor, T.N. and Hass, H. 1994. Early Devonian fungi: A blastocladalean fungus with sexual reproduction. American Journal of Botany, 81: 690-702.

Rosendahl, C.O. 1943. Some fossil fungi from Minnesota. Bulletin of the Torrey Botanical Club, 70: 126-138.

Schopf, J.W. 1968. Microflora of Bitter Springs Formation, Late Precambrian, Central Australia. Journal of Palaeontology, 42: 651-688.

Schopf, J.W. and Barghoorn, E.S. 1969. Microorganisms from the late Precambrian of South Australia. Journal of Palaeontology, 43: 111-118.

Selkirk, D.R. 1975. Tertiary fossil fungi from Kiandra, New South Wales. Proceedings of the Linnean Society of New South Wales, 100: 70-94.

Singh, C. 1971. Lower Cretaceous mircofloras of the Peace River Area, north-western Alberta. Bulletin of the Research Council of Alberta, 28: 1-542.

Staplin, F.L. 1976. Tertiary biostratigraphy, Mackenzie Delta region, Canada. Bulletin of the Canadian Petroleum Geology, 24: 117-136.

Taylor, T.N. and White, J.F.Jr. 1989. Fossil fungi (Endogonaceae) from the Triassic of Antarctica. American Journal of Botany, 76: 389-396.

Tiffney, B.H. and Barghoorn, E.S. 1974. The fossil record of the fungi. Occasional papers of the Farlow Herbarium of Cryptogamic Botany, 7: 1-42.

Tiwari, R.S. and Tripathi, A. 1995. Palynological assemblages and absolute age relationships of Intertrappean beds in Rajmahal Basin, India. Cretaceous Research, 16: 53-72.

Tripathi, A. 2001. Fungal remains from Early Cretaceous Intretrappean beds of Rajmahal Formation in Rajmahal Basin, India. Cretaceous Research, 22: 565-574.

Tyler, S.A. and Barghoorn, E.S. 1954. Occurrence of structurally preserved plants in Pre-Cambrian rocks of the Canadian Shield. Science, 119: 606-608.

van der Hammen, T. 1956. A palynological systematic nomenclature. Boletin Geologico, 4: 63-101.

Venkatachala, B.S. and Kar, R.K. 1969. Palynology of the Tertiary sediments in Kutch-2. Epiphyllous fungal remains from the borehole no. 14; The Palaeobotanist, 17: 179-183.

White, J.F., Jr. and Taylor, T.N. 1988 Triassic fungus from Antarctica with possible ascomycetous affinities. American Journal of Botany, 75: 1495-1500.

White, J.F., Jr. and Taylor, T.N. 1989. A trichomycete-like like fossil from Triassic of Antarctica Mycologia, 81: 643-646.

White, J.M. 1990. Evidence of Palaeogene sedimentation on Graham Islands, West Coast, Canada. Canadian Journal of Earth Sciences, 27: 533-538.

Williamson, W.C. 1878. On the organization of fossil plants of the Coal Measures, Part IX. Philosophical Transactions of the Royal Society of London, 169: 319-364.

Williamson, W.C. 1880. On the organization of the fossil plants of the Coal-Measures, Part X. Including an examination of the supposed radiolarians of the Carboniferous rocks. Philosophical Transactions of the Royal Society of London, 171: 493-539.

Young, F.G. and McNeil, D.H. 1984. Cenozoic stratigraphy of the Mackenzie Delta, North-west Territories. Geological Survey of Canada, Bulletin, 336: 1-63.

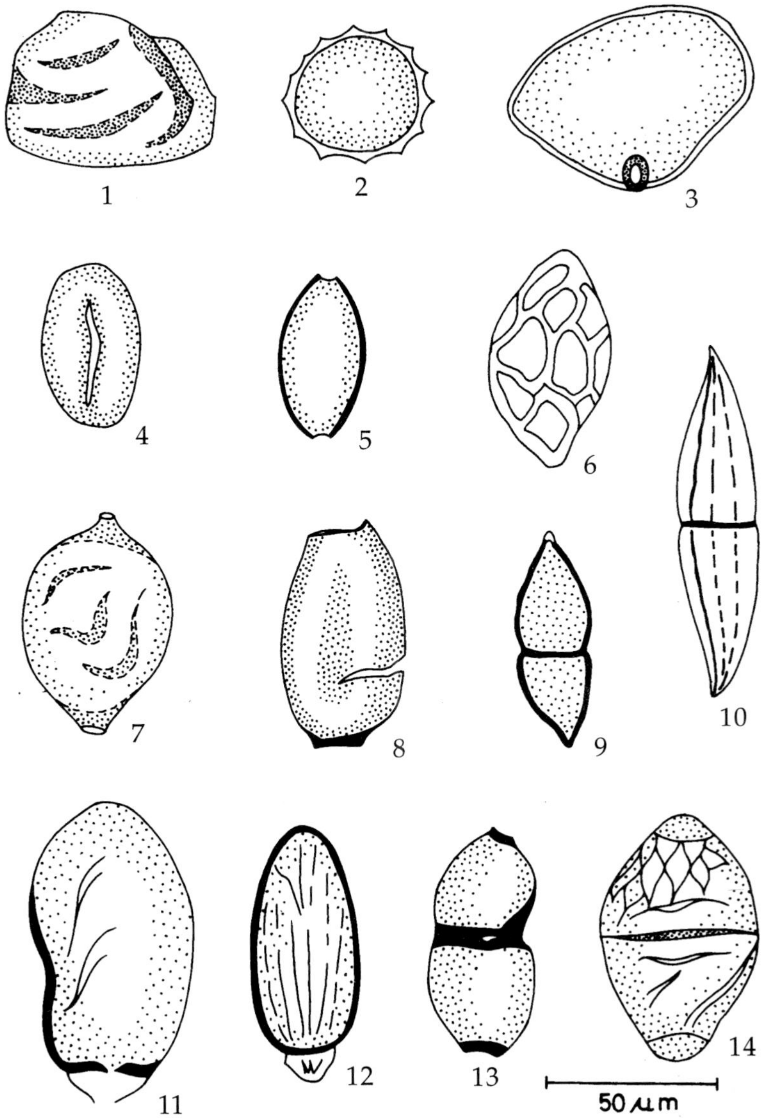

Plate 1—1. *Inapertisporites kedvesii* Elsik. 2. *Exesisporites verrucatus* Kumar. 3. *Monoporisporites psilatus* Chandra, Saxena and Settey. 4. *Haploxylonites ramanujamii* Elsik. 5. *Diporisporites psilatus* Kumar. 6. *Striadiporites irregularis* Kalgutkar. 7. *Foveodiporites conspicuous* (Ramanujam and Rao) Kalgutkar and Jansonius. 8. *Psilodiporites elongates* Varma and Rawat. 9. *Dicellaesporites obnixus* Norris. 10. *Fusiformisporites acutus* Kumar. 11. *Didymoporisporonites longus* (Kar) Kalgutkar and Jansonius. 12. *Diploneurospora tewarii* Jain and Gupta. 13. *Dyadosporites udarii* (Gupta) Kalgutkar and Jansonius. 14. *Ornasporites inequalis* Ramanujam and Rao

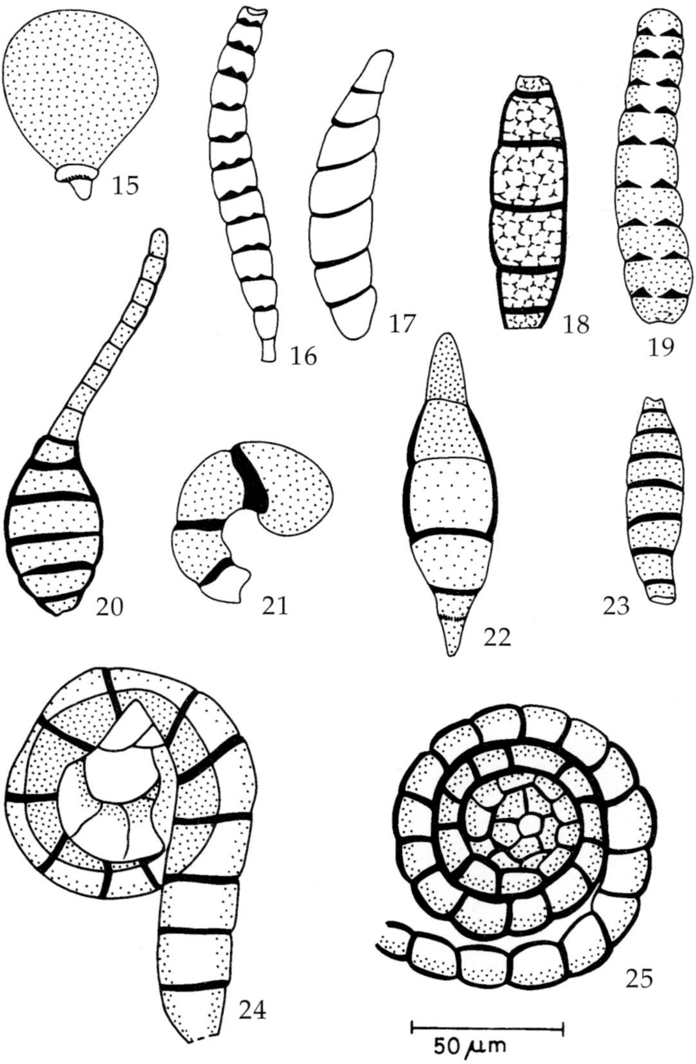

Plate 2—15. *Brachysporisporonites tenuis* Kumar. 16. *Scolicosporites scalaris* (Kalgutkar) Kalgutkar and Jansonius. 17. *Multicellites confuses* (Chandra et al.) Kalgutkar and Jansonius. 18. *Foveoletisporonites indicus* Ramanujam and Srisailam. 19. *Pluricellaesporites psilatus* Clarke. 20. *Quilonia alleppeyensis* (Ramanujam and Srisailam) Kalgutkar and Jansonius. 21. *Palaeocirrenalia oligoseptata* Ramanujam and Srisailam. 22. *Multicellaesporites denticulatus* (Ramanujam and Rao) Kalgutkar and Jansonius. 23. *Diporicellaesporites ordinates* (Sheffy and Dilcher) Kalgutkar and Jansonius. 24. *Involutisporonites chowdharyi* (Jain and Kar) Kalgutkar and Jansonius. 25. *Collegerites kutchensis* (Kar and Saxena) Jain and Kar

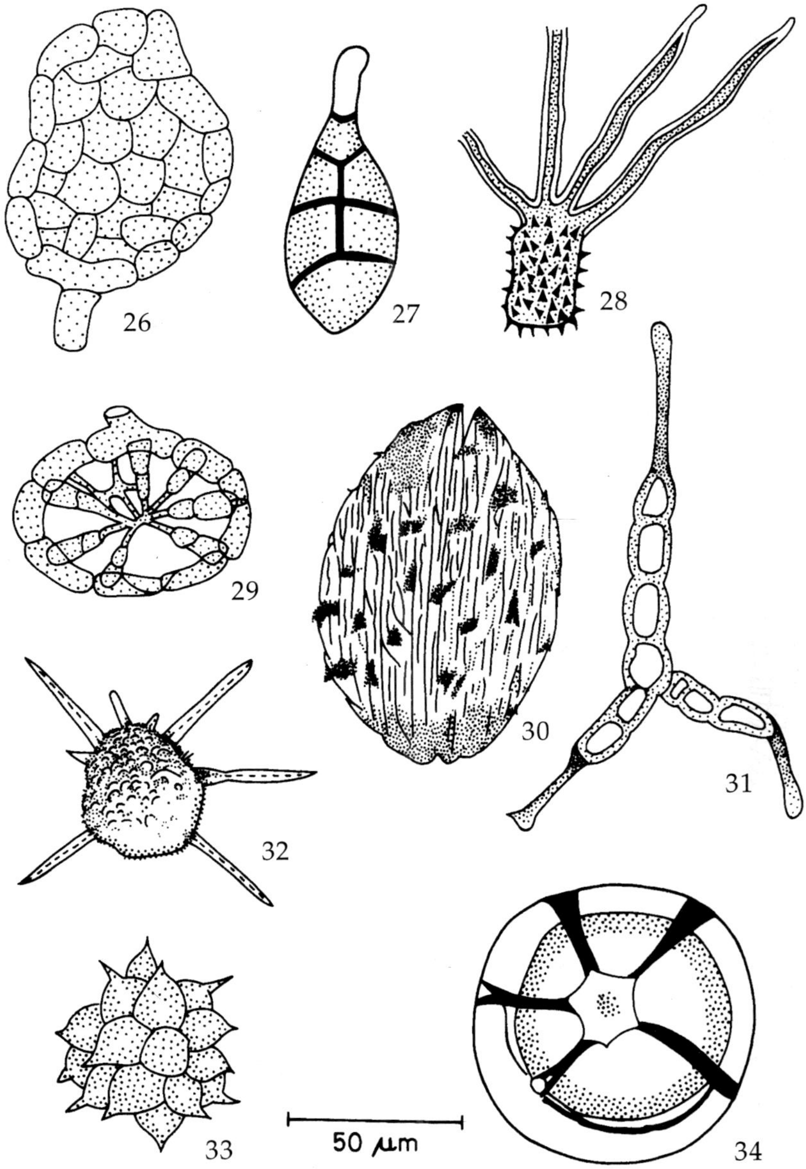

Plate 3—26. *Staphlosporonites neyveliensis* Ambwani. 27. *Polycellaesporonites bellus* Chandra et al. 28. *Frasnacritetrus conatus* Saxena and Sarkar. 29. *Pesavis tagluensis* Elsik and Jansonius. 30. *Lirasporis intergranifer* Potonie and Sah. 31. *Alleppeysporonites scabratus* Ramanujam and Rao. 32. *Appendicisporonites typicus* Saxena and Khare. 33. *Spinosporonites indicus* Saxena and Khare. 34. *Cucurbitariaceites bellus* Kar et al.

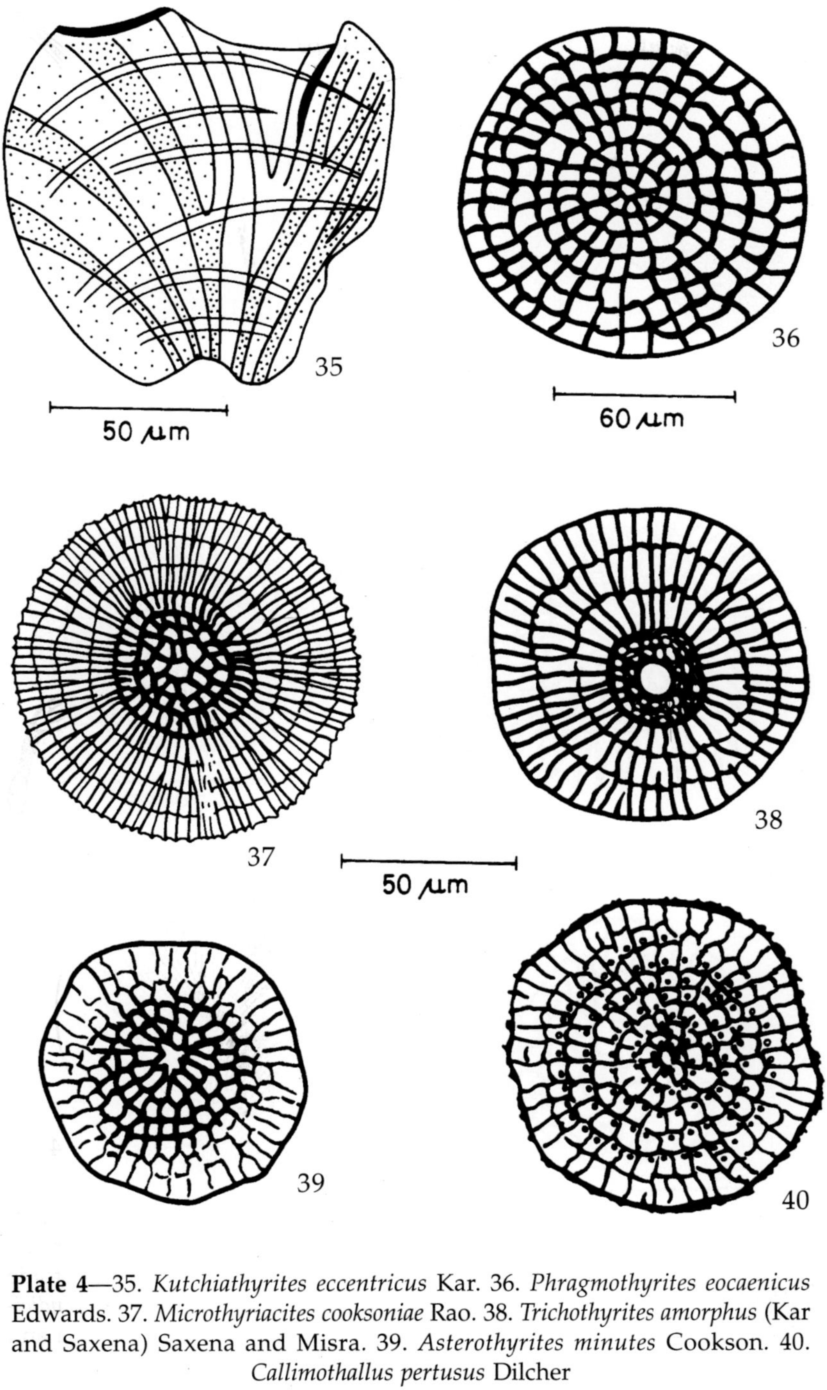

Plate 4—35. *Kutchiathyrites eccentricus* Kar. 36. *Phragmothyrites eocaenicus* Edwards. 37. *Microthyriacites cooksoniae* Rao. 38. *Trichothyrites amorphus* (Kar and Saxena) Saxena and Misra. 39. *Asterothyrites minutes* Cookson. 40. *Callimothallus pertusus* Dilcher

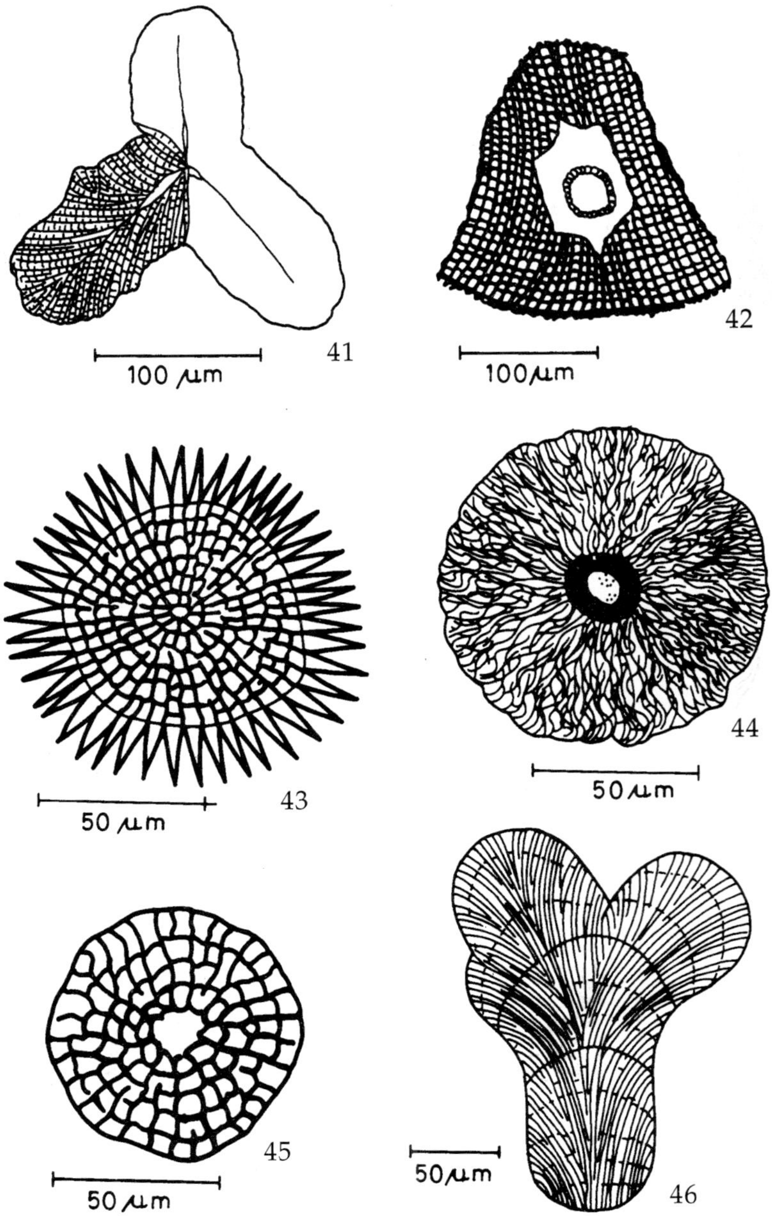

Plate 5—41. *Euthythyrites oleinites* Cookson. 42. *Brefeldiellites fructiflabellus* Dilcher. 43. *Parmathyrites indicus* Jain and Gupta. 44. *Plochmopeltinites* Cookson. 45. *Paramicrothallites* Jain and Gupta. 46. *Trichopeltinites pulcher* Cookson

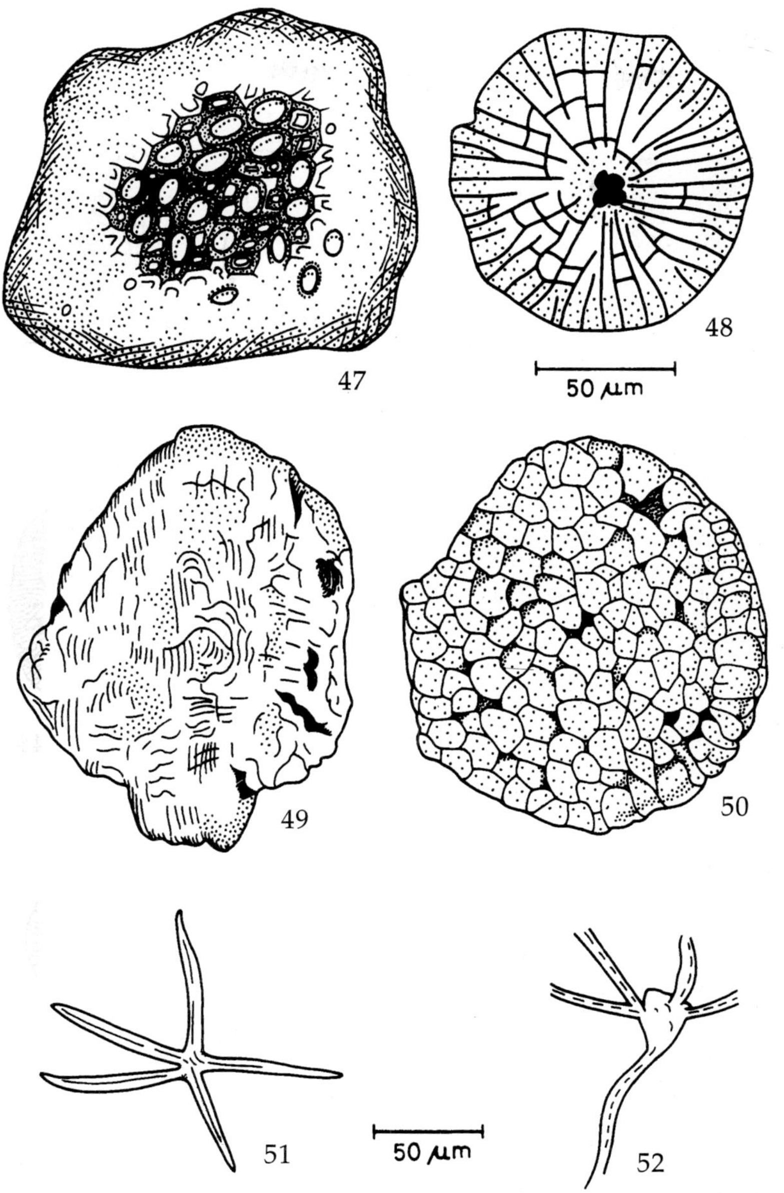

Plate 6—47. *Ratnagiriathyrites hexagonalis* Saxena and Misra. 48. *Microthallites* Dilcher. 49. *Lithopolyporales zeerabadensis* Kar et al. 50. *Kalviwadithyrites saxanae* Rao. 51. Aquatic fungi belonging to Ingoldian type with 4-5 needle-shaped arms of conidia, Kar et al. 52. Same as above with five armed globular conidia, Kar et al.

2

Fungi in the Air—Aeromycology: An Overview

S.T. Tilak
Y.M. College, Erandwane,
Bharati Vidyapeeth (Deemed University), Pune 411037, India
Correspond to: 18, Vidya Sagar Society, Near Mahesh Society,
Bibvewadi, Pune 411037, India

Abstract

India with its varied climatic conditions—temperate, tropical and coastal is distinct for aerobiological studies in general and aeromycological studies in particular. The credit goes to Cunningham who initiated studies of aerobiology in India (Cunningham, 1873). Currently, there are several centres where work on aerobiology and aeromycology is underway. The work done at these centres has been referred to at appropriate places here.

Environmental mycology or aeromycology constitutes one of the major aspects of aerobiology mainly because of the dominance of fungal spores in the ambient air. Aeromycological investigations take into account the identification of source, mode of release, dispersal, deposition, impaction and effects of impaction of fungal spores on various living systems. The fungal spores and hyphal fragments are commonly recorded in the air, and are important for the survival and subsequent continuation of generations. Many of the fungal spores have unique structures and the capacity to survive unfavourable environmental conditions.

Fungal spores form an important constituent of bioaerosol and they are often well adapted to airborne dispersal. In the course of evolution, the fungi have probably exploited the wind for their dispersal more thoroughly than any other group of organisms and consequently dominate the airspora (80%-90%).

The spores or fungal propagules are quite variable in size and shape. The spores or conidia range from 3-200 µm, most of these are about 10 µm in diameter. And they are often liberated in the air en masse and remain there for a long time.

Aeromycological studies in India have mainly been for—outdoor and indoor environments—monitoring airborne fungal spores in the atmosphere of metropolitan cities and towns, simultaneous comparative studies in urban and rural areas including air mycoflora over crop fields, quantification and biodiversity of moulds in indoor environments, experimental aspects of aeromycology, and health hazards, both to plants and human beings including animals. In the recent years more attention has been given to the allergic fungal spores suspended in various environs and causing health hazards to humans.

This chapter encompasses the knowledge gathered for aeromycology during the last two decades.

INTRODUCTION

India has the unique distinction of being one of the earliest countries where aerobiological studies in general and aeromycological studies in particular were initiated (Cunningham, 1873). Later Dr. T. Sreeramulu at Waltair in South India continued the work on aeromycology, while Drs. S.T. Tilak and A. Ramalingam carried on the work at Aurangabad, Maharashtra, and Mysore, respectively. During the last 20 years several centres in the North and North-east, South, East and West came up and have contributed towards the study of aeromycology. The work done at these centres has been referred at appropriate places here.

Environmental mycology or aeromycology constitutes one of the major aspects of aerobiology, mainly because of the dominance of fungal spores in the ambient air. In general, aeromycological investigations take into account the identification of source, mode of release, dispersal, deposition, impaction and effects of impaction of fungal spores on various living systems. The fungal spores and hyphal fragments are commonly recorded in the air and are important for the survival and subsequent continuation of generations. Many of the fungal spores are endowed with unique structures and capacity to survive under unfavourable environmental conditions and these probably account for their predominance in the air. Nearly all the spores are essentially dispersive units and their significance as gene dispersal units should not be lost.

Fungal spores form an important constituent of bioaerosol and they are often well adapted to airborne dispersal either by having tall conidiophores, that help them penetrate into/pass through the laminar boundary layer, or specialized liberation mechanisms that help them to eject forcibly through the laminar layer. In the course of evolution, the fungi have probably exploited the wind for their dispersal more thoroughly than any other group of organisms and consequently dominate the airspora (80%-90%).

The spores or fungal propagules are quite variable in size and shape. The spores or conidia range from 3-200 µm, most of these are about 10 µm in diameter. And they are often liberated in the air massively and remain there for a long time.

Aeromycological studies in India have mainly been for—1. Outdoor environment—monitoring airborne fungal spores in the atmosphere of metropolitan cities and towns including simultaneous comparative studies in urban and rural areas including air mycoflora over crop fields; 2. Indoor environment—quantification and biodiversity of moulds in indoor environments; 3. Experimental aspects of aeromycology; and 4. Health hazards, both to plants and human beings including animals due to airborne fungal spores.

In the recent years more stress is being given to allergic fungal spores suspended in various environs and causing health hazards to humans.

This chapter overviews what has been added to our knowledge of aeromycology during the last two decades after the two reviews of Tilak (1990b, 2005).

INSTRUMENTATION

Aeromycological investigations require volumetric samplers. Andersen and Burkard samplers are in use but the cost is prohibitive. Tilak air sampler is being used widely by Indian workers. Rotorod sampler, Personal sampler, Insect traps, designed and developed by Tilak also have been used by several workers. A new model of Tilak air sampler—2005, and Rotorod sampler is now commonly used.

FUNGI IN OUTDOOR ENVIRONMENT

Fungal spores and their fragments in the outdoor air have been studied all over India by various workers either to compare them from the indoor environment or to know the significant fungal types—pathogens or saprophytes (Gopi and Kumar, 1990; Datta and Jain, 1990; Roy and Trivedi, 1996; Sinha et al., 1997, 1998; Bhat and Rajasab, 1991; Ramalingam and Nair, 1994; Singh et al., 2004; Chauhan and Kulshrestha, 2004; Cholke, 2007).

Vegetable market

The atmosphere of the vegetable market has been found to contain a variety of fungal components. Vegetable and fruit markets are of varied types like completely closed, partly closed or in sheds or open. This influences the spore load. Vegetable markets have been surveyed by various workers like Sahaney and Purwar (2002), Sawane and Saoji (2004), Gaikwad (1997), Sarma and Bora (1996), Kakde and Saoji (1996, 1998a, b), Kakde et al. 2001). Shastri (1998) carried out aeromycological investigation of the vegetable market with special reference to hyphal fragments and insect parts that are known to play an important role in the spread of diseases of vegetables and fruits in the market and in the storage. It was observed that during the day time hyphal fragments contributed more to the airospora and maximum number was recorded (2145/m^3) in December and minimum in July (285/m^3). Exposed petriplates in the atmosphere of vegetable markets at Guwahati, Assam revealed various types of fungal spores in the atmosphere throughout the year. The high concentration of fungal spores was recorded during July-March and the lowest in April-June. In other months the concentration of fungi in the air varied. The important types of fungal spores recorded during July-March were *Cladosporium, Alternaria, Fusarium, Curvularia, Aspergillus, Penicillium, Rhizopus, Cercospora, Mucor* and *Colletotrichum*. Of the types isolated *Cladosporium* constituted the highest number, followed by *Aspergillus, Alternaria, Penicillium, Mucor, Rhizopus,* and *Fusarium*. It may be noted that some types of fungi which are generally associated with rotten vegetables are present in the market atmosphere throughout the year. The qualitative and quantitative variations in the airspora were dependent on the type of vegetable stored in the market place. Significance of aerobiodeteriogens in affecting or damaging the vegetables and fruits in the market as well as in the place of storage is well established. The role of airborne *Penicillia* in fruit markets has been worked out in detail. Several species of *Penicillium* like *P. chrysogenum, P. citrinium, P. cyclopium, P. digitatum, P. italicum, P. expansum and P. funiculosum* were isolated from Kalamna and Santra Markets of Nagpur (Sawane and Saoji, 2004). Different control measures of green mould, *P. digitatum* were worked out (Saoji and Sawane, 2001).

Garbage depot

Garbage depot. constitutes an outdoor environment. An aeromycological survey was undertaken by Tilak and his associates at Pune, India (More et al., 1997). Since there was no proper mechanism for further treatment of the garbage, this resulted in a huge population of microbial forms specially the saprophytic ones. The garbage depot provides an ecological niche in which a variety of forms grew luxuriantly, continuously disseminating a huge amount of sporeload in the atmosphere. There were reports of the adjoining population suffering from various health hazards. It was, therefore, decided to carry out air monitoring with an intention to study the aeromicrobiota with particular reference to aeroallergens.

Results of the survey indicated that over 90% of the adjoining population suffered from respiratory diseases, skin diseases, asthma, reddening of the eyes and other allergic disorders. However, confirmatory tests are required for establishing relevance of aerollergens and health hazards of the local affected population.

Airborne fungal spores were studied from 1st February 1995 and is in process uptil now. A total of 39 fungal spore types were identified, out of which 23 were known to be allergenic. *Alternaria, Aspergillus, Chaetomium, Cladosporium, Curvularia and Pleospora* were predominant.

During the rainy season, concentration of spores was comparatively more in the morning hours as compared to the evening. Low temperature (22°C) and high humidity (88%) show direct correlation with the concentration of spores.

A definite and positive correlation between the high counts of the aeroallergens during the rainy season and increase in allergy symptoms of the patients was also noted.

INDOOR ENVIRONMENT

In a country like India with a fairly huge number and variety of environments, including occupational ones have not received the desired attention. However, in the recent past many aerobiologists have greatly contributed to this aspect with special emphasis on biodiversity and the occupational health hazards (Santra and Chanda, 1989; Singh et al., 1990, 1995; Sumbali and Badyal, 1991; Misra and Jamil, 1991; Pandit and Singh, 1992; Verma and Chile, 1992a, b; Ghani, 1993, 1994; Ghani et al., 1993; Pugalmaran and Vittal, 1994, 1999; Singh and Singh, 1994; Nadimuthu and Vittal, 1995; Surekha and Reddy, 1996; Rafiyuddin et al., 1997; Raha and Bhattacharya, 1997; Agashe and Anuradha, 1998; Giri and Saoji, 1998, 2003; Singh and Singh, 2000; Bagwan and Meshram, 2000; Verma et al., 2000; Kulshrestha and Chauhan, 2001; Sharma and Dutta, 2001; Sahaney and Purwar, 2001; Udaya Prakash and Vittal, 2003; Misra et al., 2003; Majumdar and Bhattacharya, 2004; Singh, 2007). It is now realized that the importance of suspended fungal spores in causing allergic disorders and also in the damage to stored materials is being understood (Table 2).

Misra's et al. (2003) review has covered the fungal diversity of 32 indoor environments of occupational importance which include—agarbathi (incense sticks) factories, animal care facilities, bakeries, sugar mills, cattle sheds, cinema halls, cobbler shops and tanneries, coffee curing works, cow sheds, fire wood depots, flour mills, food storage places, fruit shops/ markets, ginnery, grain shops/storage godowns, hospital wards, hostel kitchens, industry—mechanical and textile, jute mills, libraries, matches factory, oil mills, paper mills, pig farms, poultry sheds/farms, rice mills, saw mills, lecture halls/school rooms, scientific laboratories, silk filatures, snuff depots, working environments, etc. but many more still await investigations. Udaya Prakash (2004) has published a useful manual for identification of indoor moulds.

In most of the indoor environments *Aspergillus, Penicillium* and *Cladosporium* are most commonly encountered. In general, it has been observed that well ventilated rooms have more concentration of spores in the air, while in airconditioned rooms, the concentration is low. The dwellings of allergy patients indicated the presence of more aerobio-allergens. House dust provides a constant source for airborne microbes.

The impact of airborne fungal spores in the indoor environment has been studied by numerous workers in the recent years but they are not more than 30 in number. There are more than 15 research papers on library work that have appeared. This further indicates that we follow the trend of research, rather than investigating newer and significant areas that deserve attention. Occurrence of *Histoplasma capsulatum* from a library is a significant finding as this fungus is a causal organism of *Histoplasmosis,* a very serious lung disease. Why have other investigators not found this fungus? This poses a serious question and points towards the need for further search in libraries in different parts of our country. Similarly, many places such as agarbathi (incense stick) factories, cinema halls, food storage places, ginnery, mechanical and textile industries, jute mills, pig farms, saw mills and so on have a single or a few publications on their indoor fungal flora. This strongly suggests that these environments need more surveys with a view of finding out both fungi suspended in the air and their possible correlation with the ailments of the workers of such places. Studies in relation to the hospital environment—operation theatres and dentistry wards, etc. have helped to modify the concepts in hygiene and sterilization to provide a 'clean' environment.

Recent studies of Singh (2007) have revealed the rich biodiversity of indoor environments (industrial and non-industrial) in North-East India during the last five decades. Industrial indoor workplace environments include cinema halls, saw mills, rice mills, paper mills, and bakeries. Non industrial workplaces are hospitals, poultry farms, libraries, pig farms, while stored printed paper materials include state archives, hostel kitchens, FCI grain godowns and potato storage chambers. Singh's findings are indicative of the fact that biodiversity of fungi fluctuates with place, meteorological parameters and the substrate. Many of the indoor environments showed higher concentrations of *Penicillium* and *Aspergillus* indicating a possible source of indoor contamination. He has reported of fungi inside an operating cinema hall (non-air conditioned) in Shillong (Meghalaya). Spores of Aspergilli-Penicilli were trapped throughout the period of investigation by slide exposure method contributing 82.7% of the flora. Thus, the fact that airspora inside an operating cinema hall might be one of the sources for fungal respiratory allergy to frequent cinemagoers can not be overlooked. He reported 20 different species of fungi from the indoor air of a saw mill. *Aspergillus, Cladosporium, Fusarium, Penicillium, Trichoderma*, etc. were dominantly isolated genera. His findings are in conformity with that of Misra (1988) and Misra et al. (1989).

Sheds of various domestic animals like cows, buffaloes, horses, sheep, pigs and others have evoked great interest in aeromycological research. Tilak and Pande (2004) have emphasized the role of environmental biopollution in health hazards of domestic animals. Many diseases of commercial and domestic animals are transmitted by contact or through ingestion of contaminated food. Mammalian Aspergillosis, Mycotoxicoses, facial eczema and other fungal diseases are receiving attention from veterinary pathologists and aerobiologists. The foot and mouth disease, Rinderpaste and Ephemeral fever in cattle and infectious Laryngo-tracheatis of poultry have been observed. Khillare (1990) conducted aero-mycological surveys inside the cattleshed. However, definite relationship between the airborne pathogens and disease incidence could not be established. *Aspergillus, Penicillium* and *Rhizopus* grow within tissues of various animals. It is gratifying that aeromycologists have taken an active interest in airborne diseases of veterinary animals. It would go a long way in arresting the health hazards of animals (Table 1).

Table 1. Pathogens of veterinary importance (Tilak and Pande, 2005)

Abisidia ramosa	*Histoplasma capsulatum*
Aspergillus flavus	*Histoplasma farcinosum*
Aspergillus fumigatus	*Mucor corymbifer*
Aspergillus nidulans	*Nocardia asteriodes*
Aspergillus terreus	*Rhenosporidium seeberi*
Blastomyces dermatitides	*Rhizopus equines*
Coccidioides inmitis	*Rhizopus sunus*
Cryptococcus neoformans	

Exact data about the incidence of several animal diseases are not available, nevertheless losses to the tune of several crores of rupees every year are due to the foot and mouth disease that drastically reduce the productive capacity of milking animals and working power of bullocks. *Aspergillus fumigatus* has been associated with abortion of cattle. Acute intoxication is due to the ingestion of saprophytic or phytopathogenic fungi associated with the feed of cattle.

The highly contagious skin infection of ring worm, which may infect many domestic animals and men, may be due to species of different genera viz., *Microsporum, Trichophyton, Epidermophyton,* which are able to digest and utilize keratin, the portion of horny outer layers of skin, hairs and nails. *Trichophyton* may infect skin between the toes causing athlete's foot or the skin of scalp or the hair itself. *Aspergillus, Penicillium* and *Rhizopus* are other saprophytic species that may grow within tissues of various animal skins.

Thus, it can be concluded that intramural airspora of animal shed is rich in potential pathogens.

Sharma and Dutta (2001) recovered 32 fungal forms from the indoor air of the Hindustan paper mill, Pashgoan, Silchar, Assam. Out of 17 genera identified, *Aspergillus* sp. contributed 34.57% of the total population and *Penicillum* sp. contributed 26.3% of the total spores collected from the finishing house. The highest percentage contribution was shown by *Aspergillus* (54.4%) in the finishing house whereas the highest percentage contribution in Chipper House was shown by *Penicillium* sp. (77.8%). The pulp house showed the lowest distribution of fungal forms.

Kukeraja (2007) carried out an aerobiological survey at three different places—viz., 1. The place where raw material (waste paper) was stored. 2. Paper machine plant. 3. Paper storage godown (manufactured) of a paper mill. The maximum CFUs/M3 were recorded from raw material storage places followed by paper machine plants and paper storage godowns. Dominant genera were *Aspergillus* (41.14%,) *Penicillium* (37.14%), *Curvularia* (16.5%), *Alternaria* (2.85%) and *Torula* (2.28%).

Sharma and Dutta (2001) also reported 14 fungal genera from the indoor air of bakeries in Greater Silchar, Assam. They found *Aspergillus* contributing 38.6% of the population whereas *Penicillium* contributing 20.3% of the total population. The other fungal types isolated were *Humicola* sp. (2.6%), *Curvularia* sp. (2.9%), *Cladosporium* sp. (7.5%), *Geotrichum* sp. (11.1%) *Rhizoctonia* sp. (6.1%) and *Mucor* sp. (2.6%).

Singh et al. (1990) also observed, in the air spora of a large bakery in Delhi, *Aspergillus flavus* and *A. niger* contributing to 90% of the total Aspergilli recovered. *A. flavus* was characteristic of the storage section (87%) while *A.niger* was in packing (56.3%). Peak period of occurrence was from

May to August. Most of the bakery workers suffered from one or the other respiratory disorders (breathlessness, cough, asthma, etc.) that might be due to aeromycoflora of the indoor bakery environment (Singh et al., 1994, 1998b). Rafiyuddin et al. (1997) have reported aeromycoflora of a bakery elaborating mycotoxins.

Survey of aeromycoflora of different tanneries was conducted by Chauhan and Rathore (1998). A total of 50 fungal forms were isolated. Their results are in conformity with that of Shukla and Misra (1984) and Shukla (1987). Chauhan and Rathore (1998) also investigated fungi from different types of finished leather and assessed their influence on the quality of leather. They found that fungal attack on leather reduce their quality such as—colour, odour, crackness, stiffness, and tensile strength and so on and thereby drastically reducing the commercial value of the leather and its product.

Air spora of medical wards dominantly comprises of spp. of *Aspergilli* contributing 33% of the total population whereas *Penicillium* spp. contribute 15.2% of the total population (Verma and Chile, 1992a, b; Singh et al., 1994; Govind et al., 2002; Giri and Saoji, 2003). The lowest fungal population was observed in operation theatres and inside hospitals where filtered air was allowed to enter.

The poultry farm provides a very congenial environment for fungal spores and is gaining more importance due to the outbreak of fungal diseases of birds. Microorganisms present in the air therein affect the health of birds as well as workers involved in poultry. *Aspergillus* species has been reported as the dominant one in such places by Verma and Bhandari (1992), Singh et al. (2000) Verma and Shrivatsava (2004), and Verma et al. (2006).

Recently Shrivastava (2007) investigated the poultry environment using a combination of techniques and recovered 71 fungal forms dominated by the members of Deuteromycotina (60 in number). Analysis of feed also revealed the dominance of Deuteromycotina (20 in number)

The only report in print regarding the fungi in the air of a pig farm is that of Begum et al. (2001). A total of 21 fungi were recorded by them. Dominant fungal forms were—*Aspergillus glaucus, A. niger, A. flavus, Alternaria solani, A. alternata, Cladosporium herbarum, Fusarium oxysporum, F. moniliformae*, and the species of *Helminthosporium, Trichoderma, Nigrospora, Mucor* and *Penicillium.*

Aeromycoflora of various libraries in India has been worked out by different authors (Singh et al., 1990; Singh et al., 1995; Tilak and Pande, 1997; Nadimuthu and Vittal, 1995; Saoji and Giri, 1997; Sahaney and Purwar, 2001; Atluri and Padmini, 2002; Mohammad et al., 2003; Rane and Gandhe, 2005; Majumdar and Hazara, 2005; Upadhyaya and Jain,

2005). In almost all surveys it has been found that species of *Aspergillus, Penicillium and Cladosporium* dominate the flora of fungi. These fungi in the library air, besides deteriorating paper and book binding material (Nyuksha, 1994; Majumdar et al., 2003; Sarma and Basumatary, 2004; Dhawan and Nigam, 2005; Kukeraja, 2007) also significantly affect the health of library and other paper related industry staff (Majumdar and Bhattacharya, 2004).

Godowns of grains have also received some attention by aeromycologists such, Pugalmaran and Vittal (1999) and Sharma and Dutta (2001). In general it has been found by all that various species of the genus *Aspergillus* dominate the flora of such places. Species of other genera like, *Alternaria, Cladosporium, Curvularia, Fusarium, Geotrichum, Humicola, Nigrospora, Penicillium, Torula,* and *Trichoderma* are also recovered, of course, in varying percentages.

Sick buildings

The indoor environment of residential buildings, offices, schools and other places are also now known to have fungal contaminants in significant proportions. And, more importantly airconditioned rooms are now a problem of growing concern in India as these contain a number of allergenic fungal forms. Such infested buildings are known as 'sick buildings'. A few studies have been conducted for such 'sick' buildings. Carpetted rooms are more of a problem. The coarse filters of air conditioners have more fungal spores while other layers also contain enough quantities of spores. In general, the species of *Aspergillus, Penicillium* and *Cladosporium*, etc. are commonly encountered. Allergy from such fungal forms are reported in sensitive patients. It may be noted that spores retained in the filters germinate and sporulate providing additional source of allergenic material. Since very little work has been done in India on this important aspect, hence investigations for the presence of fungi in air conditioned indoor environment would be a rewarding venture for any enthusiastic worker.

Furthermore, the information on fungi in the indoor air is still very limited both for the number of places looked into and the assessment of damages done by the fungi to human beings and the material.

Meteorological factors and air spora

The effects of temperature, humidity and rainfall on the frequency of fungal spores in the air have been worked out. Spore concentration in the atmosphere varies with the fluctuations taking place in the meteorological factors. High wind velocity increases the spore load in the air. This is particularly true for hyphal fragments. Roy and Trivedi (1996) have

Table 2. Common fungal spores recovered from indoor air of occupational environs by various workers

Alternaria	*Helminthosporium*
Aspergillus	*Hendersonula*
Basidiospores	*Heterosporium*
Beltrania	*Hysterium*
Bispora	*Nigrospora*
Botryodiplodia	*Penicillium*
Cephaliophora	*Periconia*
Cercospora	*Pithomyces*
Chaetomium	*Pleospora*
Cladosporium	Rust spores (mainly Urediniospores)
Corynespora	Smut spores
Curvularia	*Sordaria*
Dictyoarthrinium	*Spegazzinia*
Didymosphaeria	*Sporomiella*
Diplodia	*Stigmina*
Drechslera	*Tetraploa*
Epicoccum	*Torula*
Fusarium	*Trichoconis*
Ganoderma	*Trichothecium*
Harknessia	Unidentified spores

provided useful data on the role of meteorological factors and the occurrence of fungal spores in the air.

The aeromycoflora remains rich during intermittent rainy and dry periods and becomes poor during prolonged rainy and dry periods. Scarcity or absence of suitable substrate, sufficient moisture, vegetational density and diversity affect the occurrence and composition of fungal airspora. Temperature and RH have a pronounced effect on spore productivity which probably explains high spore incidence in the rainy season and low in dry periods (Shrivastava, 2007).

EXPERIMENTAL AEROMYCOLOGY

a) Terminal velocity and density

Dorycanta et al. (1994) studied the terminal velocity and spore density of *Penicillium, Cladosporium* and *Nigrospora* spores inside a room. The spore concentration of test fungi just after stirring were almost equal and high.

A difference in spore concentration was observed after one hour of undisturbed state. The air at five feet above ground (average nose level) was almost free from *Cladosporium* and *Nigrospora*, whereas the concentration of *Penicillium* spore even after one hour was quite significant. This can be explained on the basis that bulky spores (*Nigrospora*, density = 61.13×10^{-3} g/cm^3) descend freely (velocity = 2.13 cm/s) and sedimented on the ground causing the air free from spores when left undisturbed. The lighter spore (*Penicillium*, density = 2.48×10^{-3} g/m^3) descends slowly (velocity = 0.035 cm/s) and at the same time a slight disturbance makes it ascend and remain suspended in the air and become a major fungal pollutant in the indoor environment. In most of the indoor airspora studies, concentration of *Penicillium* was found to be appreciably high (2.5). Thus, the degree of pollution caused by fungal spore is largely dependent on its terminal velocity and the density of spore.

b) Splash dispersed spores

Earlier, it was believed that splash-dispersed spores do not form components of airspora. Aeromycological investigations carried out by the author have helped to clarify and modify the earlier concept that splash dispersed spores would not contribute to airspora. He established the significance of splash dispersed spores and airspora. However, it must be remembered that the effective range of splash is limited but when combined with wind velocity the range may extend considerably.

Splash dispersal is typical of slime spored fungi. However, when a splash drop with conidia falls on and glides over hydrophobic plant surface of a leaf, 'non wettable' (dry) conidia e.g., those of *Penicillium*, tend to be deposited near the moving droplet and wettable (slimy) conidia of *Verticillium* are washed away or deposited where the droplet comes to rest.

Spores of *Colletotrichum* and *Pestalotia* are produced in saucer shaped acervuli, when a thick film of water gets deposited on the acervuli, efficient splash dispersal is assured. Dispersal by splash mechanism is well known and is very common in *Fusarium, Colletotrichum, Phytophthora, Botrytis, Hemilea, Gibberella, Cercospora, Phyllosticta, Venturia,* and *Nectria*, etc. It is thus, evident that for practical purposes splash dispersed spores not only contribute to aeromicrobiota but can also be transported to a considerable distance if associated with high wind speed.

The work on splash dispersal carried out by Devi (1992) in the laboratory and in the fields for three plant pathogenic spores—*Albugo candida, Helminthosporium oryzae* and *Alternaria solani* revealed that the number of splash droplets was dependent on height, size of the incident drop and target film of the spore suspension. The maximum number of splash droplets was recorded by an incident drop of 5 mm dia falling from a height of 3.5 m with a maximum of 4500, 1630 and 1455 droplets for all the 3 test fungi. It was also observed that the size of the spores influences the frequency of their association with the splash droplet.

The smaller spores were more readily picked up from the suspension than conidia of a larger size. The larger drop of water i.e., 5 mm produced the highest number of splash droplets.

It may, therefore, be concluded that splash dispersed spores though not truly airborne, play a significant role as effective plant pathogens and human health hazards.

c) Circadian rhythms of some common air-spora components

Circadian periodicity patterns of 36 spore types have been identified by Suman et al. (1992) out of a total of 80 airspora components trapped in an aerobiological study at Hajipur (Bihar). Diurnal rhythms have been categorized as night patterns, post dawn patterns, middle day patterns and double peak patterns. Meterological and topographic factors have been recognized for varying circadian rhythms.

Spore trapping in the immediate aerial environment of spore producing structures reveals several periodic patterns, many of which are related to diurnal rhythm of alternating day and night. Spore concentrations have also been related to humidity levels in reproductive structures as also to the changing climatic cycles. Circadian periodicity studies are of great help in distinguishing broadly between the dry spora and wet spora, the former comprising the spores, whose release appears to be followed by low humidities and the latter by high humidities. Distinct circadian periodicity patterns can be recognized for different airborne spores (Agashe and Anuradha, 1996).

Results of this investigation are of immense value in evolving a disease forecasting system for the crops grown in a region.

FUNGAL SPORES AS BIOINDICATORS

Tilak (1990a) has provided a detailed account of the utility of airborne fungal spores as bioindicators. The occurrence and dominace of fungal spores in air depend on a variety of environmental factors like rainfall, humidity, temperature, wind speed and direction, etc. In several aero-mycological investigations by various workers in India, this relationship has been clearly brought out.

The superiority of biological indicators over physico- chemical factors has been advocated by many workers as it is direct method to indicate the prevailing conditions of the environment (Tilak, 1990a). Most of the spores of fungi occur dominantly during the rainy season (June-Sept.) when temperature ranges between 20-30°C and relative humidity remains 75% or above.

The interest of meteorologists would be of a different nature. Some spores would indicate the possible prevailing weather conditions while some would hint at future meterological conditions. Such types are often designated as 'markers' or 'biological indicators'. The meteorologists would naturally be interested in such markers which would help them to forecast meteorological conditions.

Apart from the plant pathogenic forms, the other fungal spores which appear regularly in abundance in the atmosphere could also be correlated with meteorological conditions.

Rhizopus Ehrenh. spores are observed almost the whole year but the concentration varies from one season to another. Extremely high concentrations of these spores indicate the rainy season (from June to September)—high rainfall, high relative humidity (above 75%) and low temperatures (20°-27°C). Moderate concentration of this spore type indicates the winter season when there are scattered rains and a relative humidity and temperature ranging between 50% to 75% and 27° to 37°C, respectively. Low concentration or the absence of the spores can be assigned to the summer season when there is no rainfall and low relative humidity (below 50%) and high temperatures (above 35°C). This spore type is known to be potentially allergenic.

Spores of *Didymella* and *Leptosphaeria* and Basidiopsores of genera like *Ganoderma* are found in abundance in the rainy season. In other seasons the number of their spores goes down.

Spores of the members of Deuteromycotina dominate the air spore population, both qualitatively and quantitatively. The spore types like *Alterneria, Cladosporium and Pyricularia* are abundant in the air during the rainy season. However, the spores of *Alternaria* are observed almost the whole year in the atmosphere. Besides being a pathogen to cause leaf spot diseases in plant leaves, this spore type is also allergenic in nature.

FUNGAL SPORES AND ALLERGY

Many aerobiologists have contributed towards this aspects (Misra, 1988; Misra et al., 1989; Tilak, 1990b; Atluri and Appana, 1990; Misra and Jamil, 1991; Verma and Chile, 1991, 1992a, b, c, d, 1993, 1997; Shrivastava and Wadhwani, 1992; Ganguly, 1992; Ghani, 1993; Agashe et al., 1994; Wadhwani, 1994; Singh et al., 1995; Singh and Singh, 1994, 1996; Jain, 1997, 1998; Shah, 1997; Singh et al., 1998a; Singh et al., 1999; Verma and

Vidyanidhi, 2001; Mishra, 2002; Singh and Kumar, 2003; Singh and Pandit, 2003; Chauhan et al., 2004; Verma and Shrivastava, 2005; Saoji, 2006; Verma et al., 2006).

Fungal spores predominate the other bioparticles in the airspora. The sources for such fungal spores are the substrates that are at the ground level. Many of the airborne fungal spores are potential allergen (Table 3). Therefore, indoor and outdoor aeromycological surveys help considerably to locate the sources of spores, their identification, concentration and seasonal variation. Thus, such information, provides basic data for the treatment of sensitive individuals suffering from an allergy. Data obtained from such a survey help to obtain spore calendar for the allergens, their avoidance and management strategies.

Table 3. Some allergenic fungal spores (Tilak and Pande, 2004)

Acremonium	*Neurospora*
Acremoniella	*Paecilomyces*
Alternaria	*Penicillium*
Arthrinium	*Phoma*
Aspergillus	*Puccinia*
Aureobasidium	*Rhodotorula*
Botrytis	*Rhizopus*
Candida	*Saccharomyces*
Drechslera	*Scopulariopsis*
Epicoccum	*Stachybotrys*
Epidermophyton	*Stemphylium*
Fusarium	*Trichocladium*
Ecliocladium	*Trichoderma*
Graphium	*Trichophyton*
Helminthosporium	*Trichothecium*
Humicola	*Ustilago*
Microsporon	*Verticillium*
Mucor	

In general, it has been established that particles of 7.5 μm in size cause Type I allergy and particles of size 5 μm cause Type III allergy and this may be applicable for patients with normal nose breathing. A patient suffering from nasal hindrance may breathe by the mouth and in that case particles up to 20 μm may be deposited in the bronchiole. *Aspergillus fumigatus*, a common airborne mould and with spores of 3 μm diameter is a good example of spores that may cause infection of type I or III allergy.

The percentage of fungal spores in the air is approximately 10 times that of pollen grains. However, by volume the pollen grains dominate. The most predominant spores are those of Deuteromycotina. Among these the spores of *Cladosporium, Alternaria, Curvularia, Helminthosporium* and *Aspergillus* are encountered in maximum concentration. But, all these viable fungal spores in the air are not allergenic, some may be harmless, while some may cause diseases of plants and animals. Out of over 200 spore types identified from airspora studies in different geoclimatological regions of India, only 38 have been reported as allergenic to human beings (Tilak, 1998).

Spores of mushrooms, toadstools, rust and smuts have also been reported to be of allergenic significance.

The threshold level of different fungal spores in evoking allergic symptoms, has been estimated for some fungi, for examples, 1000 *Alternaria* spores/m^3 of air and 3000 *Cladosporium*/m^3 of indoor air can initiate allergic reactions, however, these concentrations are encountered in multiples in the outdoor environment.

CLINICAL STUDIES

Allergenicity of fungi has been tested by a number of workers. The clinical investigations have established the significance and necessity of treatment in sensitive individuals. Hence, there is a need of trained aerobiologists and clinicians in the Indian subcontinent.

In earlier research, the relationship of airborne fungi and symptoms of allergic respiratory diseases was reported extensively, however, it should be remembered that correlations do not necessarily mean causal relationships between the composition and sporeload of indoor airborne fungal spores and the symptoms of respiratory allergies. It is well known that the inhalation of spores can induce upper and lower respiratory tract symptoms. In weak and inmmunocompromised patients, some fungal spores which are normally considered non-pathogenic may cause severe and fatal infections (Verma and Vidyanidhi, 2003).

The skin prick test for allergenicity of fungal spores was assessed in 20 poultry workers. The highest allergenicity was exhibited by *Aspergillus flavus* (27.77%) followed by *Penicillium* sp. (22.2%). Enzymes of the blood serum regarding the length of work was studied in poultry farmers with exogenous allergic alveolities. The result demonstrated a decrease in alkaline phosphates and creatinine phosphokinase in patients with the length of work ranging from 1 to 20 years. Respiratory allergenic disorders were estimated to be upto 59% in poultry workers. High concentration of

Scopulariopsis in the air of poultry can sensitize predisposed patients. ELISA showed a significant high sensitivity in upto 10% patients (Singh et al., 1998b).

Studies in the dwellings of asthmatic children

Asthma among children is becoming a significant health problem these days. Aerobiological investigations in the management of pediatric asthma was carried out in the dwellings of 15 asthmatic children patients in the city of Pune, using standard methods, by Tilak, More and Jogdand in the year 2005 under a project financed by the WHO in which Dr. Salvi from the Chest Institute, Pune, Dr. Peter Slay and Mrs. Slay from Perth University Australia and Bharti Vidyapeeth, Pune participated. Twenty five different fungal forms, some hyphal fragments, a few pollen, trichomes, algal filaments, insect scales and bacteria were observed. Qualitative and quantitative analysis of aeromycospora indicated the dominance of *Aspergillus* (25%), *Cladosporium* (20%), *Rhizopus* (10%), *Alternaria* (10%), *Penicillium* (5%) and *Fusarium* (5%) in their descending order. Other fungal forms such as *Periconia, Helminthosporium, Curvularia, Memnoniella, Pleospora,* and *Bispora,* etc. were encountered in a lesser concentration. The findings also indicated a presence of allergenic pollen grains of *Parthenium, Mangifera, Lantana, Moringa, Cocos,* and *Papaya* which were also detected in varying concentrations. Attempts were made to correlate the incidence of air borne fungal forms and pollen with the manifestation of symptoms of asthma in the pediatric patients (unpublished).

The preliminary survey clearly indicated the need of continuing this project.

AEROMYCOLOGY IN RELATION TO CROP DISEASES AND THEIR FORECASTING

Aeromycological studies over crop fields were intended mainly to find out preventive measures against pathogens leading to epiphytotics. Epidemiological surveys provide useful information to arrive at a suitable preventive method to protect plant diseases well in advance. Plant disease forecasting service in India is still in its infancy (Tilak, 1994, 1996a). This has not got the attention of governments as it deserves. The estimation of inoculum in the air forms one of the major bases of devising an efficient disease forecasting system.

Factors such as the concentration of pathogenic spore load in the air, meteorological parameters, growth stage of the crop and the appearance and spread of disease leading to epiphytotics have a close relationship among themselves. And, hence aeromycological investigations can play a significant role in the disease forecasting system and help in preventing epidemics and subsequent crop losses for different crops like ground nut (Tikka disease), Bajra—*Pennisetum typhoides* (Rust and Ergot diseases), sorghum and rice—*Oryza sativa* (Blast of rice—*Pyricularia oryzae*) that have been studied by Indian workers (Tilak and Pande, 2005).

Different crops have been studied at different locations and regions of India for fungal spores over crop fields and their relationship with varying ecological factors have also been worked out. The importance of such studies in plant disease forecasting has been emphasized for different crops such as bajra, rice, maize, pulses, groundnut, safflower, cotton, soybean, grapes, vegetables—like, carrots, tomatoes, onions, potatoes, mustard, cabbage and sugarcane, etc. (Murdhankar and Pande, 1991; Rajasab and Rao, 1992; Atluri et al., 1992; Chawda and Rajasab, 1992a, 1992b, 1994, 1997; Pande, 1994, 1995, 2001; Rajasab and Chawda, 1994; Jayrajan and Palaniswami, 1994; Shaikh and Talde, 1994; Prasad et al., 1994; Sahu, 1995, 1996; Uddin and Chakaravarty, 1995; Tilak, 1996b; Rajasab and Kallurmath, 1997; Gaikwad, 1998; Naik and Pande, 1998; Nagpurne et al., 1998; Tilak et al., 1999; Rajasab and Frederiksen, 2001; Hegde and Kulkarni, 2002; Hegde et al., 2002; Johnson and Rajasab, 2003; Jagannathan and Gaikwad, 2003; Vittal, 2004, 2005; Rajasab et al., 2004; Singh, 2006)

Studies by Singh and Doryacanta (1992) for maize diseases clearly indicate that pathogenic spores become abundant in the air 4-5 weeks before the onset of the disease. These findings are helpful as they indicate a close correlation between meterological factors, growth stages of the crop and spore load. Ramachander Rao and Tilak (1990) and Ramachander Rao (1993) have also found dominance of *Alternaria alternata* causing leaf spot of sunflower at the flowering stage associated with favourable meterological factors for the disease. Chawda and Rajasab (1994) similarly noted conidia of *Alternaria porri* causing purple blotches of onions when the crop was at the 7-8 leaf stage. Channabasavraj et al. (1994) observed that moderate temperature and more humidity increased the incidence of *Alternaria dauci* causing blight of carrots. Prasad et al. (1994) studied the aerobiology of mulberry rust.

The probable 'Phakospora' path of soybean rust was studied by Hegde et al. (2002) based on trap nursery trial by the aeroscope method and weather data, the probable path of *P. pachyrhizia* was traced out and it was concluded that Ugarkhurd is one of the foci of infection spots for rust out break in Karnataka.

Aeromycological investigations on groundnut fields by Tilak (1996a) have clearly brought out the close correlation between the spore load of pathogens in the air, growth stages of crop, meteorological factors and disease incidence and its subsequent spread. Tikka disease of groundnut could be forecasted 15-20 days in advance, considering the meteorological factors and growth stages of the crop. Similalrly, investigations for rust of groundnut have also revealed that the warning system can be of great utility because pathogenic spore load appears in the air 18-20 days in advance before the first appearance of the disease (Pande, 2001).

Aeromycological investigations on the bajra crop have indicated that in case of rust disease of bajra, moderate temperature and high humid conditions are favourable for disease incidence and spread. Regarding the ergot disease of bajra, the relative humidity of 70% or above and the temperature between 25 and 30°C promoted the liberation of ascospores into the atmosphere, which directly results in the increase of the primary inoculum and hence the increase in disease incidence (Tilak and Babu, 1984; Tilak and Pande, 1989).

The incidence of downy mildew of sorghum can be predicted on the basis of aerobiological and epidemiological studies.

Rice blast pathogen—*Pyricularia oryzae* is a menacing problem particularly for Tamil Nadu and Kerala. Abundance of conidia (inoculum), the influence of environmental factors and resistance or susceptibility of hosts are important factors responsible for the onset of the disease. The reproductive propagules—the conidia persisting over secondary hosts develop fast under high humid conditions (80-95%), low temperature (23-25°C) associated with rainfall of a few mm and creat a spore load sufficient to cause blast disease in rice. The presence of conidia and their load can be estimated 3-5 days in advance through an aeromycological study that can provide a useful forecasting system of the disease.

Hence a specific forecasting model for different crops and pathogens would be of use for a geographical area in forecasting diseases of wider significance.

It would be of interest to note that in modern aerospora diagnosis one should reach the molecular way to understand not only the taxa, genus, species but also virulence and frequency of the type of pathogen if at all this information is to be used for disease management. The interaction of obnoxious gases with airspora also needs to be investigated. In the Indian subcontinent with a variety of crops, different geographical regions and varied climatic zones, the prediction system and its utility has a much wider scope.

What needs to be done: There are very few publications (Tilak, 1982, 1989a, b and Udaya Prakash, 2004) of text books and manuals for identification of airborne bio-components in India. Any efforts in this direction would be welcome. There are still a large number of dwellings and workplaces where we need to search for the presence of fungi, particularly for those that may incite allergenic reactions. A collaborative work with physicians practising for respiratory diseases would be fruitful in combatting many allergenic/respiratory diseases. Hence there is a need for a collaborative programme at an all India level.

References

Agashe, S.N. and Anuradha, H.G. 1996. Studies in circadian periodicity of fungal spores of Bangalore city. Indian J. Aerobiol, 8: 17-23.

Agashe, S.N. and Anuradha, H.G. 1998. Aeromycological studies of a library in Bangalore. Indian J. Aerobiol, 11: 24-26.

Agashe, S.N. Philip, E. and Anuradha, H.G. 1994. Aerobiological analysis of intramural environment and atopic allergy in Bangalore (India), pp. 169-176. In: Aerobiologia-Proc. 5th International Conference. S.N. Agashe (ed.), Oxford & IBH Publishing Co. Pvt. Ltd., New Delhi, India.

Atluri, J.B. and Appanna, N. 1990. A survey of airborne allergenic fungal spores at Vijaywada. J. Environ. Biol, 11: 321-329.

Atluri, J.B. and Padmini, V. 2002. Aeromycoflora of Andhra University Library. Indian J. Aerobiol, 15: 47-50.

Atluri, J.B., Nageshwara Rao, G. and Rao, S.S. 1992. The incidence of airborne fungal spores over certain pulse crops. Frontier Botanist, 4 & 6: 5-12.

Bagwan, N.B. and Meshram, B.M. 2000. Relationship between fungal air-spora of apple storehouses and incidence of post-harvest fruit rots. Indian J. Aerobiol, 13: 29-32.

Begum, J., Ahmad, K. and Bora, K.N. 2001. A study of mycoflora in the indoor air of a Pig farm at Khanapara (Assam). Indian J. Aerobiol, 14: 36-38.

Bhat, M.M. and Rajasab, A.H. 1991. Incidence of airborne fungal spores at two different sites in Gulbarga during 1986. Indian J. Aerobiol, 4: 1-6.

Channabasavraj, G.A. and Reddy, N. 1994. Aeromycoflora over a carrot field with special reference to airborne inoculum of *Alternaria dauci*, pp. 243-249. In: Aerobiology-Proc. 5 International Conference, Bangalore. S.N. Agashe (ed.) Oxford & IBH Publishing Co. Pvt. Ltd., New Delhi, India.

Chauhan, S. and Rathore, D.S. 1998. Role of aeromycoflora in the deterioration of finished leather, pp. 125-146. In: Environment and Aerobiology. A.K. Jain (ed.) Research Periodicals and Book Publishing House, Texas, U.S.A.

Chauhan, S.V.S. and Kulshrestha, A. 2004. Comparative incidence of fungal spores of two gardens of Agra city. Indian J. Aerobiol, 17-12-16.

Chauhan, S.V.S., Kulshreshta, A. and Goyal, R. 2004. Airborne fungal spora of Agra city with special reference to allergic significance. Indian J. Aerobiol, 17: 17-22.

Chawda, H.T. and Rajasab, A.H. 1992a. Epidemiology of Anthracnose of onion. I. Stages of infection spots and conidial production. Indian J. Aerobiol, Special Vol. 121-125.

Chawda, H.T. and Rajasab, A.H. 1992b. Epidemiological investigations on

Anthracnose and purple blotch of onion (*Allium cepa* L.). Onion Newsl. Tropics, 4: 65-66.

Chawda, H.T. and Rajasab, A.H. 1994. Aerobiology of *Alternaria porri* and its relation to purple blotch disease in onion. Indian J. Mycol. Plant Pathol, 24: 41-45.

Chawda, H.T. and Rajasab, A.H. 1997. Day to day variations in the incidence of *Alternaria porri* conidia over a purple blotch infected onion field, pp. 221-228. In: Aerobiology-Proc. 5th International. Conference. S.N. Agashe (ed.) Oxford & IBH Publishing Co. Pvt. Ltd., New Delhi, India.

Cholke, P.B. 2007. Extramural and intramural aerobiological studies with reference to Pimpri-Chinchwas area. Ph.D. Thesis, University of Pune, Pune, India.

Cunningham, D.D. 1873. Microscopic Examination of Air. Govt. Printer, Calcutta, India.

Datta, T.R. and Jain, A.K. 1990. Aerobiological studies at Gwalior fungal spores in a textile mill area. Bionature, 11: 77-81.

Devi, K.K. 1992. Aerobiological Studies at Imphal. Ph.D. Thesis, Manipur University, Manipur, India.

Dhawan, S. and Nigam, N. 2005. Fungal diversity in paper materials and its role in biodeterioration, pp. 93-113. In: Fungal Diversity and Biotechnology. M.K. Rai and S.K. Deshmukh (eds.). Scientific Publishers, Jodhpur, India.

Dorycanta, H., Amita, Y., Shabitri, and Singh, I.N. 1994. Terminal Velocity—factors determining indoor environmental pollution caused by fungal spores. Front. Bot, Special Vol. 86-89.

Gaikwad, Y.B. 1997. Vegetable Aerobiology & Disease, pp. 155-163. In: Environment & Aerobiology. A.K. Jain (ed.) Research Periodicals & Book Publishing House, Texas, USA.

Gaikwad, Y.B. 1998. Aerobiological approach to aerolate mildew disease of cotton. Indian. J. Aerobiol, 11: 33-37.

Ganguly, M. 1992. Studies on incidence of *Alternaria* in the atmosphere of Bangalore. A major fungal allergen. Indian J. Aerobiol, 6: 30-35.

Ghani, T. 1993. Studies of Aeromycoflora of Bhagalpur. Ph.D. Thesis, Bhagalpur University, Bhagalpur, India.

Ghani, T. 1994. Incidence of air-borne fungi inside different scientific laboratories at Bhagalpur. Indian J. Aerobiol, 7: 8-12.

Ghani, T., Kumar, S. and Kale, S.R. 1993. Aeromycoflora inside and outside a college library. Intl. J. Environ. Hlth. Res, 3: 49-58.

Giri, S.K. and Saoji, A.A. 1998. Microbial pollution of air in intramural environment of Nagpur city, residential quarters and food storage place. Botanique, 2: 27-32.

Giri, S.K. and Saoji, A.A. 2003. Aerobiological studies in intramural-environment of a hospital. Botanique, 12: 68-73.

Gopi, T.V. and Kumar, K.P. 1990. Comparative analysis of airspora of two urban localities in Kerala. Indian J. Aerobiol, 3: 39-44.

Govind, K., Varghese, J.P., Prakash Kumar, R., Gokulapalan, C. and Ravindran, P. 2002. Comparative study of culturable fungi in an asthmatic patient's bedroom and inside a hospital ward. Indian J. Aerobial, 15: 52-58.

Hegde, G.M. and Kulkarni, S. 2002. Aerobiology and disease prediction models of soybean rust caused by *Phakospora pachyrhizi*. Indian J. Aerobiol. 15:1-14.

Hegde, G.M., Kulkarni, S. and Anahosur, K.H. 2002. Probable "*Phakospora*" path of soybean rust. Indian J. Aerobiol, 15: 15-19.

Jagannath, S. and Gaikwad, Y.B. 2003. A preliminary observation of incidence of *Alternaria* leaf spot disease over Safflower fields in Maharastra. Indian J. Aerobiol, 16: 51-56.

Jain, A.K. 1997. Some aerobiological studies in relation to allergic incidence at Gwalior (India), pp. 69-74. In: The Aerobiology. S.N. Agashe (ed.) Oxford & IBH Publishing Co. Pvt. Ltd., New Delhi, India.

Jain, A.K. 1998. Airborne fungi and hypersensitivity among allergenic patients at Gwalior, pp. 251-265. In: Environment and Aerobiology. A.K. Jain (ed.) Research Periodicals and Book Publishing House, Texas, USA.

Jayarajan, R. and Palaniswami, A. 1994. Epidemiology and forecasting of rice blast disease, pp. 195-201. In: Advances in Mycology & Aerobiology: Prof. S.T. Tilak Comm. Vol. (ed. U.K. Talde). Today and Tomorrow's Printers and Publishers, New Delhi, India.

Johnson, N. and Rajasab, A.H. 2003. Prevalence of *Claviceps sorghi* on sorghum in Karnataka, India. International Sorghum News Letter, 44: 93-95.

Kakde, U.B. and Saoji, A.A. 1996. Study of air borne fungal spores in vegetable/ fruit market environment. Intl. J. Mendel, 12: 17-18.

Kakde, U.B. and Saoji, A.A. 1998a. Airborne fungal spores in the market environment of Nagpur city. University J. Sci, 8: 45-48.

Kakde, U.B and Saoji, A.A. 1998b. Study of mycoflora associated with four different places. Botanique, Special Vol. 11: 45-51.

Kakde, U.B., Saoji, A.A. and Kakde, H.U. 2001. Seasonal variation of fungal propagules in a fruit market environment of Nagpur. Aerobiologia, 17: 177-182.

Khillare, S.S. 1990. Aerobiological investigations at Ahmednagar. Ph.D. Thesis, Marathwada University, Aurangabad, India.

Kukeraja, S.G. 2007. A Detailed Study of Paper Deterioration by Cellulolytic Aeromycoflora. Ph.D. Thesis, Nagpur University, Nagpur, India.

Kulshrestha, A. and Chauhan, S.V.S. 2001. Aeromycoflora of some hospitals of Agra city. Indian J. Aerobiol, 14: 33-35.

Majumdar, M.R. and Bhattacharya, K. 2004. Measurement of indoor fungal contaminants causing allergy among the workers of paper related industries of West Bengal, India. Indoor & Bull. Environment. Sage Publications, U.K., 13: 189-197.

Majumdar, M.R. and Hazara, S. 2005. Assessment of fungal contaminants in the libraries of Presidency College, Kolkata. Indian J. Aerobiol, 18: 1-5.

Majumdar, M.R., Gangopadhya, S. and Chanda, S. 2003. Cellulytic activity of some promising fungi in selected biodeterioration studies. J. Mycopathol. Res, 41: 27-33.

Misra, J.K. 1988. Fungi in the Environs of Occupational Importance and Their Possible Role in Causing Allergenic Diseases Among Workers. A final report submitted to the Department of Environment, Govt. of India, New Delhi, India.

Mishra,V.K. 2002. Immunopharmacological studies of fungal extracts in guinea pigs. National Conference on Fungal Diversity and Biotechnology. K.V. Pendharkar College, Dombivali (East), Abstract, 34.

Misra, J.K and Jamil, Zafar. 1991. Fungi in the indoor environments of flour mills in Lucknow. Grana, 30: 398-403.

Misra, J.K., Tewari, Rashmi and Jamil, Z. 1989. Occurrence, distribution and phenology of indoor fungal flora of Saw mills in Lucknow and their allergenic significance. Biol. Mem, 15: 73-78.

Misra, J.K., Hashim, I. Bahoray, S.N. and Mishra, V.K. 2003. Fungal diversity of indoor environment of occupational importance in India. An overview, pp. 717-740. In: Frontiers of Fungal Diversity in India (Prof. Kamal Festschrift). G.P. Rao, C. Manoharachary, D.J. Bhat, R.C. Rajak and T.N. Lakhanpal. (eds.) International Book Distributing Co., Lucknow, India.

Mohammad, S., Johnson, S.N. and Rajasab, A.H. 2003. Aeromycoflora of Gulbarga University Library. Indian J. Aerobiol, 16: 28-30.

More, T.N., Bharati, S.K., Ghatpande, P.S. and Shinde, P.B. 1997. Survey of aerobiopollutants in the garbage depot. Bharati Vidapeeth Univ. Journal, 1: 21-25.

Murdhankar, S.V. and Pande, B.N. 1991. Aerobiological and epidemiological approach to ground nut rust. Indian J. Aerobiol, 4: 19-22.

Nadimuthu, N. and Vittal, B.P.R. 1995. A volumetric survey of culturable molds in a library. Indian J. Aerobiol, 8: 12-16.

Nagpurne, S.D., Patil, T.M. and Gaikwad, Y.B. 1998. Vegetable aerobiology and disease, pp. 154-170. In: Environment & Aerobiology. A.K. Jain (ed.) Research Periodicals and Book Publishing Co., Houston, Texas, USA.

Naik, M.N. and Pande, B.N. 1998. Aerobiological survey over sunflower fields at Aurangabad, pp. 285-292. In: Aerobiology, Proc. 5th Intl. Conference, Bangalore. S.N. Agashe (ed.) Oxford & IBH Publishing Co. Pvt. Ltd., New Delhi, India.

Nyuksha, Y.P. 1994. The biodeteriotation of paper and books, pp. 1-88. In: Recent Advances in Biodeteriotation. K.L. Garg and K.G. Mukerji. (eds.) Naya Prakashan, Calcutta, India.

Pande, B.N. 1994. Air monitoring of two spp. of *Cercospora* pathogenic to groundnut. Dr. B.A.M.U.I., 27: 26-29.

Pande, B.N. 1995. Epidemiology of jowar rust. Indian Bot. Reporter, 14: 138-148.

Pande, B.N. 2001. Incidence of airborne microbial components over groundnut fields, pp. 128-135. In: Frontiers in Fungal Biotechnology and Plant Pathology. C. Manoharachary (ed.). Allied Publishers Limited, New Delhi, India.

Pandit, T.R. and Singh, A.B. 1992. Prevalance of airborne fungi in a sugar factory environment. Indian J. Aerobiol, 5: 145-152.

Prasad, K.V., Jadhav, Dayakar and Sullia, S.B. 1994. Aerobiology of *Peridiospora mori* causing mulberry rust, pp. 261-272. In: Aerobiology-Proc. 5th International Conf., Bangalore, S.N. Agashe (ed.) Oxford & IBH Publishing Co. Pvt. Ltd., New Delhi, India.

Pugalmaran, M. and Vittal, B.P.R. 1994. Aeromycological survey of indoor environment in leather godowns, pp. 143-148. In: Aerobiology-Proc. 5th International Conference. S.N. Agashe (ed.). Oxford & IBH Publishing Co. Pvt. Ltd., New Delhi, India.

Pugalmaran, M. and Vittal, B.P.R. 1999. Fungal diversity in the indoor and outdoor environments and dust of grain storage godowns. Indian J. Aerobiol, 12: 24-29.

Rafiyuddin, M., Girisham, S. and Reddy, S.M. 1997. Aeromycoflora of bakery elaborating mycotoxins—A preliminary study. Indian J. Aerobiol, 10: 9-12.

Raha, Subrata and Bhattacharya, K. 1997. Aeromycoflora of residential area at two distinct biozones of West Bengal, India. Indian J. Aerobiol, 10: 38-42.

Rajasab, A.H. and Vasanth Rao, C. 1992. *Aspergillus niger* Van Teigh, the casual organism of black mould of onion. Indian J. Aerobiol, Special Vol. 127-132.

Rajasab, A.H. and Chawda, H.T. 1994. Dispersal of conidia of *Colletotrichum gloeosporioides* by rain and the development of Anthracnose on onion. Grana, 33: 162-165.

Rajasab, A.H. and Kallurmath, R.S. 1997. Airborne incidence of *Aspergillus niger* populations and their pathogenic ability on onion (*Allium cepa* L.) and garlic (*Allium sativum* L.) pp. 181-188. In: Environment & Aerobiology. A.K. Jain (ed.) Research Periodicals and Book Publishing House, Houston, Texas, USA.

Rajasab, A.H. and Frederiksen, R.A. 2001. Prevalence of *Sorghum* diseases in Karnataka and Maharashtra States, India during 2001 rainy season. International Sorghum News Letter, 42: 40-44.

Rajasab, A.H., Hosmani, Narayan and Kandan, Ganesh. 2004. Biology and Epidemiology of *Phoma sorghina*—the grain mold pathogen of *Sorghum*. Indian J. Mycol. Plant Pathol, 34: 606-608.

Ramachander Rao, K.S. 1993. *Alternaria* Leaf Spot of Sunflower I: Epidemiology and airborne conidia. Indian J. Aerobiol, 6: 45-48.

Ramachander Rao, K.S. and Tilak, S.T. 1990. Meteorology and spore load of *Alternaria* over a sunflower field. Oil Crops News Letter, 7: 34-38.

Ramalingam, C.G.K. and Nair, J. 1994. Comparative study of the airpora at an urban complex in Hyderabad over a period and its significance. Geophytology, 23: 343-347.

Rane, A.M. and Gandhe, R.N. 2005. Air and dust mycoflora of a library at Jalgaon. Indian J. Aerobiol, 18: 95-101.

Roy, M. and Trivedi, R.N. 1996. Effects of meteorological variables on the incidence of fungal spores in the atmosphere in Patna, pp. 122-142. In: Dimensions of Safe Environment. R.N. Trivedi and M. Roy (eds.). Annual Publications Pvt. Ltd., Delhi, India.

Sahaney, M. and Purwar, A. 2002. Incidence of fungal airspora in the market area of Allahabad. Indian J. Aerobiol, 15: 32-45.

Sahaney, M. and Purwar, A. 2001. Aeromycoflora inside a library of Allahabad University. Indian J. Aerobiol, 14: 20-23.

Sahu, S.K. 1995. A comparative study on aeromycoflora of *Onion cepa* L. Plant Flora & Fauna, 35: 38.

Sahu, S.K. 1996. Fungal association of leaf surface of some pulse crops in relation to aeromycoflora. Indian J. Aerobiol, 9: 23-28.

Santra, S.C. and Chanda, S. 1989. Airborne fungal flora in indoor environments of the Calcutta metropolis, India. Grana, 28: 141-145.

Saoji, A.A. 2006. *Aspergillus*: An important aeroallergen, pp. 537-544. In: Advancing Frontiers of Ecological Researches in India. A.K. Kandya and A. Gupta (eds.). Bishen Singh Mahendra Pal Singh Publishers and Distributors, Dehradun, India.

Saoji, A.A. and Giri, S.K. 1997. Concentration of aeroallergenic fungal spores in intramural environment of a Nagpur city hospital ward and library, pp. 211-218. Proc. 5th Intl. Conf. Aerobiol, Bangalore 1994.

Saoji, A.A. and Sawane, A.M. 2001. Control of post harvest green mould of citrus fruits: A comparative study. Indian J. Aerobiol, 14: 16-19.

Sarma, G.C. and Bora, K.N. 1996. Spore content of the atmosphere of a vegetable market at Gauhati, Assam, pp. 165-170. In: Environment & Aerobiology. A. K. Jain (ed.) Research Periodicals & Book Publishing House, Texas, USA.

Sarma, R.B. and Basumatary, S. 2004. Airborne fungal spores in a library and their association with paper and leather materials. Indian J. Aerobiol, 17: 38-43.

Sawane, A.M. and Saoji, A.A. 2004. Airborne *Penicillium* in fruit markets: one of the post harvest decay agent. Indian J. Aerobiol, 17: 25-31.

Shah, A. 1997. Allergic bronchopulmonary Aspergillosis: A clinical profile. Indian J. Aerobiol, 10: 7-8.

Shaikh, S.K. and Talde, U.K. 1994. Airborne fungal spores observed over *Vitis vinifera* L. Indian J. Aerobiol, 7: 28-31.

Sharma, D. and Dutta, B.K. 2001. Observations on the aeromycoflora of indoor working environments of Greater Silchar. Indian J. Aerobiol, 14: 8-15.

Shastri, S.D. 1998. Atmospheric concentration of hyphal fragments and insect parts inside a vegetable market, pp. 212-219. In: Environment and Aerobiology. A.K. Jain (ed.). Research Periodicals and Book Publishing Co., Houston, Texas, USA.

Shrivastava, P. 2007. Clinical and Immunological Study of Some Airborne Allergic Fungi in Poultry Workers' Environment of Jabalpur. Ph.D. Thesis, R.D. University, Jabalpur, India.

Shrivastava, A.K. and Wadhwani, K. 1992. Dispersion and allergenic manifestations of *Alternaria* airspora. Grana, 31: 61-66.

Shukla, P. 1987. Aeromycoflora of Cobbler Shops and Tanneries in Uttar Pradesh and Their Allergenic Potentialities. Ph.D. Thesis, University of Garhwal, Srinagar, U.P.

Shukla, P. and Misra, J.K. 1984. Fungi in the environs of occupational importance: I. Cobbler Sop. Bio Mem, 9: 95-97

Singh, N.I. 2006. Aerobiology, epidemiology and forecasting of fungal diseases found in certain crops of North-Eastern India. Indian J. Aerobiol, 19: 12-18.

Singh, N.I. 2007. Airborne mould biodiversity of certain indoor workplace Environments in (North-Eastern) India. Indian J. Aerobiol, 20: 9-16.

Singh, I.N. and Dorycanta, H. 1992. Aerobiology and crop disease in Manipur VIII. Fungal airpsora over a maize field in Senapati District. Indian J. Aerobiol, Special Vol. 141-144

Singh, A. and Singh, A.B. 1994. Airborne fungi in a bakery and the prevalence of respiratory disfunction among workers. Grana, 33: 349-358

Singh, A. and Singh, A.B. 1996. *Aspergillus* as an important environmental risk factor among suspectible individuals. Indian J. Aerobiol, 9: 43.

Singh, W.M and Singh, I.N. 2000. Airborne fungi from indoor air of a poultry farm at Lamangdong, Bishnupur Dist. Manipur. Indian J. Aerobiol, 13: 20-24.

Singh, A.B. and Kumar, P. 2003. Aeroallergens in clinical practice of allergy in India. An overview. Ann. Agric. Environ. Med, 10: 131-136.

Singh, A.B. and Pandit, T. 2003. Aerial Fungal Biodiversity and Allergic Diseases, pp. 741-768. In: Frontiers of Fungal Diversity in India (Prof. Kamal Festschrift). G.P. Rao, C. Manoharachari, D.J. Bhat, R.C. Rajak, T.N. Lakhanpal (eds.). International Book Distributing Co., Lucknow, India.

Singh, A.B., Chatterji, M., Singh, B.P. and Gangal, S.V. 1990. Airborne viable fungi in a library before and after agitation of books. Indian J. Aerobiol, 3: 32-38.

Singh, A., Singh, A.B., Bhatnagar, A.K. and Gangal, S.V. 1990. Relevance of Aspergilli in a bakery. Indian J. Aerobiol, 3: 15-21

Singh, A., Gangal, S.V. and Singh, A.B. 1994. Airborne fungi in the hospitals of metropolitan Delhi. Aerobiologia, 10: 11-21.

Singh, A., Ganguli, M. and Singh, A.B. 1995. Fungal spores are important components of library air. Aerobiologia, 11: 231-237.

Singh, A.B., Gupta, S.K., Pareira, B,M.J. and Prakash D. 1995. Sensititzation to *Ganoderma lucidum* in respiratory allergy patients in India. J. Clin. Expt. Allergy, 25: 440-447.

Singh, A., Singh, A.B. and Pandit, T. 1995. Prevalence of culturable and non-culturable airborne fungi in a grain storage in Delhi. Aerobiology, 11: 172-182.

Singh, A., Bhatnagar, A.K. and Singh, A.B. 1998a. *Scopulariopsis brevicaulis*. A potential environment allergen in poultries. pp. 11-124. Proceeding 7th National Conf. on Aerobiology Environment, A.K. Jain (ed.).

Singh, A., Prakash, D. and Singh, A.B. 1998b. Sensitization to different species of *Aspergillus* in bakery workers and general population. Asian Pacific J. Allergy and Immunol, 16: 5-15.

Singh, A., Singh, A.B. and Pandit, T. 1999. Respiratory disorders among occupational workers in India. A cross sectional epidemiological study. Ann. Agri. Environ. Medicine, 6: 1-11.

Singh, W.M., Singh, Kh.R., Sharma, M.H. and Singh, N.I. 2000. Airborne fungi from indoor air of a poultry farm at Lamangdong, Bishnupur district, Manipur. Indian J. Aerobiol, 13: 20-24.

Singh, D., Dutta, B.K. and Singh, A.B. 2004. Studies on seasonal and annual variations in the atmospheric pollen and fungal spores of Greater Silchar, Assam. Indian J. Aerobiol, 17: 1-11.

Sinha, A., Singh, M.K. and Raj Kumar. 1997. Aerofungi—an important atmospheric biopollutant at Jamshedpur. Indian J. Aerobiol, 11: 19-23.

Sinha, S.S., Mishra, K.B. and Singh, K.P. 1998. Aerobiology in Bihar—A review, pp. 59-68. In: Environment and Aerobiology. A.K. Jain (ed.) Research Periodicals and Book Publishing House, Houston, Texas, USA.

Suman, S.N., Roy, P.M. and Singh, S.P. 1992. Circadian rhythms in some common airspora components of Hajipur atmosphere. J. Mendel, 9: 153-154.

Sumbali, G. and Badyal, K. 1991. Relationship between fungal airspora of fruit shops and incidence of fruit rots. Indian Phytopath, 44: 214-218.

Surekha, M. and Reddy, S.M. 1996. Aeromycoflora of poultry shed in relation to

incidence of mycotoxigenic fungi. J. Palynol, 32: 135-141.
Tilak, S.T. 1982. Aerobiology. Vaijayanti Prakashan, Aurangabad, India.
Tilak, S.T. 1989a. Atlas of Airborne Pollen grains and Fungal Spores. Vaijayanti Prakashan, Aurangabad, India.
Tilak, S.T. 1989b. Aerobiology. Satyajeet Prakashan, Pune, India.
Tilak, S.T. 1990a. Airborne spores as bioindicators, pp. 227-236. In: Prof. G.P. Agrawal Festschrift. Today and Tomorrow's Printer, New Delhi, India.
Tilak, S.T. 1990b. Fungal spores and allergy. Currents Perspective in Palynological Research, Silver Jubilee Vol. J. Palynology, 91: 369-386.
Tilak, S.T. 1994. Aerobiology and cereal crop diseases, pp. 329-354. In: Rev. Trop. Pathol. S.P. Raychaudhary and J.P. Verma (eds.) Today and Tomorrow's Printer and Publishers, New Delhi, India.
Tilak, S.T. 1996a. Plant disease forecasting, pp. 235-263. In: IPM System in Agriculture. R.K. Upadhyay, K.G. Mukerjee and R.C. Rajak (eds.), Aditya Books (Pvt.) Ltd., New Delhi, India.
Tilak, S.T. 1996b. Relevance of Aerobiology to Phytopathology, pp. 225-231. Proc. Nat. Symposium-Microbial Biotechnology. S.M. Reddy (ed.). Scientific Publishers, Jodhpur, India.
Tilak, S.T. 1998. Aeroallergen and Human Health. A Report to Ministry of Environment. Govt. of India, New Delhi, 142 pp.
Tilak, S.T. 2005. Memoirs of an Aerobiologist. India J. Aerobiol, 18: 57-59.
Tilak, S.T. and Babu, M. 1984. Aerobiology and epidemology of certain diseases of bajra. Indian J. Aerobiol, 7: 77-83.
Tilak, S.T. and Pillai, S.G. 1988. Fungi in a library: an aerobiological survey. Indian J. Aerobiol, 1: 92-94
Tilak, S.T. and Pande, B.N. 1989. Aerobiology and epidemiology of Bajra diseases, pp. 79-96. In: Environmental Ecology. S.T. Tilak (ed.), Today and Tomorrow's Printers and Publishers, New Delhi, India.
Tilak, S.T. and Pande, B.N. 1997. Biodeteriogens inside a library, pp. 177-189. In: Aerobiology-Proc. 5th International Conf., Bangalore 1994. S.N. Agashe (ed.). Oxford & IBH Publishing Co. Pvt. Ltd., New Delhi, India.
Tilak, S.T. and Pande, B.N. 2004. Role of environmental biopollution in health hazards of domestic animals, pp. 270-283. In: Environmental Pollution and Management. P.C. Tiwari (ed.). Aaviskar Publications, Jaipur, India.
Tilak, S.T. and Pande, B.N. 2005. Current trends in aeromycological research in fungi, pp. 481-510. In: Fungal Diversity and Biotechnology. M.K. Rai and S.K. Deshmukh (eds.), Scientific Publishers, Jodhpur, India.
Tilak, S.T. Naik, M.N. and Pande, B.N. 1999. Aeromycological studies over groundnut fields at Aurangabad. 10th Nat. Conf. on Aerobiol and its Appl, India, 51 p.
Udaya Prakash N.K. and Vittal, B.P.R. 2003. Aerometrical study on thermophilous fungi in a farm house, Chennai. Czech Mycol, 55: 253–259.
Udaya Prakash, N.K. 2004. Indoor Molds—Isolation and Identification. Author's own publication, Chennai, India.
Uddin, A. and Chakravarty, R. 1995. Airspora measure in a paddy field in West Bengal. Grana, 34: 345-349.
Upadhyaya, H. and Jain, A.K. 2005. Fungal flora inside the library environment

of Jiwaji University, Gwalior. Indian J. Aerobiol, 18: 6-11.
Verma, K.S. and Chile, S. 1991. Allergenic fungal spores in certain environments at Jabalpur. J. Mendel, 8: 39-42.
Verma, K.S. and Bhandari, S. 1992. Incidence of *Aspergillus* inside poultry sheds at Jabalpur. Indian J. Aerobiol, 5: 46-47.
Verma, K.S. and Chile, S. 1992a. The airspora of hospital wards in relation to bronchial allergy. J. Palynology, 28: 99-106.
Verma, K.S. and Chile, S. 1992b. Fungi in the Medical College of the Jabalpur city and the allergenic behaviour of some species. J. Indian Bot. Soc, 71: 247-249.
Verma, K.S. and Chile, S. 1992c. Fungi in the medical college of the Jabalpur city and the allergenic behaviour of some species. J. Indian Bot. Soc, 4: 13-18.
Verma, K.S. and Chile, S. 1992d. The airspora of a hospital ward in relation to bronchial allergy. J. Palynol, 28: 99-106.
Verma, K.S. and Chile, S. 1993. Environmental distribution of clinically important fungal spores at various locations of Jabalpur. J. Econ. Tax. Bot, 17: 385-390.
Verma, K.S. and Chile, S. 1997. Allergenicity of certain airborne fungal spores at Jabalpur with reference to selected fungi in etiology of respiratory allergy. Res. J. Indian Bot. Soc, 231-232.
Verma, K.S. and Shrivastava, P. 2004. Airborne fungi of poultry environment. Advances in Pollen Spore Research, 22: 93-94.
Verma, K.S. and Shrivatsava, P. 2005. Assessment of mould allergy in poultry workers with special emphasis to fungal concentration in their working areas. Viselesana, 9: 65- 70.
Verma, K.S. and Vidyanidhi, G. 2001. Aero Allergenic Moulds. In: Contemporary Trends in Biological Sciences. S.C. Pathak Festschrift. R.D. University, Jabalpur, India.
Verma, K.S. and Vidyanidhi, G. 2003. Importance of IgG level diagnosis for immune therapy in allergic patients. Viselesana, 8: 57-62.
Verma, K.S., Maini, H., Shrivatsava, P. and Verma, Pratibha. 2006. Evaluating poultry workers susceptibility against fungal aeroallergens using Andersenn and Rotorod sampler. J. Phytol. Res, 19: 15-18.
Verma, S., Sen, S. and Kaur, M. 2000. Survey of microbial aeroallergens at Solan-Himachal Pradesh. Indian J. Aerobiol, 13: 10-13.
Vittal, B.P.R. 2004. Indian contributions to aerobiology of fungal plant pathogens. An overview. In: Biotechnological Approaches for the Integrated Management of Crop Diseases. Mayee et al. (eds.), Daya Publication, New Delhi, India.
Vittal, B.P.R. 2005. Progress of aerobiology in India during the last quarter century —An overview. Indian J. Aerobiol, 18: 60-68.
Wadhwani, K. 1994. Air fungal spore and working environment. Some medical aspects, pp. 271-277. In: Advances in Mycology and Aerobiology. Dr. S.T. Tilak, Commemoration Volume. U.K. Talde (ed.). Today and Tomorrow's Printers and Publishers, New Delhi, India.

3

Fungi in Saline Water Bodies with Special Attention to the Hypersaline Dead Sea Mycobiota

Asya S. Buchalo[1], Solomon P. Wasser[2] and Eviatar Nevo[3]

[1]N.G. Kholodny Institute of Botany,
National Academy of Sciences of Ukraine,
2 Tereshchenkivska St., 01601 Kiev, Ukraine

[2]Institute of Evolution and Department of Evolutionary and Environmental Biology, Faculty of Science and Science Education, University of Haifa, Mt. Carmel, Haifa 31905, Israel

[3]Institute of Evolution, University of Haifa, Mt. Carmel, Haifa 31905, Israel

Abstract

The number of fungi documented to date in saline water bodies is about 1500 species including 444 spp. belonging to the higher obligate marine mycota (Ascomycota–81%; Basidiomycota–2%; Mitosporic fungi–17%). Marine fungi were isolated from water and a wide variety of substrates: wood, sediments, mud, algae, corals, mangrove leaves, and other ligno-cellulose substrata. Fungi isolated from sediments are usually fast-growing species (e.g. *Aspergillus, Penicillium*), which inhabit soils also. Some marine fungi are cosmopolitan. Many fungal species isolated from salt water bodies showed high osmotolerance and remained viable under hypersaline conditions. During 1995-2006, 70 filamentous species were isolated, from one of the most saline lakes on earth, the Dead Sea water (340-350 g salt per liter) that belonged to 26 genera, 10 orders and 3 divisions (Oomycota, Zygomycota, and Ascomycota). The Dead Sea is located in the Syrian-African rift valley on the border between

Israel and Jordan. Many species were isolated irregularly over space and time. The low similarity between species richness at different localities indicated that most of the diversity observed is periodic and they are not the common inhabitant of the sea. The species *Aspergillus versicolor, Eurotium amstelodami, E. herbariorum, Cladosporium cladosporioides*, and *C. sphaero-spermum* had the highest spatial and temporal frequency of occurrence. These species probably form a stable core of the community. In saline water bodies, fungi evolve a number of metabolic strategies, which allow them to adapt to extreme environmental conditions: to tolerate salt stress; to use a wide variety of carbon substrates including inorganic carbon that are present at very low concentrations; to grow not only aerobically, but also microaerophilically or anaerobically. It seems scientifically unjustified to classify fungi isolated from salt water bodies as 'true' and 'occasional' simply because the latter are also distributed in other habitats. It is possible to conclude that the ecological situation in salt water bodies corresponds to the nutritional and ecological demands of fungal organisms. Saline as well as hypersaline aquatories may be considered econiches for fungi including habitats other than marine.

1. Distribution and taxonomy of fungi in saline water bodies

Water bodies represent about 90% of the biosphere volume on Earth. Most are characterized by a high concentration of salts, mainly sodium chloride. Many fungi adapt to the presence of salt even up to quite high concentrations. A large group of obligate and facultative halophilic marine fungi inhabit saline and brackish water environments. Marine mycology has evolved as a specialized field since 1940, and its studies have contributed to our knowledge of the taxonomy, ecology, and physiology of marine fungi (Sparrow, 1934; Barghoorn and Linder, 1944; Johnson and Sparrow, 1961; Hughes, 1975; Waguri, 1976; Kohlmeyer and Kohlmeyer, 1979; Artemchuk, 1981; Moss, 1986; Booth and Kenkel, 1986; Austin, 1988; Kohlmeyer and Volkmann-Kohlmeyer, 1991; Hyde and Pointing, 2000). Marine fungi are not taxonomically defined, but are characterized ecologically and physiologically. There is no consensus today as to the definition of the term 'marine fungi'. According to Kohlmeyer and Kohlmeyer (1979), obligate marine fungi are those that grow and sporulate exclusively in marine or estuarine habitats; facultative marine fungi are those living in a freshwater or terrestrial milieu, but are able to grow and possibly also sporulate in the marine environment. At the 7th International Marine and Freshwater Mycology Symposium held in Hong Kong in 1999 it was proposed to use terms such as maritime, halotolerant, estuarine, resident, and native in addition to marine fungi. A further suggestion was that the ability to germinate and to form mycelium under natural marine conditions should be used as criterion for the definition of a marine fungus. Today, the majority of fungi classified as obligate marine fungi are parasites on plants and animals, as symbionts in lichenoid association with algae, and as saprobes on dead organic matter of plant or animal origin. In their life cycle, these fungi are associated with organisms that constantly live in oceans and estuaries. However, most marine-occurring fungi are facultative halophiles that grow in freshwater or terrestrial environments in addition to the marine environment. Such not 'truly marine fungi' represent a very large group, mainly of Ascomycota and Mitosporic fungi as well as Oomycota, Zygomycota, and Basidiomycota. The most abundant representatives of this group belong to the genera *Aspergillus, Penicillium, Cladosporium, Chaetomium, Acremonium, Alternaria, Fusarium, Mucor, Absidia, Rhizopus*, etc. Many authors have recorded the presence of these fungi in the different geographic regions of the world; in sediments of ocean, mud, submerged wood, and other plant residues (Cronin and Post, 1977; Artemchuk, 1981; Austin, 1988; Hyde and Pointing, 2000).

The latest estimate of the number of marine fungi is 1500 species, not including those from lichens (Hyde et al., 1998). In *Marine Mycology* (Hyde and Pointing, 2000), a list of 444 species of fungi belonging to the higher marine mycota is given (Table 1), and the dominance of Ascomycota among the groups of marine fungi is documented (81%). The rest are Basidiomycota (about 2%) and Mitosporic fungi (17%).

Table 1. Numbers of higher marine fungi (according to Hyde and Pointing, 2000)

Group	*Genera*	*Species*
Ascomycetes	177	360
Basidiomycetes	7	10
Ceolomycetes	23	28
Hyphomycetes	28	46
Total	235	444

The predominance of sexual Acomycetes in saltwater bodies can be explained by analogy with soil fungi. These, in contrast to asexual fungi, are mainly stress-selected, occupying environments with various stress levels that exclude many asexual species (Dix and Webster, 1995; Grishkan et al., 2002). Recent molecular studies have shown that marine ascomycetes originated from terrestrial ancestors (Hyde and Pointing, 2000). Taxonomic groups of obligate marine fungi are presented in table (Table 2).

Species diversity of marine fungi is controlled by an amalgam of interacting factors: effect of habitats, availability of substrate for colonization,

Table 2. Filamentous fungi recovered from the marine environment (according to Austin, 1988)

Sub-division	*Class*	*Family*	*Genus*
Ascomycotina	Discomycetes	Orbiliaceae	*Orbilia*
	Plectomycetes	Eurotiaceae	*Amylocarpus, Eiona*
	Pyrenomycetes	Laboulbeniaceae	*Laboulbenia*
		Physosporellaceae	*Buergenerula*
		Spathulosporaceae	*Spathulospora*
		Diaporthaceae	*Gnomonia*
		Halosphaeriaceae	*Aniptodera, Bathyascus, Carbosphaerella, Ceriosporopsis, Chadefaudia, Corollospora, Haligena, Halosarpheia, Halosphaeria, Lignincola, Lindra, Lulworthia, Nais, Nautosphaeria, Trailia.*
		Hypocreaceae	*Halonectria, Heleococcum, Hydronectria, Nectriella*
		Polystigmataceae	*Haloguignardia, Phycomelaina*
		Sordariaceae	*Biconiosporella, Zopfiella*
		Sphaeriaceae	*Abyssomyces, Chaetosphaeria, Pontogeneia*
		Verrucariaceae	*Pharcidia, Turgidosculum*
	Incertae sedis		*Oceanitis, Ophiobolus, Savoryella Torpedospora*
	Loculoascomycetes	Herpotrichiellaceae	*Herpotrichiella*
		Mycoporaceae	*Leiophloea*
		Mycosphaerellaceae	*Didymella, Mycosphaerella*

Contd.

Table 2 continued

Sub-division	*Class*	*Family*	*Genus*
		Patellariaceae	*Bankegyia, Kymadiscus*
		Pleosporaceae	*Didymosphaeria, Halotthia, Helicascus, Keissleriella Leptosphaeria, Manglicola, Massarina, Microthelia, Paraliomyces, Phaeosphaeria, Pleospora, Pontoporeia, Thalassoascus, Trematosphaeria*
	Incertae sedis		*Crinigera, Orcadia, Sphaerulina*
Basidiomycotina	Gasteromycetes	Melanogastraceae	*Nia*
	Hymenomycetes	Corticiaceae	*Digitatispora*
		Incertae sedis	*Halocyphina*
	Teliomycetes	Tilletiaceae	*Melanotaenium*
Deuteromycotina	Hyphomycetes	Agronomycetaceae	*Papulaspora,*
		Moniliaceae	*Blodgettia, Botryopphilalophora Clavatospora, Varicosporina*
		Dematiaceae	*Asteromyces, Cirrenalia, Cladosporium, Clavariopsis, Cremasteria, Dendryphiella, Dictoyosporium, Drechslera, Humicola, Monodictys, Orbimyces, Periconia, Sporidesmium, Stemphylium, Trichocladium, Zalerion*
		Tuberculariaceae	*Allescheriella, Tubercularia*
	Coelomycetes	Excipulaceae	*Dinemasporium*
		Melanconiaceae	*Sphaceloma*
		Sphaerioidaceae	*Ascochyta, Ascochytula, Camarosporium, Coniothyrium, Cytospora, Diplodia, Macrophoma, Phialophorophoma, Phoma, Rhabdospora, Robillarda, Septoria, Stagonospora*

temperature, salinity, inhibition competitions, dissolved organic nutrients, hydrogen ion concentration, osmotic effects, oxygen availability, pollutants, vertical zonation, hydrostatic pressure, tidal amplitude, light, etc. (Jones, 2000).

Marine fungi grow on a wide variety of substrates from wood to sediments, mud, soils, sand, algae, corals, calcareous tubes, decaying mangrove leaves, and more. Members of Halosphaeriaceae predominate on wood in the open ocean and Loculoascomycetes in mangrove habitats (Hyde, 1989; Jones, 2000). Many marine fungi also sporulate on sand grains and hard material such as coral (Jones and Mitchell, 1996). Arenicolous fungi are generally found on sand associated with wood from which they derive their nutrients.

Marine fungi can live within a wide temperature range, with optimum conditions for most species 12-25°C; some species are thermotolerant or psychrophilic. Temperature plays a major role in the geographical distribution of marine fungi particularly with the species that are typically tropical (*Antennospora quadricornuta* and *Halosarpheia ratnagiriensis)*, temperate (*Ceriosporopsis trullifera* and *Ondiniella torquata*), and arctic (*Spathulospora antartica* and *Thraustochytrium antarticum*) while others are cosmopolitan (*Ceriosporopsis halima* and *Lignincola laevis*). Collections of some fungi, such as *Digitospora marina*, during the winter months when water temperatures are below 10°C, indicate their seasonality (Jones, 2000). Yeasts and filamentous fungi are able not only to survive, but live actively and complete their life cycle under deep sea conditions and possibly play an important role in this habitat (Zaunstöck et al., 1994; Lorenz and Molitoris, 1997). The authors showed that some yeasts and filamentous fungi (*Aspergillus niger*) that tolerate pressures 20-40 Mpa can also grow and germinate after incubation at 10 Mpa (about 100 atmospheres, corresponding to the pressure at a depth of 1 km). *Aspergillus ustus* and *Graphium* sp., isolated from a depth of 860 and 965 m, were able to germinate, grow, and sporulate at a pressure of 100 bar (Raghukumar and Raghukumar, 1998).

Some marine fungi are cosmopolitan (Jones and Mitchell, 1996) and are more or less common in temperate and tropical seas. In most cases, the occurrence of a fungus in a particular habitat is related to water temperature and availability of substrates. Biogeographically, the marine fungi can be divided into two major groups: pan temperate and pan tropical. In the regions between the tropics and subtropics, the composition of the mycobiota depends on water temperature and salinity rather than on air temperature. Most marine fungi have been obtained from substrates containing lignocellulose: mostly soft rot and white rot degraders of wood. Peroxidase and laccase activity was detected in some isolates (Pointing et al., 1998). In contrast, exposed shores or depauperate habitats support few fungi.

Little is known about the role of marine fungi in sediments and in the decaying or dead animal parts. When fungi are isolated from marine sediments, typical fast-growing weedy species (e.g., *Aspergillus, Penicillium*) are usually recovered.

Extreme habitats generally sustain low species diversity, and only a few fungi are able to grow in saturated NaCl solutions. At a still lower water potential, the active mycobiota is dominated by species of *Aspergillus* and *Penicillium* (Griffin and Luard, 1979). Some species of the genera *Polypaecilium* and *Basipetospora,* which are commonly encountered on salted fish, can be cultured in saturated salt solutions (Wheeler et al., 1988).

It was shown that in *Penicillium ochrochloron, Emericella nidulans,* and in some other species of filamentous fungi the internal Na^+ concentration increased at high salinity, and Na^+ became as prevalent as potassium; the intracellular K^+/Na^+ ratio is about one (Gadd et al., 1984; Beever and Laracy, 1986). It has become widely accepted that yeasts and fungi may be halotolerant but not obligatory halophilic. In some cases, they show better growth under conditions of moderate to high salt. *Torulopsis candida,* a yeast isolated from Arabian Gulf seawater, grew best in 2 M NaCl at 37°C. Radwan et al. (1984) found several strains of *Penicillium* that grew optimally in 9-10% salt and one strain of *P. notatum* that grew optimally in 10-15% NaCl. Andrews and Pitt (1987) described several xerophilic fungi that grew better on NaCl than on sugars including *Polypaecilium pisce, Exophiala werneckii,* and *Aspergillus wentii* (Table 3). However, there was no obligatory requirement for NaCl.

Some degree of halotolerance may be associated with temperature dependence. True salt dependence, or halophily, has been recorded in the yeast *Metschnikowia bicuspidata* var. *australis,* a parasite of the brine shrimp (*Artemia salina*) that lives in salt ponds with 10-12% NaCl (Phaff and Starmer, 1980). These authors reported that it only grew on medium supplemented with 10-12% NaCl. It is the only yeast reported with an obligatory salt requirement. Spencer et al. (1964) reported isolating a similar yeast (*M. kamienski*) from brine shrimp.

Jennings (1986) concludes that a combination of factors enables these fungi to grow in the sea; they can tolerate concentrations of ions present in seawater and prefer the alkaline pH of seawater.

Yeasts and fungi have also been documented from salt marshes of elevated salinity (Abdel-Hafez et al., 1977). Radwan et al. (1984) isolated four strains of *Penicillium* that grew optimally in about 10% NaCl with maximum salt tolerances ranging up to 15-30% salt (Table 3). Most samples collected were associated with the root zones of higher plants. The

Table 3. Salt tolerances of osmophilic fungi (according to Javor, 1989)

Fungus	*Maximum salt tolerance*	*Source*
Aspergillus spp.	≥ 5%	Desert soil
Aspergillus spp.	20-25%	Marine
A. repens	≥ 25%	Marine
A. ochraceus	Saturated	–
A. wentii	Saturated	Salted fish
A. penicilloides	Saturated	Salted fish
Basipetospora halophila	Saturated	Salted fish
Cladosporium sp.	36%	Great Salt Lake
Drechslera spp.	5%	Desert soil
Eurotium repens	25%	Salted fish
Exophiala werneckii	Saturated	Salted fish
Fusarium spp.	20-25%	Marine
Glomus fasciculatum	≥ 13%	Mycorrhizae
G. mosseae	≥ 11%	Mycorrhizae
G. etunicatus	≥ 10%	Mycorrhizae
Penicillium spp.	20-25%	Marine
P. notatum	25-30%	Desert soil
Polypaecilium pisce	25%	Salted fish
Ulocladium spp.	5%	Desert soil
Wallemia sebi	Saturated	Bread

numerically most common taxa cited were *Aspergillus* and *Penicillium* although several other genera were also reported.

Osmotolerant yeasts and fungi have also been directly isolated from seawater (Onishi, 1963; Norkrans, 1966; Pal et al., 1979). Fifty-six strains were isolated with varying degrees of salt tolerance. One-third of the isolates had a maximum tolerance of 10-15% NaCl, one-fifth of the strains tolerated 20-25% NaCl, and *Aspergillus repens* tolerated ≥ 25% NaCl (Javor, 1989). Ross and Morris (1962) isolated 10 species of yeasts from seawater, including *Debaryomyces kloeckeri* and *D. subglobosus* with a tolerance of 22-24% NaCl. The other isolated strains tolerated 9-22% NaCl. Species of filamentous fungi belonging to 30 genera of Mucorales, Ascomycota, and Mitosporic fungi were isolated from the sediments of the Black Sea (Artemchuk, 1981).

2. How do we define marine fungi?

The question is whether it is proper to divide fungi isolated from salt water bodies as obligate (true) and facultative (occasional), which are distributed in salt water bodies but also in soils and other habitats. The study of marine fungi has progressed along two research lines. The first most widely accepted approach was Kohlmeyer's direct observation method (Kohlmeyer and Kohlmeyer, 1979). This method permits observation of the fruiting stage of fungi inhabiting woody tissues of mangroves and driftwood and does not use any culture techniques. The taxonomy of marine fungi, mainly based on the works of Kohlmeyer and Kohlmeyer (1979) and of Kohlmeyer (1986), was compiled by Austin (1988) and that is presented here (Table 2). The second approach involves culture work with fungi that can be isolated from sediments, beaches, or under mangrove stands. These studies dealt only with filamentous fungi and Zygomycotina. As might be expected, the species of fungi reported by this method include none of those reported by Kohlmeyer, and these were not regarded as typical marine fungi. This was probably the reason why species that were isolated, thus far, from the Dead Sea, using the cultural method (Buchalo et al., 1998a, b; Molitoris et al., 2000; Buchalo, 2003), are excluded from the list of marine fungi (Hyde and Pointing, 2000), pending the answer to the question of how to define 'marine fungi'. The opinion of Ritchie (1954), rejecting *Aspergillus* as a marine form because it does not occur exclusively in the sea, is like rejecting the typhoid bacillus as a pathogen because it can live in water. One might, with equal justification, not accept *Aspergillus* as a terrestrial genus because it can be found in the ocean. A somewhat similar situation was obtained in soil mycology in the early part of the 20th century when the idea of a fungal biota of the soil was dubiously received until the presence of living hyphae in soil was clearly demonstrated. Sparrow (1943) emphasized that no clear distinctions exist between aquatic, amphibious, and terrestrial fungi. There is certainly active obligate and facultative mycobiota in the sea. Although some of the species are halophiles that can utilize different metabolic strategies and grow anaerobically, many are able to thrive and reproduce in salt water or out of it.

Marine environments represent a complex ecosystem with great variation in many parameters. To determine the true fungal diversity in salt water bodies new exploratory studies have been undertaken, in such habitats as rhizosphere of mangrove trees, soils and mud of coastal beaches, hypersaline waters, marine salterns, and mortalities of marine animals. Also, new isolation media and procedures need to be developed (Jones, 2000). It seems scientifically unjustified to classify fungal organisms isolated from marine and other salt aquatories as 'true' and 'occasional' simply because the latter are also distributed in other habitats besides marine habitats. The ecological situation in salt water bodies are met by the physiological and ecological demands of fungal organisms, and these water bodies can be considered a habitat available for fungi.

3. Adaptation of fungi to stressful living conditions in saline water bodies

There are some stress factors, in addition to salinity, that support fungal life in saline water bodies. Fungi continue to be regarded by many as exclusively aerobic organisms, but today there is evidence that strict anaerobiosis is widespread in fungi. It was shown that representatives of the genera *Geotrichum*, *Fusarium*, *Mucor*, etc., which have also been recorded in saline water bodies, can grow in a nitrogen atmosphere where they ferment glucose anaerobically (Wainwright, 1988). According to Javor (1989), the majority of osmophilic fungi grow facultatively as anaerobes, although they grow most rapidly under aerobic conditions. It is likely that anaerobic growth will allow certain fungi to resume aerobic growth directly as hyphae and mycelium, rather than having to go through the relatively long process of germination from spores (Wainwright and Killham, 1993). It was shown that at least some fungi are capable of growing anaerobically. Data exist on the mycelial growth of the soil fungus *Fusarium solani* both aerobically and anaerobically (Wainwright et al., 1993). Our investigation of fungi in saline paddy soils in the southern regions of the Ukraine showed that in water-logged conditions many representatives of soil mycobiota can grow under low tension of oxygen and maintain viability under prolonged anaerobic conditions. Although the total number of fungal isolations has increased, members of the genera *Penicillium*, *Aspergillus*, *Fusarium*, *Acremonium*, and *Cladosporium*, which are common in saline water bodies, were relatively more abundant (Buchalo, 1978, 1984).

Fungi are metabolically diverse organisms, which should be capable of growing through mechanisms other than strict heterotrophy. Such diverse metabolic strategies should allow certain species to grow in habitats lacking large amounts of available carbon. It appears that fungi are facultative oligo-carbotrophs. They can be grown using carbon-free media (Parkinson et al., 1990), yet they also grow sparsely when carbon is limited, but more luxuriously when large amounts of carbon become available. Many species of fungi can oxidize reduced forms of elements including sulphur, nitrogen, and manganese. Presumably they can grow chemolitho-autotrophically by fixing CO_2 and gaining energy from such oxidation reactions (Wainwright, 1988). Saline water bodies are generally considered to contain insufficient carbon to allow the continuous growth of fungi. Fungi are thought to subsist on endogenous metabolism or to lie dormant in these environments as spores. It was shown (Wainwright, 1988) that fungi may be capable of growing as oligotrophs, chemolithoheterotrophs or even as chemolitho-autotrophs. They participate widely in mineral cycling and are not restricted to the role of decomposer. It was shown that *Fusarium oxysporum* assimilated $^{14}CO_2$ under strict oligotrophic conditions; most of the assimilated ^{14}C was present in the wall/wall-associated membranes, and the CO_2 did not involve the Calvin cycle (Parkinson et al., 1990).

Fungi are undoubtedly very versatile heterotrophs, capable of efficiently scavenging nutrients from solution and from the atmosphere while augmenting their carbon supplies with CO_2. They may also be capable of gaining additional energy from processes such as sulphur oxidation and nitrification, or H_2 oxidation (Wainwright, 1988; Parkinson et al., 1990). Growing in this way, and using a number of substrates at the same time, there seems to be no reason why fungi should not grow in saline water bodies even in the absence of large amounts of available organic carbon substrates. Extraordinary adaptation to extreme stress factors is illustrated by the cosmopolitan soil fungus *Cladosporium cladosporioides*, which was also isolated from water, sediments, and wood of the Dead Sea and other hypersaline aquatic habitats (Javor, 1989; Buchalo et al., 2000 a, b; Kis-Papo et al., 2001, 2003 a, b). According to Zhdanova and Vasilvskaya (1982, 1988) *C. cladosporioides* is highly resistant to γ-radiation (up to 9.000 Gray), to hyper- and hypoxia (microaerophilic), and to a low concentration of carbon in nutrient media. This species is also able to consume carbon from air CO_2 and $NaHCO_3$. High resistance to different stress factors was also observed by the above authors for melanin-pigmented species belonging to the genera *Stemphylium*, *Aureobasidium*, *Stachybotrys*, etc., which were also recorded in the Dead Sea and other salt water bodies (Austin, 1988; Javor, 1989; Buchalo et al., 2000a, b; Kis-Papo et al., 2001, 2003a, b).

The myth that fungi are solely aerobic heterotrophs, whose only role in nature is the degradation of plant remains, has finally been dispelled (Mirocha and Devay, 1971; Wainwright, 1988). While the role of fungi in the carbon cycle remains of paramount importance, it is clear that they can participate more widely in nutrient cycling than generally recognized, and that they have evolved a number of metabolic strategies other than strict heterotrophy to allow them to adapt to various environmental conditions including extreme ones. Under extreme conditions of saline water environments, fungi will enjoy an advantage by being able to tolerate salt stress; to grow oligocarbotrophically; to use a wide variety of carbon substrates that are present at very low concentrations; to utilize substrates that contain inorganic carbon, e.g., CO and CO_2; to utilize metabolic strategies, e.g. chemolithoheterotrophy, or even possibly chemolitho-autotrophy; to grow aerobically, microaerophilically, or anaerobically, depending on the oxidation status of the habitat. In the natural environment a mixture of many substrates is often present at growth-limiting concentrations, and it might therefore be expected that, during evolution, microorganisms have been under a constant pressure to develop and improve their ability to utilize more than one substrate simultaneously (Kuenen, 1986). Such a strategy might have led to the development of metabolically highly versatile organisms. Fungi isolated from the Dead Sea can, without a doubt, be considered extremophiles, i.e., organisms living under extreme conditions at the edge of life (Madigan and Marrs, 1997; Buchalo et al., 1998a, b; Molitoris et al., 2000). These organisms do not merely tolerate their extreme living conditions, which are detrimental to most organisms. Remarkably, they indeed do best in these extreme habitats and in many cases require one or more extremes for reproduction. In other words, they have evolved unique adaptive evolutionary strategies to withstand these extreme living conditions, thereby becoming narrow extreme specialists. It is possible to conclude that the ecological situation in salt water bodies corresponds to the nutritional and ecological demands of fungal organisms, and saline aquatories may be considered an econiche available for fungi.

4. Adaptation of Dead Sea mycobiota to extreme environments

The Dead Sea is located in the Syrian-African rift valley, on the border between Israel and Jordan (Fig. 1). The Dead Sea is a harsh environment, a hypersaline (340-350 g salt per liter) desert lake even for those microorganisms best adapted to life at high salt concentrations. Among the hypersaline lakes, the Dead Sea is unique because of its peculiar Ca-chloride composition [i.e. $Ca/(SO_4+HCO_3) > 1$]. In most salt lakes worldwide Na^+ is the dominant cation, and the concentrations of divalent cations are relatively low (Javor, 1989; Oren, 2002). In the Dead Sea, divalent cations dominate (presently $Mg^{2+} + Ca^{2+} = 2.33$ M, as compared to $Na^+ + K^+ =$ 1.79 M, see also Table 3). Cl^- and Br^- are the dominant anions (99% and 1% of the anion sum, respectively), and concentrations of SO_4^{2-} and HCO_3^- are very low. Don Juan Lake (Antarctica) reportedly has an even higher salt concentration than the Dead Sea with 477 g/l total dissolved salts, most of it being $CaCl_2$. However, it probably does not support microbial life (for discussions see Javor, 1989 and Oren, 2002, 2003). The Dead Sea waters are slightly acidic, while most other hypersaline lakes have a neutral or alkaline pH.

The peculiar ionic composition of Dead Sea water, with its high concentration of the divalent cations magnesium and calcium, is highly inhibitory even to the most halophilic and halotolerant microorganisms. Since the first reports on the existence of an indigenous microbiota in the Dead Sea (Wilkansky, 1936; Elazari-Volcani, 1940; Volcani, 1944),

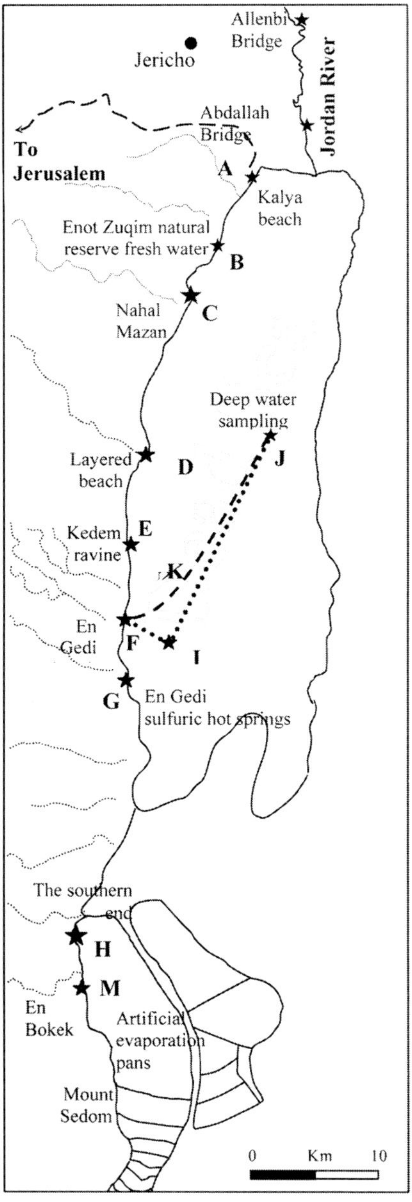

Fig. 1. The Dead Sea map, showing microfungi isolation sites. A depth profile of the water column was made in point J, the deepest part (–305 m) of the lake

knowledge of the microbiology of the lake has greatly increased. The main primary producer in the lake is the green alga *Dunaliella* (Oren, 1988, 1997, 2000, 2003). Development of *Dunaliella* blooms is followed by massive growth of halophilic *Archea* of the family *Halobacteriaceae.* Taking into consideration that the Dead Sea shore is sparsely vegetated, it contains only low amounts of organic material, derived partially from wood and other plant material entering with the periodic influx of freshwater from the Jordan River. Although prokaryotic microorganisms were considered to be the only decomposers in the Dead Sea, it has recently been suggested that fungi, long neglected as a component of the food web in the Dead Sea and in other hypersaline environments too, may also play a role (Buchalo et al., 1998a, b, 2000a, b; Molitoris et al., 1998; Kurchenko et al., 1998; Kis-Papo et al., 2001, 2003a, b; Wasser et al., 2003).

The first isolation of a fungus from the Dead Sea water column was reported by Kritzman (1973), who isolated an osmophilic yeast from the lake that grew in a medium containing 15% glucose + 12% salt. No further details were given, and unfortunately no cultures were preserved.

Between 1995 and 2006, a variety of fungi were isolated from the Dead Sea, from surface water at the shoreline and in the centre of the lake, as well as from deepwater samples. To date, 70 species belonging to 26 genera have been found, most of them belonging to the Ascomycota (66 spp.), but Oomycota (1 sp.) and Zygomycota (3 spp.) have also been encountered (Buchalo et al., 1998a; Kis-Papo et al., 2001, 2003a; Wasser et al., 2003). The predominance of Ascomycota in the Dead Sea as well as in other hypersaline water bodies is in accordance with the hypothesis that sexuality increases with the increase of saline drought stress (Grishkan et al., 2002; Wasser et al., 2003; Kis-Papo et al., 2003a, b). Most species identified were common soil fungi. Of the species isolated from the waters of the Dead Sea, *Aspergillus niger*, *A. terreus*, *Cladosporium cladosporioides*, *Thielavia terricola*, *Ulocladium chartarum*, *U. atrum*, *Penicillium brevicompactum*, *P. westlingii*, and *Sporothrix guttuliformis* had also been isolated from the soil around the lake including hypersaline desert soils and soils collected from the oases of Enot Zuqim and En Gedi (Steiman et al., 1995, 1997; Guiraud et al., 1995; Volz and Wasser, 1995; Volz et al., 1996; Kis-Papo et al., 2001, 2003a, b; Wasser et al., 2003).

Taxonomic analysis of Dead Sea mycobiota

No filamentous fungi had been recorded in the hypersaline waters of the Dead Sea prior to the discovery by Buchalo et al. (1998a). The first three species of filamentous fungi were isolated from the surface water samples: *Gymnascella marismortui* described as new to science, *Ulocladium chlamydosporum*, and *Penicillium westlingii*. Later, more species of filamentous fungi from the Dead Sea were discovered (Buchalo et al., 1998b, 1999, 2000a, b). Twenty-six species representing 13 genera of Zygomycota, Ascomycota (teleomorphic and anamorphic) were isolated from the Dead Sea till the end of 2000. Kis-Papo and the associates published 38 species of filamentous fungi isolated from the water samples from the surface to a depth of 304 m in the center of the sea (Kis-Papo et al., 2001).

At present, microfungal biota of the Dead Sea contains a total of 70 filamentous species (Wasser et al., 2003) and includes representatives from almost all of the main taxonomic groups (Table 4, Fig. 2).

The majority of Dead Sea fungi (66 species–94%) belong to the division **Ascomycota,** but only 12 of these species were found to have a teleomorphic (sexual) stage in their life cycle, and those species are—*Gymnascella marismortui*, a new fungal species, endemic to the Dead Sea, and three species of the strongly osmophilic genus *Eurotium*. All other ascomycete species were represented only by their anamorphs, and they are included in Ascomycota according to the modern fungal system. Among ascomycete microfungi, representatives of the order Eurotiales prevail (33 species, including the genera *Aspergillus* and *Penicillium*). The main species diversity is present in the genera *Aspergillus* (19 species including *Emericella*

Table 4. Systematic diversity of Dead Sea mycobiota

Division	*Order or class*	*Number of taxa*	
		Genera	*Species*
Oomycota	Pythiales	1	1
Zygomycota	Mucorales	2	3
Ascomycota	Dothideales	1	1
	Eurotiales	5	33
	Hypocreales	4	7
	Mycosphaerellales	1	5
	Onygenales	2	2
	Ophiostomales	1	1
	Pleosporales	5	9
	Sordariales	2	6
	Incertae sedis	2	2
	Total	23	66
Total		26	70

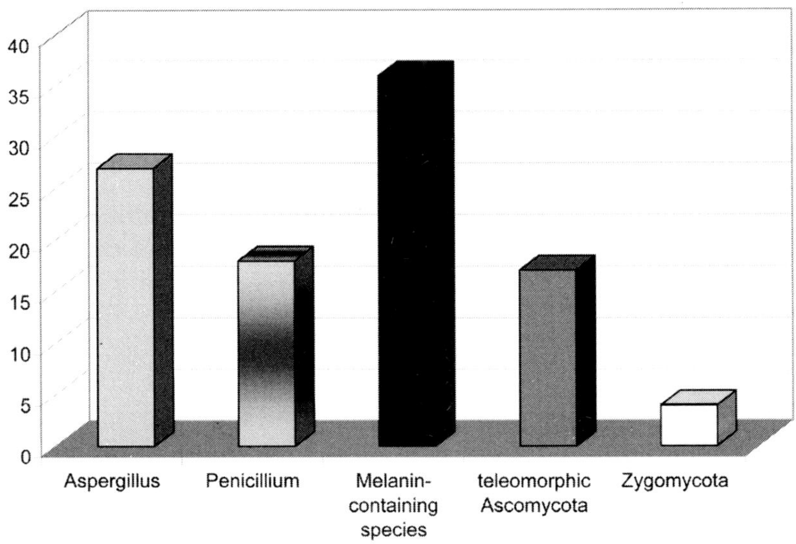

Fig. 2. Main groups of microfungi (% of general species number) in Dead Sea mycobiota (Wasser et al., 2003)

and *Eurotium* anamorphs), *Penicillium* (13), *Chaetomium* (5), and *Cladosporium* (5).

Zygomycota is represented by 3 species (4%). **Oomycota** rank last in species diversity, with only one species of the genus *Pythium*.

Among fungi isolated from the Dead Sea, 23% of the genera are common soil organisms, and 25% were also reported from the water of the Black Sea: *Absidia, Alternaria, Aspergillus, Chaetomium, Cladosporium, Penicillium, Stachybotrys*, and *Aureobasidium*. Representatives of the genera *Aspergillus, Penicillium, Chaetomium*, and *Cladosporium* were abundant in both water bodies. In contrast, species of the genera *Fusarium* and *Mucor*, which were abundant in the Black Sea (Artemchuk, 1981), have not been found in the Dead Sea.

One isolate, described as a new species, *Gymnascella marismortui* (Ascomycota), is a true halophile that grows well on agar media containing 50% Dead Sea water. Optimal growth was observed on agar media containing 10-30% (by volume) of Dead Sea water or in 0.5-2 M NaCl. Further investigations of this species showed that spores of *G. marismortui* remained viable for 4 weeks in undiluted and 12 weeks in 80% diluted Dead Sea water (Kis-Papo et al. 2003a, b). According to these authors (Kis-Papo et al., 2003a, b) not only spores of *Aspergillus versicolor* and *Chaetomium globosum* remained viable in undiluted Dead Sea water during 12 weeks of experiment, but also mycelium of these species lived for 8 weeks. Other isolated halotolerant fungi from the Dead Sea were able to grow in 50% Dead Sea water media: *Ulocladium chlamydosporum* (growing best at 3-15% NaCl at 26°C) and *Penicillium westlingii* (Buchalo et al., 1998a, b, 1999, 2000a, b; Molitoris et al., 1998).

Many fungi do not sporulate on freshly collected material but require a period of incubation. Prasannarai and Sridhar (1997) have shown that 70% of the fungi produced fruit bodies in incubation for 6 months, while others appeared after 12-18 months incubation (*Corollospora* sp. and *Dactylospora haliotrepha*). This is, therefore, a factor that must be taken into account while estimating the biodiversity of fungi on materials collected from the sea, especially when examining submerged and drift wood. Nine species of Ascomycota and Mitosporic fungi were isolated in the Dead Sea from pieces of submerged wooden constructions using sterile incubation in the moist chamber. In this case, it was anticipated finding living hypha directly on the surface of submerged wood. Species such as *Chaetomium aureum, Ch. nigricolor, Cladosporium cladosporioides*, and *Eurotium amstelodami* were also isolated from the Dead Sea water and from wood. *Ch. aureum, E. herbariorum*, and *Paecilomyces variotii* were discovered both on wood and in mud (Buchalo et al., 2000a). *Aspergillus ustus, Cladosporium cladosporioides, Chaetomium aureum, Ch. nigricolor, Eurotium herbariorum, Penicillium glabrum* (= *P. frequentans* Westl.) *Paecilomyces variotii*, and *Talaromyces stipitatum* (= *Penicillium stipitatum* Thom), isolated from wood, were also recorded in Israel from soils near the Dead Sea (Steiman et al., 1995; Kis-Papo et al., 2001, 2003b; Volz et al., 2001; Grishkan et al., 2003). It, therefore, was concluded that high salt tolerance of some fungi, which were in soil near the Dead Sea, as well as in water from this lake, may be the result of their adaptation to extreme environments (Kis-Papo et al. (2003a, b)). So the assumption was made that fungal mycelium can develop in the lake during periods of massive inflow of freshwater.

Yeasts and fungi are well suited to life in natural habitats of high salt content (Javor, 1989), and their lack of representation in the literature may not reflect their inability to colonize these environments, but rather the relatively little effort that has been spent on finding them. True halophily may occur among yeasts and fungi, but most studies show that there is no absolute requirement for NaCl among salt-tolerant osmophilic strains (Adler, 1996). Effects of increased salt concentration on fungal physiology are explained in terms of the effect of a more general factor, i.e. the water potential of the environment. The limiting value of salinity for fungal growth is not absolute but depends on nutrition, temperature, and the nature of the water potential adjusting solute. The physiology of adaptation of fungal ion transport and osmoregulation of salt-tolerant fungi to a concentrated environment has been reviewed by Blomberg and Adler (1993), Clipson and Jennings (1993), Tokuoka (1993), Mager and Varela (1993), and Adler (1996). According to Adler (1996) the response of fungi to salt stress involves the integrated function of diverse capacities. The intracellular physiology is protected from the external salinity primarily by an effective exclusion of Na^+ and by a compensatory accumulation of polyols, mainly glycerol, to achieve an internal environment that is suitable for enzyme function and growth under salt stress (Jin and Nevo, 2003).

Colonial growth on Petri plates as an indication of adaptation to the extreme conditions of the Dead Sea was tested using isolates obtained from the Dead Sea (*Gymnascella marismortui*, *Ulocladium chlamydosporum*, and *Penicillium westlingii*) at salinities ranging from 0 to 26% and temperatures from 15 to 35°C (Molitoris et al., 2000). The ascomycete *Gymnascella marismortui* strains isolated from this habitat (Buchalo et al., 1998a, b) are obligately halophilic, as they did not grow in freshwater medium and showed better growth with increasing salinity. The ionic composition of the medium had little effect on growth. *Gymnascella marismortui*, therefore, might be well adapted to life in the Dead Sea. The ascomycete *Penicillium westlingii*, a cosmopolitan soil fungus also isolated from the Dead Sea, was relatively indifferent in its response of colonial growth to salinity and ionic composition of the medium and cultivation temperature. This fungus, therefore, is halotolerant and thermotolerant and may thus be expected to be able to grow and propagate in the Dead Sea at sites of locally lowered salinity.

It was shown (Molitoris and Schaumann, 1986) that the group of marine isolates, which are normally terrestrial fungi and facultative marine as well as obligate marine fungi, possesses nitrate reductase activity in contrast to the investigated group of terrestrial fungi, which have the smallest percentage of nitrate-reductase-producing strains. These authors suggest that the group of marine isolates may, therefore, occupy an intermediate position between the obligate marine and the terrestrial fungi. The data obtained offer a deeper insight into the role of facultative marine fungi in the ecosystem of saline water bodies.

Physiological activities providing energy and metabolites for growth and other processes require the enzymes to be active under the prevailing conditions. A number of investigations have been made on enzyme activities of marine fungi (Molitoris and Schaumann, 1986; Rau and Molitoris, 1991; Schimpfhauser and Molitoris, 1991; Rohrmann and Molitoris, 1992). Hardly any data are available on enzyme activities of filamentous fungi in hypersaline habitats. Activities of enzymes involved in degradation and use of organic material were tested in fungal isolates of *Gymnascella marismortui*, *Ulocladium chlamydosporum* and *Penicillium westlingii* from the Dead Sea (Kurchenko et al., 1998; Molitoris et al., 2000). Enzymes are: amylase for degradation of starch (present in many plant tissues); caseinase as an enzyme degrading protein of plant or animal origin; cellulase, the enzyme splitting cellulose, the major component of wood, representing the ubiquitous substrate for fungi, also in marine habitats; and urease, an enzyme involved in the use of urea of animal origin as a nitrogen source. Generally, all four enzymes could be produced by all strains, at all temperatures and salinities, but enzyme production decreased with increasing temperature and salinity. In a few cases (*Gymnascella*: amylase, cellulase, urease; *Ulocladium*: amylase, caseinase) optimal enzyme production was observed at intermediate salinities and temperatures. Only for *Penicillium westlingii* did temperature and salinity have little or no influence on enzyme production. Since all strains investigated could grow and produce the enzymes at higher ranges of salinities and temperatures tested, they probably represented inhabitants of the hypersaline Dead Sea.

Physiological activity of fungi can conveniently be assayed according to decolourization of synthetic dyes. Such dyes have been widely used as model compounds to monitor the self-cleaning capacity of waters. Fungi have already been shown to be promising degraders of such dyes (Vyas and Molitoris, 1995). Among isolates from the Dead Sea, the best degraders were *Emericella nidulans*, *Ulocladium chlamydosporum*, and *Aspergillus phoenicis*. The ascomycete *Gymnascella marismortui* did not degrade any dye. Generally, dye-degradation capacity decreased with increasing salinity (Rawal et al., 1998).

While *Ulocladium* seems less adapted to temperatures and salinities of the Dead Sea, the *Gymnascella* strains seem much better adapted and may represent authentic members of this habitat. The cosmopolitan *Penicillium westlingii* strains show little growth differences, and they may inhabit the Dead Sea. Pitt (1975) showed that the genera *Aspergillus* and *Penicillium* have many species generally tolerant to low water ability. It is too early to claim on the basis of the available data that fungi are an important component of the heterotrophic community in the Dead Sea. However, the presence of halophilic and halotolerant fungi in the sea and their ability to proliferate in nutrient media containing a high (50% and higher) content of Dead Sea water, suggests that the potential for fungal activity in the sea does exist (Kis-Papo et al., 2003a, b). It was shown (Rau and Molitoris, 1991) that all marine fungi tested produced nitrate reductase (Na-R). Nitrogen often constitutes a limiting factor for growth of fungi. Natural seawater contains considerable amounts of nitrogen in the form of nitrate (up to 600 mg/L). Marine fungi that possess Na-R can use nitrate as a nitrogen source. Possession of Na-R would, therefore, constitute an important selective advantage explaining why all marine fungi investigated so far contain this enzyme.

There is also an important practical aspect to the knowledge of the mycobiota of salt inhabitants. Today, many products of the cosmetic industry include marine salt and mud, which may contain potentially hazardous species of fungi. Fifteen species were isolated representing seven genera of Zygomycota, Ascomycota, and Mitosporic fungi from Dead Sea mud. Of the isolated fungal cultures, 75-80% were identified as *Penicillium* and *Aspergillus* species; the other isolates were identified as species belonging to the genera *Mucor, Chaetomium, Cladosporium, Paecilomyces, Eurotium*, etc. Representatives of all these genera are known as a potential risk factor causing human diseases, and many of them can be classified as hazardous (Washburn, 1994; Vismer and Marasas, 1998).

The Dead Sea is a potentially an excellent model for studies of evolution under extreme environments and is an important gene pool for genetic engineering in agriculture in future. A MAPK gene from the Dead Sea fungus *Eurotium herbariorum* confers stress tolerance to lithium salt and freezing-thawing, indicating prospects for saline agriculture (Jin et al., 2005).

References

Abdel-Hafez, S.I., Moubasher, A.H., and Abdel-Fattah H.M. 1977. Studies on myco-flora of salt marshes in Egypt. IV. Osmophilic fungi. *Mycopathologia,* 63: 143-151.

Adler, L. 1996. Fungi and salt stress, pp. 217-234. In: Fungi and environmental change. J.C. Frankland, N. Magan, and M.G. Gadd, (eds.), Cambridge University Press, Cambridge, UK.

Andrews, S. and Pitt, J.I. 1987. Further studies on the water relation of xerophilic fungi, including some halophiles. J. Gen. Microbiol, 133: 233-238.

Artemchuk, N.Y. 1981. The mycoflora of the Seas of the USSR. Nauka, Moscow, (In Russian).

Austin, B. 1988. Marine Microbiology. Cambridge University Press, Cambridge, UK.

Barghoorn, E.S. and Linder, D.H. 1944. Marine fungi: their taxonomy and biology. Farlowia, 1: 395-467.

Beever, R.E. and Laracy, E.P. 1986. Osmotic adjustment in the filamentous fungus *Aspergillus nidulans*. J. Bacteriol, 168: 1358-1365.

Blomberg, A. and Adler, L. 1993. Tolerance of fungi to NaCl.pp. 209-232. In: Stress Tolerance of Fungi. D.H. Jennings (ed.), Marcel Dekker, New York, USA.

Booth, T. and Kenkel, N. 1986. Ecological studies of lignicolous marine fungi: a distribution model based on ordination and classification, pp. 297-310. In: The Biology of Marine Fungi. S.T. Moss (ed.), Cambridge University Press, Cambridge, UK.

Buchalo, A. 1984. Micromycetes of paddy soils, pp. 84-92. In: Micromycetes of Soils. V.J. Bilay (ed.), Naukova Dumka Press, Kiev (In Russian).

Buchalo, A.S. 1978. The influence of water-logging upon the paddy soil mycoflora in the Ukrane. Zentralbl. Bacteriol. Parasitenkund. Infectionskrankh. Hygiene II Abt., Bd. 133: 421-425.

Buchalo, A.S. 2003. Fungi in saline water bodies, pp. 141-161. In: Fungal Life in the Dead Sea. E. Nevo, A. Oren, and S.P. Wasser (eds.), A.R.G. Gantner Verlag K.-G., Ruggell, Switzerland.

Buchalo, A.S., Nevo, E., Wasser, S.P., Oren, A. and Molitoris H.P. 1998a. Fungal life in the extremely hypersaline water of the Dead Sea: First records. Proc. R. Soc. Lond. B, 265: 1461-1465.

Buchalo, A.S., Nevo, E., Wasser, S.P., Oren, A., Molitoris, H.P., and Volz, P.A. 1998b. Diversity of fungal life in the extremely hypersaline water of the Dead Sea, p. 180. In: Abstracts of the 6th Intern. Mycol. Congr., Jerusalem, Israel.

Buchalo, A.S., Wasser, S.P., Molitoris, H.P., Volz, P.A., Kurchenko, I., Lauer, I. and Rawal, B. 1999. Species diversity and biology of fungi isolated from the Dead Sea, pp. 283-300. In: Evolutionary Theory and Processes: Modern Perspectives. Papers in honour of Eviatar Nevo. S.P. Wasser (ed.). Kluwer Academic Publishers, Dordrecht, The Netherlands.

Buchalo, A.S., Nevo, E., Wasser, S.P., Oren, A., Molitoris, H.P. and Volz, P.A. 2000a. Fungi discovered in the Dead Sea. Mycol. Res, 104: 132-133.

Buchalo, A.S., Nevo, E., Wasser, S.P., and Volz, P.A. 2000b. Newly discovered halophilic fungi in the Dead Sea, pp. 239-252. In: Journey to Diverse Microbial Worlds. J. Seckbach (ed.), Kluwer Academic Publishers, Dordrecht, The Netherlands.

Clipson, N.J.W. and Jennings, D.H. 1993. *Dendryphiella salina* and *Debaryomyces hansenii*: Models for ecophysiological adaptation to salinity by fungi that grow in the sea. Can. J. Bot, 70: 2097-2105.

Cronin, A.E. and Post, F.J. 1977. Reports of a dematiaceous hyphomycete from the Great Salt Lake, Utah. Mycologia, 69: 846-847.

Dix, N.J. and Webster, J. 1995. Fungal ecology. Chapman and Hall, London. UK.

Elazari-Volcani, B. 1940. Studies on the microflora of the Dead Sea. Ph.D. Thesis. The Hebrew University of Jerusalem, Israel (In Hebrew).

Gadd, G.M., Chudek, J.A., Foster, R. and Reed, R.H. 1984. The osmotic responses of *Penicillium ochrochloron*: Internal solute levels in response to copper and salt stress. J. Gen. Microbiol, 130: 1969-1975.

Griffin, D.M. and Luard, E.J. 1979. Water stress and microbial ecology. pp. 49-63. In: Strategies of Microbial Life in Extreme Environments. M. Shilo (ed.), Vergal Chemie, Weinheim, Germany.

Grishkan, I., Nevo, E., and Wasser, S.P. 2003. Soil micromycete diversity in the hypersaline Dead Sea coastal area, Israel. Mycol. Progr, 2: 19-28.

Grishkan, J., Korol, A.B., Nevo, E. and Wasser, S.P. 2002. Ecological stress and sex evolution in soil microfungi. Proc. R. Soc. Lond. B, 270: 13-18.

Guiraud, P., Steiman, R., Seigle-Murandi, F., and Sage, L. 1995. Mycoflora of soil around the Dead Sea. II. Deuteromycetes (except *Aspergillus* and *Penicillium*). Syst. Appl. Microbiol, 18: 318-322.

Hughes, G.C. 1975. Studies of fungi in oceans and estuaries since 1961. I. Lignicolous, caulicolous and foliicolous species. Oceanography Mar. Biol. Ann. Rev, 13: 69-180.

Hyde, K.D. 1989. Vertical zonation of intertidal mangrove fungi, pp. 302-306. In: Recent Advances in Microbial Ecology T. Hattori, Y. Ishida, Y. Maruyama, R.Y. Morita, and A. Uchida, (eds.), Japan Sci. Soc. Press, Tokyo, Japan.

Hyde, K.D. and Pointing, S.B. 2000. Marine mycology: A practical approach. Fungal Diversity Press, Hong Kong.

Hyde, K.D., Jones, E.B.G., Leano, E., Pointing, S.B., Poonyth, A.D. and Vrijmoed, L.L.P. 1998. Role of fungi in marine ecosystems. Biodivers. Conserv, 7: 1147-1161.

Javor, B. 1989. Hypersaline environments: Microbiology and biogeochemistry. Springer-Verlag, Berlin, Germany.

Jennings, D.H. 1986. Fungal growth in the sea, pp. 1-10. In: The Biology of Marine Fungi S.T. Moss (ed.), Cambridge University Press, Cambridge, UK.

Jin, Y. and Nevo, E. 2003. Osmoadaptation strategies of the Dead Sea fungi to hypersaline stress, pp. 162-200. In: Fungal Life in the Dead Sea. E. Novo, A. Oren and S.P. Wasser (eds.) A.R.G. Gantner Verlag K.-G., Ruggell, Switzerland.

Jin, Y., S. Weining, and Nevo, E. 2005. A MAPK gene from Dead Sea fungus confers stress tolerance to lithium salt and freezing-thawing: prospects for saline agriculture. Proc. Natl. Acad. Sci., USA 102 (52): 18992-18997.

Johnson, T.W. and Sparrow, F.K. 1961. Fungi in oceans and estuaries. Cramer, Weinheim, Germany.

Jones, E.B.G. 2000. Marine fungi: some factors influencing biodiversity. Fungal Diversity, 4: 53-73.

Jones, E.B.G. and Mitchell, J.I. 1996. Biodiversity of marine fungi, pp. 31-42. In: Biodiversity: International Biodiversity Seminar A. Cimerman and N. Gunde-Cimerman (eds.), National Inst. Chemistry and Slovenia National Commission for UNESCO, Ljubljana, Slovenia.

Keunen, J.G. 1986. Mixed substrates and mixed cultures. Antonie van Leeuwenhoek, 47: 187-189.

Kis-Papo, T., Grishkan, J., Oren A., Wasser, S.P., and Nevo, E. 2001. Spatio-temporal diversity of filamentous fungi in the hypersaline Dead Sea. Mycol. Res, 105(6): 749-750.

Kis-Papo, T., Oren, A., Wasser, S.P., and Nevo, E. 2003a. Survival of filamentous fungi in hypersaline Dead Sea water. Microb. Ecol, 45: 183-190.

Kis-Papo, T., Oren, A., Wasser, S.P., and Nevo, E. 2003b. Physiological adaptatiom of filamentous fungi to the Dead Sea: survival of spores and vegetative cells, pp. 293-313. In: Fungal Life in the Dead See. E. Nevo, A. Oren, and S.P. Wasser (eds.), A.R.G. Gantner Verlag K.-G., Ruggell, Switzerland.

Kohlmeyer, J. 1986. *Ascocratera manglicola* gen. et sp. nov. and key to the marine Loculoascomycetes on mangroves. Can. J. Bot, 64: 3036-3042.

Kohlmeyer, J., and Kohlmeyer, E. 1979. Marine mycology. The higher fungi. Acad. Press, New York, USA.

Kohlmeyer, J., and Volkmann-Kohlmeyer, B. 1991. Illustrated key to the filamentous fungi. Bot. Mar, 34: 1-61.

Kritzman, G. 1973. Observations on the microorganisms in the Dead Sea. M.Sc. Thesis. The Hebrew University in Jerusalem, Israel (In Hebrew).

Kurchenko, I., Lauer I., Buchalo, A.S., Oren, A., Nevo, E., Wasser, S.P., and Molitoris, H.P. 1998. Effect of salinities and temperatures on enzyme activities of fungi from the Dead Sea. In: Abstracts of the 6th Intern. Mycol. Congr., Jerusalem, Israel: 169.

Lorenz, R. and Molitoris, H.P. 1997. Cultivation of fungi under simulated deep sea conditions. Mycol. Res, 101: 1355-1365.

Madigan, M.T. and Marrs, B.L. 1997. Extremophiles. Scient. Amer.: 66-71.

Mager, W.H. and Varela, J.C.S. 1993. Osmostress response of the yeast *Saccharomyces*. Mol. Microbiol, 10: 253-8.

Mirocha, C.J. and Devay, J.E. 1971. Growth of fungus in an inorganic medium. Can. J. Microbiol, 17: 1373-1378.

Molitoris, H.P. and Schaumann, K. 1986. Physiology of marine fungi. A screening program for marine fungi, pp. 35-47. In: The Biology of Marine Fungi. S.T. Moss (ed.), Cambridge University Press, Cambridge, UK.

Molitoris, H.P., Lauer, I., Buchalo, A.S., Oren, A., Nevo, E., and Wasser, S.P. 1998. Growth of fungi from the Dead Sea at different salinities and temperatures. In: Abstracts of the 6th Intern. Mycol. Congr., Jerusalem, Israel: 169.

Molitoris, H.P., Buchalo, S., Kurchenko, I., Nevo, E., Rawal, S., Wasser, S.P., and Oren, A. 2000. Physiological diversity of the first filamentous fungi isolated from the hypersaline Dead Sea. Fungal Diversity, 5: 55-70.

Moss, S.T. 1986. The Biology of marine fungi. Cambridge University Press, Cambridge, UK.

Norkrans, B. 1966. Studies on marine occurring yeasts: Growth related to pH, NaCl concentration and temperature. Arch. Microbiol, 54: 374-392.

Onishi, H. 1963. Osmophilic yeasts. Advances in Food Research, 12: 53-94.

Oren, A. 1988. The microbial ecology of the Dead Sea, pp. 193-229. In: Advances in Microbial Ecology, Vol. 10, K.C. Marshall (ed.), Plenum, New York, USA.

Oren, A. 1997. Microbiological studies in the Dead Sea: 1892-1992, pp. 205-213. In: The Dead Sea—The Lake and its Setting. T. Niemi, Z. Ben-Avraham, and J.R. Gat (eds.), Oxford University Press, New York, USA.

Oren, A. 2000. Bilogical processes in the Dead Sea as influenced by short-term and long-term salinity changes. Arch. Hydrobiol. Spec. Issues: Advanc. Limnol, 55: 531-542.

Oren, A. 2002. Halophilic Microorganisms and their Environments. Kluwer Academic Publishers, Dordrecht, The Netherlands.

Oren, A. 2003. Physical and chemical limnology of the Dead Sea, pp. 45-67. In: Fungal Life in the Dead Sea. E. Nevo, A. Oren, And S.P. Wasser (eds.), A.R.G. Gantner Verlag K.-G., Ruggell, Switzerland.

Pal, H.S., Khosrawi, L., and Bond, C. 1979. Halophilic n-alkane utilizing yeast from the Arabian Gulf. Biotechn. Lett, 1: 445-450.

Parkinson, S.M., Killham, K., and Wainwright, M. 1990. Assimilation of $^{14}CO_2$ by *Fusarium oxysporum* grown under oligotrophic conditions. Mycol. Res, 94(7): 959-964.

Phaff, H.J. and Starmer, W.T. 1980. Specificity of natural habitats for yeasts and yeast-like organisms, pp. 79-102. In: Biology and Activities of Yeasts. F.A. Skinner, S.M. Passmore, and R.R. Davenport (eds.), Acad. Press, New York, USA.

Pitt, J.I. 1975. Xerophilic fungi and the spoilage of foods of plant origin. In: Water Relations of Foods R.B. Duckworth (ed.), Acad. Press, London, UK.

Pointing, S.B., Vrijmoed, L.L.P., and Jones, E.B.G. 1998. A qualitative assessment of lignocellulose degrading enzyme activity in marine fungi. Bot. Mar, 41: 293-298.

Prasannarai, K. and Sridhar, K.R. 1997. Effect of incubation period of driftwood on the occurrence of marine fungi. Ind. J. Mar. Sci, 26: 380-382.

Radwan, S.S., El-Essawy, A.A., and Helal, G.A. 1984. Salinity-loving fungi in Egyptian soils. 1. Numbers, identities, and halophilism. Zentralbl. Microbiol, 139: 435-440.

Raghukumar, C. and Raghukumar, S. 1998. Barotolerance of fungi isolated from deep-sea sediments of the Indian Ocean. Aquat. Microb. Ecol, 15: 153-163.

Rau, R. and Molitoris, H.P. 1991. Determination and significance of nitrate reductase in marine fungi. Kieler Meeresforschungen, Sonderheft, 8: 369-375.

Rawal, B.S., Kurchenko, I., Buchalo, A.S., Wasser, S.P., Oren, A., Nevo, E., and Moli-toris, H.P. 1998. (Abstracts), 6th Intern. Mycol. Congr., Jerusalem, August, 23-28, p. 106, Israel.

Ritchie, D. 1954. A fungus flora of the sea. Science, 3119: 578-579.

Rohrmann, S. and Molitoris, H.P. 1992. Screening for wood decay enzymes in marine fungi. Can. J. Bot, 70, 2116-2123.

Ross, S.S. and Morris, E.O. 1962. Effect of sodium chloride on the growth of certain yeasts of marine origin. J. Sci. Food Agricult, 13: 467-475.

Schimpfhauser, G. and Molitoris, H.P. 1991. Enzyme activities of monokaryotic and dikaryotic strains of the marine basidiomycete *Nia vibrissa*. Kieler Meeresfor-schungen, Sonderheft, 8: 361-368.

Sparrow, F.K. 1934. Observation on marine Phycomycetes collected in Denmark. Dansk. Bot. Ark, 8, 1: 1-24.

Sparrow, F.K. 1943. Aquatic Phycomycetes. Univesity of Michigan Press, Ann Arbor, USA.

Spencer, J.F.T., Phaff, J.H. and Gardner, N.R. 1964. *Metschnikowia kamienski* sp. n., a yeast associated with brine shrimp. J. Bacteriol, 88: 758-762.

Steiman, R., Guiraud, P., Sage, L., Seigle-Murandi, F. and Lafond, J.L. 1995. Mycoflora of soil around the Dead Sea. I. Ascomycetes (including *Aspergillus* and *Penicillium*), Basidiomycetes, Zygomecetes. Syst. Appl. Microbiol, 18: 310-317.

Steiman, R., Guiraud, P., Sage, L., and Seigle-Murandi, F. 1997. Soil mycoflora from the Dead Sea oases of Ein Gedi and Einot Zuqim (Israel). Antonie van Leeuwenhoek, 72: 261-270.

Tokuoka, K. 1993. Sugar- and salt-tolerant yeasts. J. Appl. Bacteriol, 74: 101-110.

Vismer, H.F. and Marasas, W.F.O. 1998. How dangerous are the fungi that we work with? 6th Int Mycol. Congr., Jerusalem, 1998, 75. In: Abstracts of the 6th Intern. Mycol. Congr., Jerusalem, Israel: 180.

Volcani, B. 1944. The microorganisms of the Dead Sea, pp. 71-85. In: Papers collected to commemorate the 70th anniversary of Dr. Chaim Weizmann, Daniel Sieff Research Institute, Rehovoth, Israel.

Volz, P.A. and Wasser, S.P. 1995. Soil micromycetes of selected areas of Israel. Isr. J. Pl. Sci, 43: 281-290.

Volz, P.A., Wasser, S.P., and Nevo, E. 1996. New and rare soil micromycetes for the biota of Israel. Ukr. Bot. Zhurn, 53: 51-57.

Volz, P.A., Ellanskaya, I.A., Grishkan, I., Wasser, S.P., and Nevo, E. 2001. Soil Microfungi of Israel. A.R.G. Gantner Verlag K.-G., Ruggell, Switzerland.

Vyas, B.R.M. and Molitoris, H.P. 1995. Involvement of an extracellular H_2O_2-dependent ligninolytic activity of the white-rot fungus *Pleurotus ostreatus* in the decolorization of Remazol Brilliant Blue R. Appl. Environm. Microbiol, 61: 3919-3927.

Waguri, O. 1976. Isolation of microorganisms from salt lakes in the Dry Valley, Antarctica, and their living environment. Antarctic Record 57: 80-96.

Wainwright, M. 1988. Metabolic diversity of fungi in relation to growth and mineral cycling in soil: A review. Trans. Br. Mycol. Soc. 90(2): 159-170.

Wainwright, M., Ali, J.A., and Killham, K. 1993. Aerobic growth of fungal mycelium from soil particles into nutrient-free silica gel. Mycol. Res, 98(7): 761-762.

Washburn, R.G. 1994. Opportunistic mold infections, pp. 147-158. In: The Mycota. K. Esser and P.A. Bemke (eds.) Springer-Verlag, Berlin, Germany.

Wasser, S.P., Grishkan, I., Kis-Papo, T., Buchalo, A.S., Volz, P.A., Gunde-Cimerman, N., Zalar P., and Nevo, E. 2003. Species diversiry of the Dead Sea fungi, pp. 203-270. In: Fungal Life in the Dead Sea. E. Nevo, A. Oren, and S.P. Wasser

(eds.), A.R.G. Gantner Verlag K.-G., Ruggell, Switzerland.
Wheeler, K., Hocking, A. and Pitt, J.I. 1988. Influence of temperature on the water relations of *Polypaecilum pisce* and *Basipetospora halophila*, two halophilic fungi. J. Gen. Microbiol, 133: 2255-2261.
Wilkansky, B. 1936. Life in the Dead Sea. Nature, 138: 467.
Zaunstöck, B., Lorenz, R., and Molitoris, H.P. 1994. Fungal spore germination under simulated deep sea conditions. In: Abstracts of the 7th Intern. Congr. Mycol. Division, Prague, Czech Republic, July 3-8: 334.
Zhdanova, N.N. and Vasilevskaya, A.J. 1982. Extremal ecology of fungi in nature and experiment. Kiev: Naukova Dumka Press, Kiev (in Russian).
Zhdanova, N.N. and Vasilevskaya, A.J. 1988. Melanincontaining fungi in experimental conditions. Naukova Dumka Press, Kiev (in Russian).

4

Filamentous Fungi in the Marine Environment: Chemical Ecology

Michio Namikoshi and Jin-Zhong Xu[1]
Tohoku Pharmaceutical University, Komatsushima,
Aoba-ku, Sendai 981-8558, Japan
E-mail: mnami@tohoku-pharm.ac.jp
[1]Present address: Pharmaceutical Informatics Institute,
College of Pharmaceutical Sciences, Zhejiang University,
Hangzhou, Zhejiang 310058, P.R. China

Abstract

Marine fungi (including marine-derived strains) are rich sources of biologically active secondary metabolites and interesting research objects for chemical ecological studies. This chapter describes chemical ecology of filamentous fungi emphasizing the defensive functions of secondary metabolites. New antimicrobial and cytotoxic metabolites of fungi isolated from marine organisms that may have ecological roles in protection mechanisms of marine organisms and the results on a probable implication of filamentous fungi in the indirect chemical protection mechanism of marine sponges are described.

A possible role of antimicrobial metabolites produced by fungi isolated from marine organisms is suggested to be a negative cue against fouling, settlement, colonization, adhesion, and infection by potentially harmful microorganisms. Cytotoxic metabolites may be deterrents against larvae and potential predators.

Our study demonstrated the possibility that fungi associated with marine sponges may participate in the indirect chemical defense system of the hosts in providing antibacterial metabolites. Bioassay-guided isolation afforded three new and five known antibacterial metabolites against a marine bacterium *Ruegeria atlantica*, a common fouling species. Penicillic acid, a well known mycotoxin first reported in 1936, may be a ubiquitous antibiotic of marine fungi because this antibiotic strongly inhibited the growth of *R. atlantica* with high selectivity from terrestrial bacteria and the producing fungi have been isolated from a wide area of the North Pacific Ocean.

Chemical ecological studies on marine fungi and their association with the hosts provide a better understanding of the importance of fungi in the marine environment. Chemical relationships between fungi and their host organisms are very interesting studies not only in understanding the marine fungal ecology but also in searching for new biologically active metabolites.

INTRODUCTION

Marine fungi are recognized as a prolific source of biologically active secondary metabolites. A number of reviews describing the importance and attractiveness of marine fungal metabolites have been published (Kobayashi and Ishibashi, 1993; Davidson, 1995; Liberra and Lindequist, 1995; Bernan et al., 1997; Pietra, 1997; Biabani and Laatsch, 1998; Verbist et al., 2000; Jensen and Fenical, 2000, 2002; Proksch et al., 2002; Bugni and Ireland, 2004; Namikoshi, 2006). Potential of marine fungi in providing biologically active metabolites and new compounds is greater than that of terrestrial counter parts (Cuomo et al., 1995; Namikoshi et al., 2000; Namikoshi, 2006). The structure of the first new metabolite, leptosphaerin, from marine fungus was reported in 1986 (Pallenberg and White, 1986; Schiehser et al., 1986; Rollin, 1987; White et al., 1989), and 437 new metabolites from 149 strains have been reported in 182 scientific journals up to June, 2004 (Namikoshi, 2006). These new compounds were mostly obtained in the course of studies on biologically active metabolites for searching new drug candidates and their leads, and there were very few reports describing ecological significance of marine fungi and their metabolites. The ability to produce new secondary metabolites by marine fungi may be acquired by the tolerance or adaptation to the marine environment such as salinity and high pressure and by interactions with marine micro- and macroorganisms such as symbiosis, commensalism, infection, and so on. Therefore, the marine fungal metabolites may have ecological effects on the producing organisms, for the hosts, or to the other organisms.

Marine fungi are not of a taxonomic but of an ecological group. Kohlmeyer and Kohlmeyer (1979) proposed the ecological definition of marine fungi, that is, "*Obligate marine fungi* are those that grow and sporulate exclusively in a marine or estuarine habitat; *facultative marine fungi* are those from freshwater or terrestrial milieus able to grow (and possibly also to sporulate) in the marine environment". This definition is accepted by taxonomists (Kohlmeyer and Volkmann-Kohlmeyer, 2003). Although such a definition of marine mycology is important to study the ecological roles of biologically active metabolites, many compounds have been isolated from marine isolates that are taxonomically similar or identical to terrestrial fungi such as *Aspergillus* and *Penicillium*. It is not clear if these fungi are active in the marine environment or if dormant spores or hyphae are being isolated. These fungi are known as marine-derived strains.

Interestingly, a species of common genus, *Aspergillus sydowii*, was isolated as a sea fan pathogen and has been proved to cause an infectious disease against sea fans (Geiser et al., 1998). Other fungi that are the same genera found in the terrestrial environment have been known to cause diseases in marine animals and plants (Rand, 2000; Hyde et al., 1998). Jenkins et al. (1998b) suggested that fungal secondary metabolites are involved in the pathogenesis of marine plants. It can be hypothesized that morphologically identical marine isolates with terrestrial strains were physiologically adapted to the marine environment and acquired the ability to produce new secondary metabolites for surviving at their habitats (Höller et al., 2000). Therefore, some of the ubiquitous species are also active in the marine environment. Infection of pathogenic and non-pathogenic fungi may be an important ecological event that may lead evolutional changes and intimate associations such as symbiosis and commensalism, because mutualistic associations are mostly started by incidental infections (Bernan, 2001).

Marine fungi have an important role in the decomposition of dead plants and animals and their excretions and discharged parts, as in the terrestrial environment (Hyde et al., 1998; Pointing et al., 2000). As decomposers, marine fungi are especially important in the late stages of the decomposition process in recycling nutrients back to the marine ecosystem.

Marine fungi grow on a wide rage of substrates from surfaces and inner tissues of marine organisms to sediments, mud, and sand. Marine grasses and mangroves are rich sources of obligate and facultative endophytic fungi (Leaño, 2002; Kumaresan et al., 2002; Alva et al., 2002; Abdel-Wahab and El-Sharouny, 2002; Lintott and Lintott, 2002). The role of marine fungi in mangrove habitats has been studied (Leaño, 2002; Kumaresan et al., 2002; Abdel-Wahab and El-Sharouny, 2002). There are ecological relationships between the plants and the fungi. Marine fungi play important roles not only in the decomposition process but also in the association with marine plants and animals (Hyde et al., 1998; Kolmeyer and Volkmann-Kohlmeyer, 2003; Bugni and Ireland, 2004).

Symbiotic relationships between marine fungi and other organisms are observed in the forms of lichens, mycophycobioses, and mycorrhizas (Hyde et al., 1998). Marine lichens mostly grow on the surfaces of rocks in the intertidal zones. Mycophycobiosis are mutualistic relationships between marine fungi and macroalgae. The association of halophytic plants with arbuscular mycorrhizas has been observed in the marine and estuarine environments. Marine invertebrates and macroalgae sometimes possess antifungal substances to prevent infection and adhesion of fungi. Nevertheless, filamentous fungi can be isolated from the surfaces and even inner tissues of the macroalgae. Some of the associations between marine organisms and fungi may be the results of ecological functions. Kohlmeyer and Volkmann-Kohlmeyer (2003) have introduced autochthonous coral-inhabiting ascomycetes that have fruited in the natural habitat. Fungal hyphae have been observed in the interior of living corals and soft corals (Kendrick et al., 1982; Le Campion-Alsumard et al., 1995). There should be chemical mediations in these intimate relationships. The chemical ecological studies on marine fungi are, however, rather rare.

This chapter describes chemical ecology of filamentous fungi emphasinzing the defensive functions of secondary metabolites. New antimicrobial and cytotoxic metabolites of fungi isolated from marine organisms that may have ecological roles in protection mechanisms of marine organisms and the results on a probable implication of filamentous fungi in the indirect chemical protection mechanism of marine sponges are described.

THREATS IN THE MARINE ENVIRONMENT

The marine environment is quite different from the terrestrial environment such as salinity, high pressure, and darkness. Marine organisms are always exposed to dangers from other organisms, such as predator-prey interactions, spatial and nutritional competitions, settlements of larvae and zoospores, and infections of parasitic and pathogenic organisms (McClintock and Baker, 2001). Therefore, these organisms may have physical, behavioral, and chemical protection mechanisms (Harper et al., 2001). Moreover, a great number of bacteria, fungi, microalgae, and viruses exist in the marine environment and cause infectious diseases against marine animals and plants (Correa, 1997). As marine invertebrates and plants do not have cell-based immune systems, it is possible to presume that these animals and plants developed chemical protection mechanisms utilizing secondary metabolites (McClintock and Baker, 2001).

These threats which exist in the marine environment will be the driving force for the evolution of secondary metabolisms in marine organisms (Bernan, 2001). Secondary metabolites play important roles in the defense mechanisms, and associations of microorganisms in the production systems of the hosts that have become distinct (Gil-Turnes et al., 1989, Gil-turnes and Fenical, 1992; Lopanik et al., 2004a, 2004b).

Commensal marine bacteria, inhabited surfaces, tissues, and interior spaces of other organisms, have recently been revealed to produce biologically active secondary metabolites that have been obtained from the extracts of whole organisms (Piel et al., 2004; Hildebrand et al., 2004; Schmidt et al., 2005). The association of microorganisms will potentially be cooperative in the production of biologically active metabolites. Evidences showed that bacteria living on surfaces of marine organisms chemically protect their hosts from threats (Armstrong et al., 2001; Steinberg et al., 2001, 2002; Kjelleberg and Steinberg, 2002). These host organisms may not have an ability to produce defense substances by themselves.

In the following sections it will be seen if fungi participate in the chemical protection mechanisms of the hosts with fungal secondary metabolites.

New biologically active secondary metabolites of fungi isolated from marine organisms

Natural products have evolved under the pressure of natural selection to bring about specific receptors (Williams et al., 1989). If target receptors are common among the marine and terrestrial organisms, biological activities discovered for marine natural products can be explained. Therefore fungal metabolites possessing antimicrobial and cytotoxic activities may have ecological roles against harmful microorganisms and potential predators.

NEW ANTIMICROBIAL METABOLITES

A number of antimicrobial secondary metabolites have been obtained from marine-derived fungi in the search of new medicinal agents and their lead compounds. Although the ecological significance of these antimicrobial metabolites is uncertain, a few papers have focused on the growth inhibitory activities against marine bacteria and microalgae. Antimicrobial metabolites of fungi isolated from marine organisms may act as antiinfection, antifouling, anticolonization, and antiadhesion cues against potentially harmful marine microorganisms.

This section describes new antimicrobial metabolites produced by fungi isolated from healthy marine organisms (Table 1). The structures of selected metabolites are shown in Figure 1.

From alga

Among the antimicrobial metabolites, pestalone (**1**) is the most interesting compound. Pestalone was obtained from *Pestalotia* sp. CNL-365 isolated from the surface of the brown alga *Rosenvingea* sp. in the Bahamas (Cueto

Table 1. New antimicrobial metabolites of fungi isolated from marine organisms

Fungus	*Compound*	*Activity**	*Reference*
From alga			
Fusarium sp.	Halymecin A	A, B (G+)	Chen et al., 1996
Pestalotia sp.	Pestalone	B (G+)	Cueto et al., 2001
Trichoderma virens	Trichodermamide B	Y, B (G+)	Garo et al., 2003
Varicosporina ramulosa	Dihydrocolletodiols	F	Höller et al., 1999a
unidentified fungus	Solanapyrones E–F	A	Jenkins et al., 1998a
unidentified fungus	Seragakinone A	B (G+)	Shigemori et al., 1999
unidentified fungus	M-3	F	Byun et al., 2003
From seagrass and mangrove			
Fusarium sp.	Sansalvamide A	V	Belofsky et al., 1999
Scytalidium sp.	Halovirs A–E	V	Rowley et al., 2003
Hypoxylon oceanicum	15G256's	F	Abbanat et al., 1998; Schlingmann et al., 1998, 2002
From sponge			
Aspergillus ostianus	Chloro-asperlactones	B (G+, –)	Namikoshi et al., 2003
Aspergillus versicolor	Aspergillitine	B (G+, –)	Lin et al., 2003
Cladosporium sp.	Cladospolide D	F	Zhang et al., 2001
Cladosporium herbarum	Acetyl Sumiki's acid	B (G+)	Jadulco et al., 2001
Coniothyrium sp.	Phenylbutanone	F	Höller et al., 1999b
Curvularia lunata	Lunatin	B (G+, –)	Jadulco et al., 2002
Emericella variecolor	Varixanthone	B (G+, –)	Malmstrøm et al., 2002a

Contd.

Table 1 continued

Fungus	*Compound*	*Activity**	*Reference*
Exophiala pisciphila	Exophilin A	B (G+)	Doshida et al., 1996
Microsphaeropsis sp.	Microsphaeropsisin	F	Höller et al., 1999b
Penicillium citrinum	Isocyclocitrinols	B (G+)	Amagata et al., 2003b
Penicillium cf. Montanense	Xestodecalactone B	Y	Edrada et al., 2002
Phoma sp.	YM-202204	F, Y	Nagai et al., 2002
unidentified fungus	Hirsutanol A	B (G+)	Wang et al., 1998
unidentified fungus	Secocurvularin	B (G+)	Abrell et al., 1996
From ascidian			
Acremonium sp.	Fumiquinazolines H, I	Y	Belofsky et al., 2000
Aspergillus niger	Yanuthones	Y, B (G+, –)	Bugni et al., 2000
From soft coral			
unidentified fungus	Spiroxins A–E	B (G+)	McDonald et al., 1999
From jellyfish & mollusc			
Emericella unguis	Guisinol	B (G+)	Nielsen et al., 1999
Paraphaeosphaeria sp.	Unguisins A, B	B (G+)	Malmstrøm, 1999
	Modiolides A, B	M, F	Tsuda et al., 2003

*Biological activity: A—antialgal; B—antibacterial (G+ —Gram-positive; G– —Gram-negative); F—antifungal; H—antihemoflagellate; M—antimycobacterial; V—antiviral; Y—antiyeast.

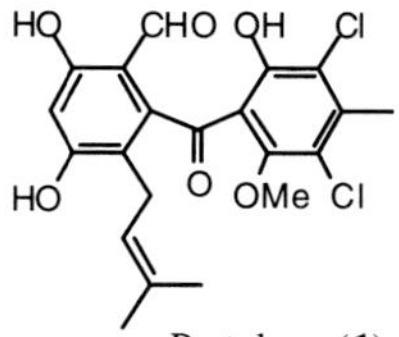

Pestalone (**1**)

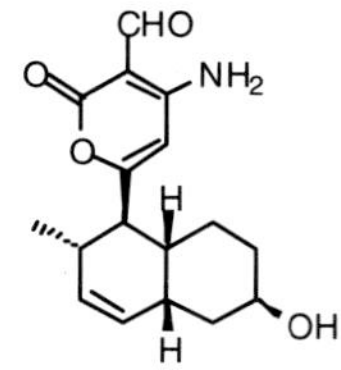

Solanapyrone E (**2**)

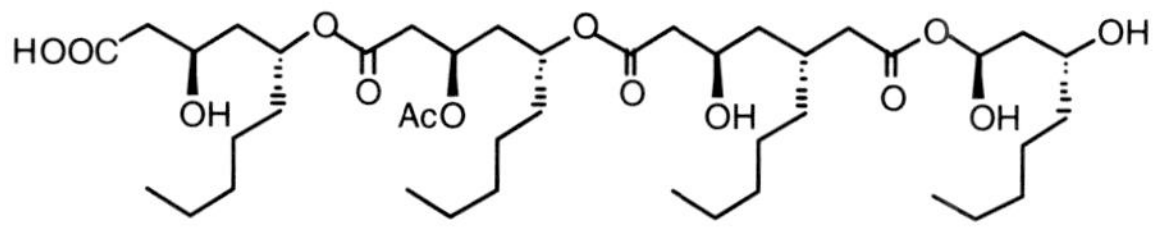

Halymecin A (**3**)

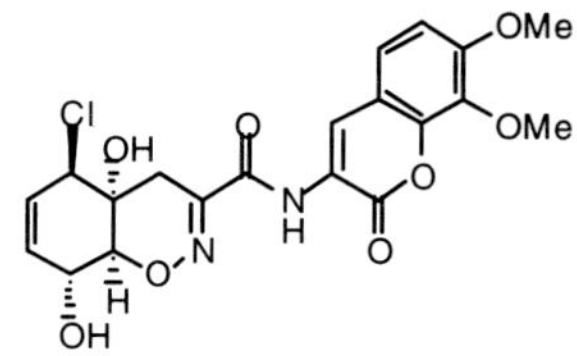

Trichodermamide B (**5**)

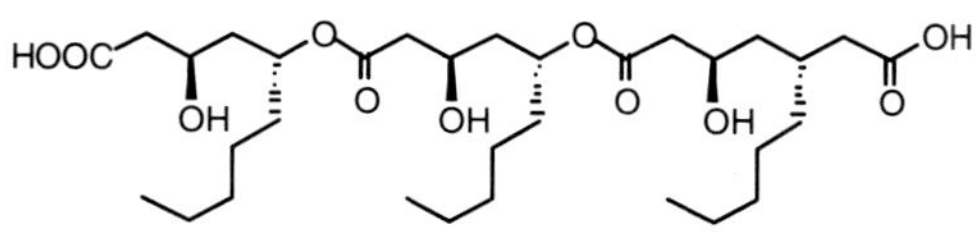

Exophilin A (**4**)

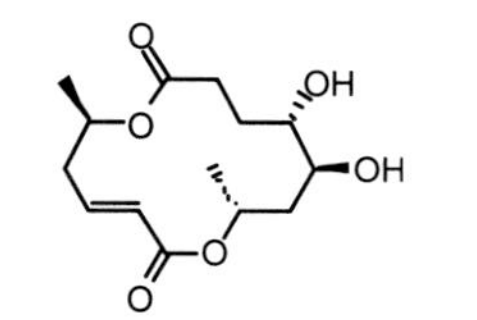

9,10-Dihydrocolletodiol (**6**)

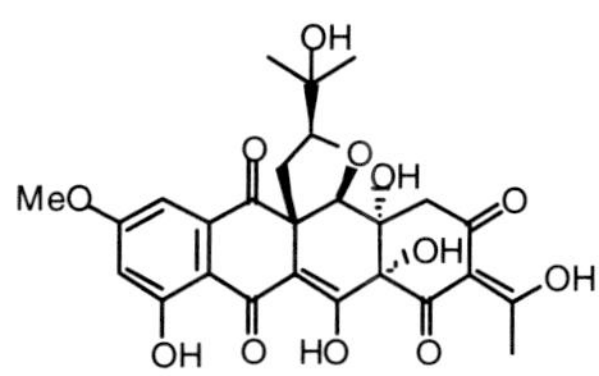

Seragakinone A (**7**)

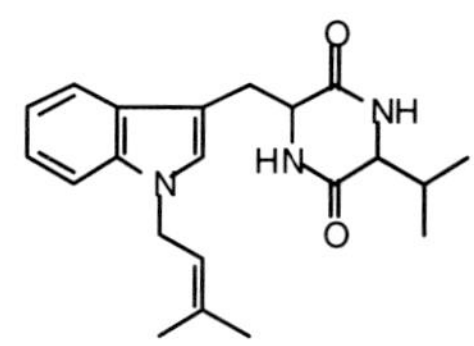

M-3 (**8**)

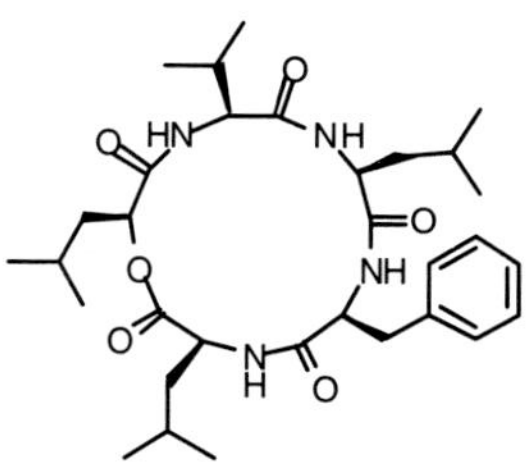

Sansalvamide A (**9**)

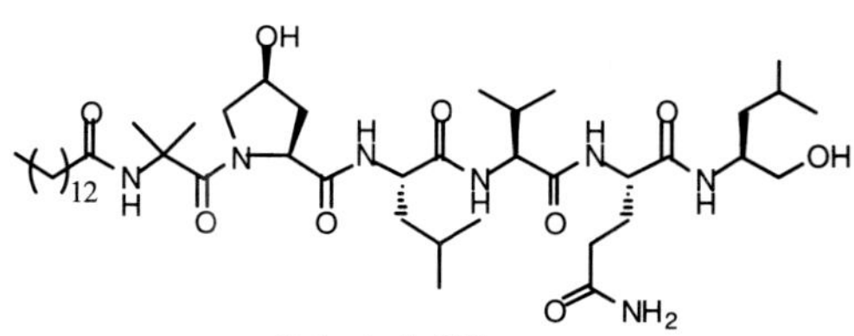

Halovir A (**10**)

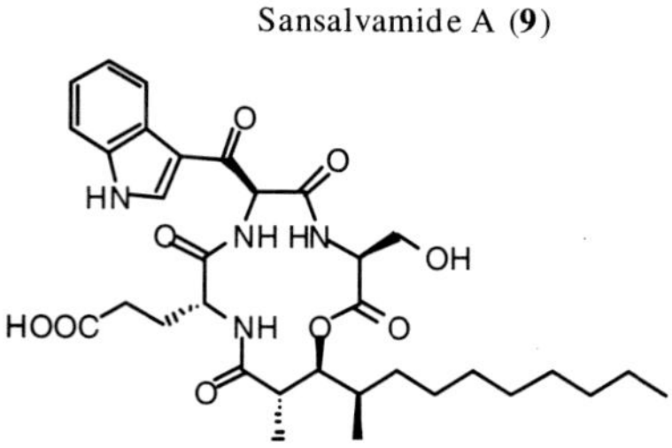

15G256γ (**11**)

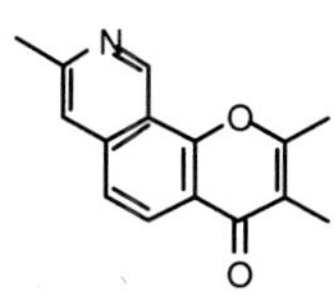

Aspergillitine (**12**)

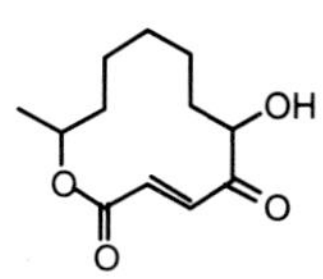

Cladospolide D (**13**)

Fig. 1. Structures of antimicrobial metabolites of fungi isolated from marine organisms

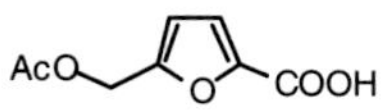

Acetyl Sumiki's Acid (**14**)

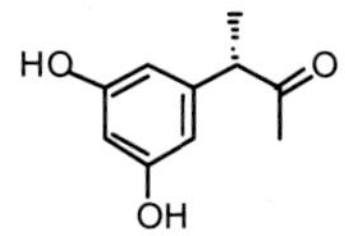

(3,5-Dihydroxyphenyl)butan-2-one (**15**)

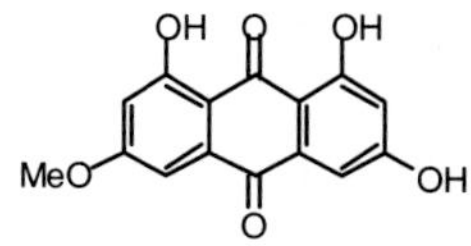

Lunatin (**16**)

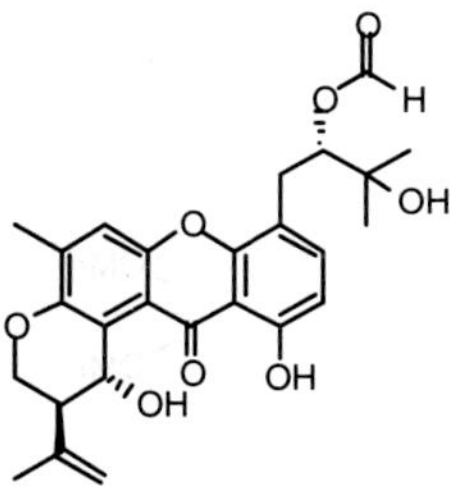

Varixanthone (**17**)

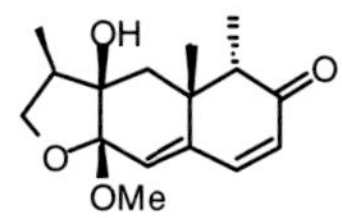

Microsphaeropsisin (**18**)

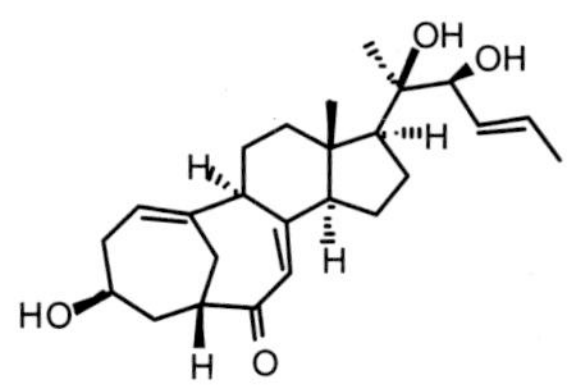

Isocyclocitrinol A (**19**)

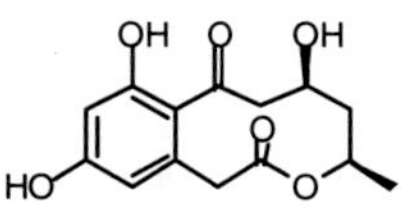

Xestodecalactone B (**20**)

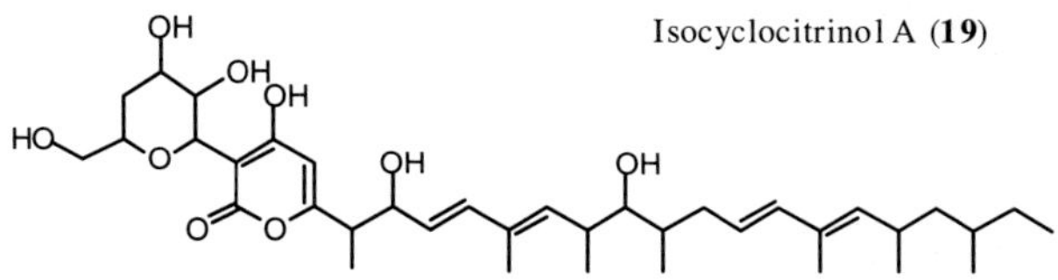

YM-202204 (**21**)

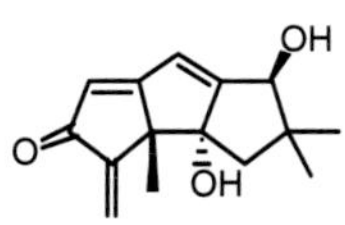

Hirsutanol A (**22**)

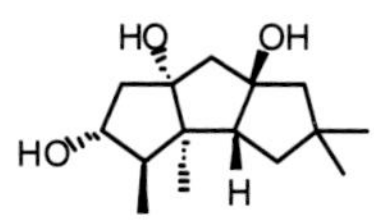

Ent-Gloeosteretriol (**23**)

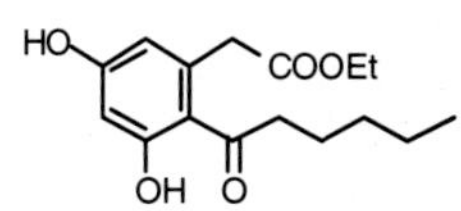

Secocurvularin (**24**)

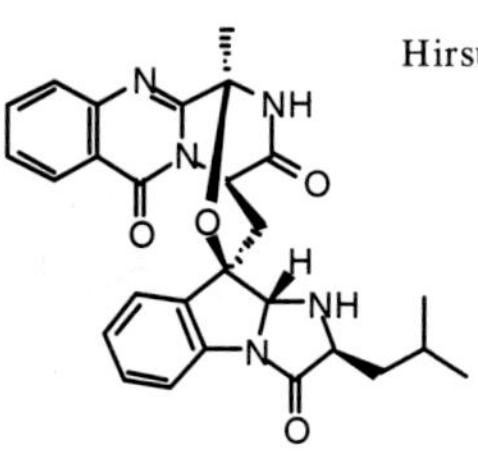

Fumiquinazoline H (**25**)

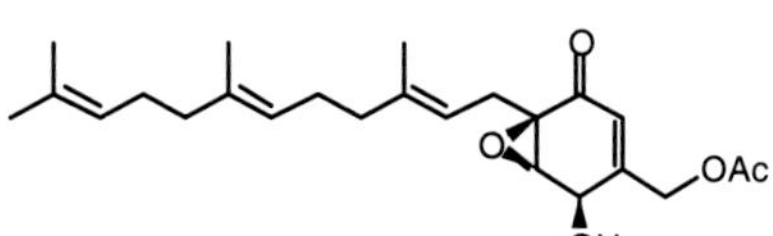

Yanuthone A (**26**)

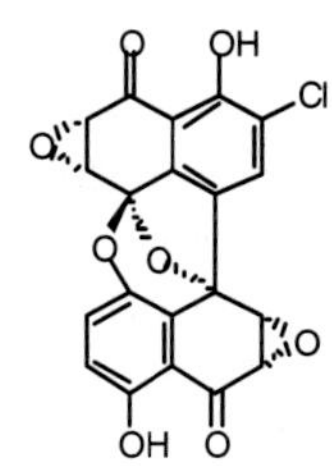

Spiroxin A (**27**)

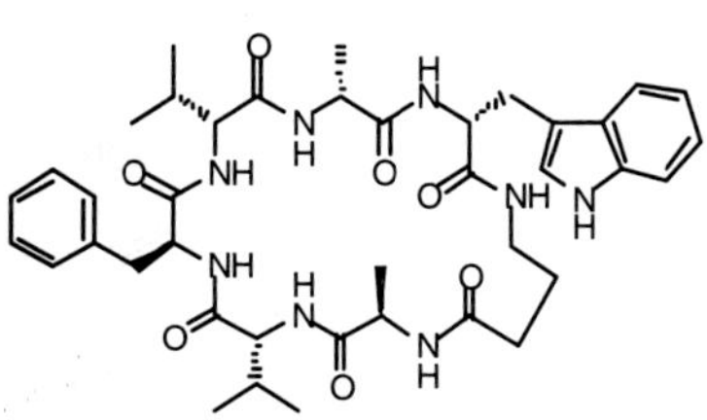

Unguisin A (**28**)

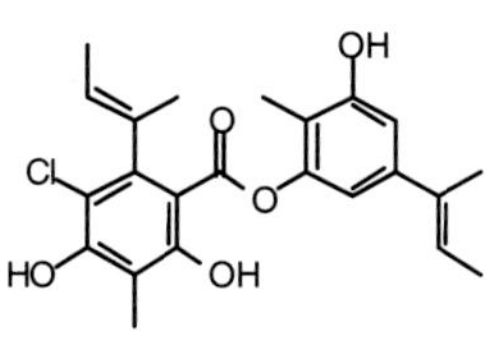

Guisinol (**29**)

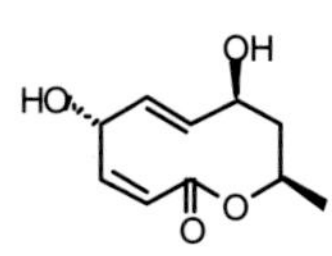

Modiolide A (**30**)

Fig. 1 continued. Structures of antimicrobial metabolites of fungi isolated from marine organisms

et al., 2001). Interestingly, this compound was produced only when this strain was cultured in the presence of a marine bacterium. This fungus or the marine bacterium alone did not produce pestalone, and the production of this compound in very low yield was induced by the addition of ethanol (1% v/v) in the culture medium after 24 h of fermentation. These facts suggested that pestalone may be produced by the fungus when a bacterial challenge occurs on the surface of the brown alga. This compound exhibited potent antibacterial activity against methicillin-resistant *Staphylococcus aureus* (MRSA) and vancomycin-resistant *Enterococcus faecium* (VREF) with MIC of 37 and 78 ng/mL, respectively. Pestalone also showed moderate *in vitro* cytotoxicity (GI_{50} = 6.0 μM) in the NCI's human tumor cell line screen.

An unidentified fungus CNC-159 was isolated from the surface of the green alga *Halimeda monile* in the Bahamas. This fungus produced solanapyrones E (**2**), F, and G (Jenkins et al. 1998a). Solanapyrones E and F inhibited the growth of a marine unicellular green alga *Dunaliella* sp. (15 and 25% inhibition at 200 μg/mL, respectively). *Fusarium* sp. FE-71-1 isolated from the alga *Halumenia dilatata* at Palau gave halymecins A (**3**), B, and C (Chen et al., 1996). Halymecin A showed growth inhibition against the diatom *Skeletonema costatum* (MIC = 4 μg/mL), antibacterial activity, and cytotoxicity. The structure of halymecin A resembled that of exophilin A (**4**), which was obtained as an antibacterial metabolite of *Exophiala pisciphila* NI10102 isolated from the sponge *Mycale adhaerens* (Doshida et al., 1996). Therefore, exophilin A may have an antialgal property. These antialgal metabolites may be produced to play an ecological role in the association of marine fungi with algae and sponges.

Two strains of *Trichoderma virens* (CNK266 and CNL910) gave trichodermamides A and B (**5**). Strains CNK266 and CNL910 were isolated from the green alga *Halimeda* sp. and the ascidian *Didemnum molle*, respectively, in Papua New Guinea (Garo et al., 2003). Trichodermamide B showed significant cytotoxicity and moderate antimicrobial activities against amphotericin-resistant *Candida albicans*, MRSA, and VREF with MIC values of ca. 15 μg/mL.

Varicosporina ramulosa 195-31 isolated from the brown alga *Cytoseira* sp. in Spain afforded (6*R*,11*S*,12*S*,14*R*)-9,10-dihydrocolletodiol (**6**) and (6*R*,11*R*,12*R*,14*R*)-9,10-dihydrocolletodiol, which exhibited weak antifungal activity against *Eurotium repens* (2 and 1 mm inhibition zone at 50 μg/disc, respectively) (Höller et al., 1999a).

An unidentified fungus K063 isolated from the red alga *Ceratodictyon spongiosum* in Okinawa produced seragakinone A (**7**) and dictyonamides A and B (Shigemori et al., 1999, Komatsu et al., 2001). Seragakinone A inhibited the growth of bacteria *S. aureus*, *Micrococcus luteus*, *Corynebacterium xerosis* and *Bacillus subtilis* (MIC = 10, 20, 20, and 41 μg/mL, respectively).

An unidentified fungus M-3 was isolated from the red alga *Porphyra yezoensis* in Chiba, Japan and gave an antifungal compound M-3 (**8**) (Byun et al., 2003). This compound exhibited the potent activity against *Pyricularia oryzae* (MIC = 0.36 μM).

From seagrass and mangrove

Fusarium sp. CNL 292 isolated from the surface of the seagrass *Halodule wrightii* in the Bahamas yielded a cytotoxic compound sansalvamide A (**9**) (Belofsky et al., 1999). Sansalvamide A selectively inhibited the topoisomerase of the poxvirus *Molluscum contagiosum* (MCV) (Hwang et al., 1999).

Scytalidium sp. CNL240 isolated from the Caribbean seagrass *Halodule wrightii* produced five anti-HIV compounds, halovirs A (**10**)–E (Rowley et al., 2003). Halovirs A–E directly inactivated the viruses HSV-1 and HSV-2.

A mangrove fungus *Hypoxylon oceanicum* LL-15G256 isolated from the *Kandelia candel* wood at Shenzen, China afforded five new antifungal compounds, 15G256γ (**11**), 15G256δ, 15G256ε, 15G256ι, and 15G256ω and five new non-active compounds, 15G256α-2, 15G256β-2, 15G256ν, 15G256ο, and 15G256π (Abbanat et al., 1998, Schlingmann et al., 1998, 2002). These compounds showed the antifungal activity against a phytopathogenic fungus *Neurospora crassa* through inhibition of cell wall biosynthesis. 15G256γ also inhibited the growth of pathogenic dermatophytic fungi and yeast with MIC of 2–16 μg/mL.

From marine sponge

Aspergillus ostianus TUF 01F313 isolated from an unidentified sponge at Pohnpei gave three chlorinated compounds **60–62** (Figure 3) (Namikoshi et al., 2003). These compounds exhibited the antibacterial activity against the marine bacterium *Ruegeria atlantica* besides terrestrial bacteria *S. aureus* and *Escherichia coli* (details are mentioned in the section 4.2).

Aspergillus versicolor HBI-2 was isolated from the sponge *Xestospongia exigua* collected at Bali, Indonesia and afforded an antibacterial metabolite, aspergillitine (**12**). (Lin et al., 2003). Aspergillitine had moderate antibacterial activity against *B. subtilis* (7 mm inhibition zone at 5 μg).

Cladosporium sp. FT-0012 isolated from an unidentified sponge in Pohnpei produced an antifungal metabolite, cladospolide D (**13**), which inhibited the growth of fungi, *Mucor racemosus* (IC_{50} = 0.15 μg/mL) and *P. oryzae* (IC_{50} = 29 μg/mL) (Zhang et al., 2001).

Cladosporium herbarum isolated from the sponge *Callyspongia aerizusa* collected in Indonesia yielded acetyl Sumiki's acid (**14**), pandangolides 3 and 4, and herbaric acid (Jadulco et al., 2001, 2002). Compound **14** showed antibacterial activity against *B. subtilis* and *S. aureus* (7 mm inhibition zones at 5 μg/disc).

Coniothyrium sp. 193H77 was isolated from the Caribbean sponge *Ectyoplasia ferox* and gave (3*S*)-(3,5-dihydroxyphenyl)butan-2-one (**15**) and 2-(1(*E*)-propenyl)-octa-4(*E*),6(*Z*)-diene-1,2-diol (Höller et al. 1999b). Compound **15** exhibited weak antifungal activity at the 50 μg/disc level against *Ustilago violacea* and *Mycotypha microspora*.

Curvularia lunata isolated from the Indonesian sponge *Niphates olemda* afforded lunatin (**16**), which inhibited the growth of *B. subtilis*, *S. aureus*, and *E. coli* (Jadulco et al., 2002).

The Caribbean sponge-derived *Emericella variecolor* M75-2 produced varixanthone (**17**), varitriol, dihydroterrein, and varioxirane (Malmstrøm et al., 2002a). Varixanthone exhibited antibacterial activity against *E. coli*, *Proteus* sp., *B. subtilis* and *S. aureus* (MIC = 12.5 μg/mL) and toward lower potency to *E. faecalis* (MIC = 50 μg/mL).

Microsphaeropsis sp. H5-50 was isolated from the sponge *Myxilla incrustans* in Germany and gave microsphaeropsisin (**18**), which showed weak antifungal activity at the 50 μg/disc level against *Eurotium repens* and *U. violacea* (Höller et al., 1999b).

Penicillium citrinum 991084 isolated from the sponge *Axinella* sp. in Papua, New Guinea yielded two novel steroids, isocyclocitrinol A (**19**) and 22-acetylisocyclocitrinol A (Amagata et al., 2003b). These compounds had modest antibacterial activity against *Staphylococcus epidermidis* (MIC = 100 μg/mL each) and *Enterococcus durans* (MIC = 100 μg/mL each).

Penicillium cf. *montanense* HBI-3/D was isolated from the sponge *Xestospongia exigua* collected at Bali and produced xestodecalactones A, B (**20**), and C (Edrada et al., 2002). Xestodecalactone B was active against *C. albicans* (7 mm inhibition zone at 20 μM).

Phoma sp. Q60596 isolated from the Okinawan sponge *Halichondria japonica* gave YM-202204 (**21**) and a known analog (Nagai et al. 2002). YM-202204 demonstrated potent antifungal activities against *C. albicans* (IC_{80} = 6.25 μg/mL), *Cryptococcus neoformans* (IC_{80} = 1.56 μg/mL), *Aspergillus fumigatus* (IC_{80} = 12.5 μg/mL), and *Saccharomyces cerevisiae* (IC_{80} = 1.56 μg/mL).

An unidentified fungus 95-1005C isolated from the sponge *Haliclona* sp. collected at North Sulawesi, Indonesia afforded hirsutanols A (**22**)–C and *ent*-gloeosteretriol (**23**) (Wang et al., 1998). Compounds **22** and **23** were reported to show mild antibacterial activity against *B. subtilis* (no quantitative data).

An unidentified fungus 951014 was isolated from the Indonesian sponge *Spirastrella vagabunda* and gave secocurvularin (**24**), which showed a modest antibiotic activity against *B. subtilis* at 200 μg/disc (Abrell et al., 1996).

From ascidian

Acremonium sp. CNC 890 isolated from the surface of the tunicate *Ecteinascidia turbinata* at the Bahamas produced fumiquinazolines H (**25**) and I and oxepinamides A, B, and C (Belofsky et al., 2000). Fumiquinazolines H and I showed very weak antifungal activity against *C. albicans* at 500 μg/mL.

Aspergillus niger F97S11 isolated from the orange tunicate *Aplidium* sp. in Fiji afforded eight new antibacterial metabolites, yanuthones A (**26**)–E, 1-hydroxyyanuthone A, 1-hydroxyyanuthone C, and 22-deacetylyanuthone A (Bugni et al., 2000). These compounds exhibited antimicrobial activity (62–250 μg/disc) against *S. aureus*, MRSA, VREF, *C. albicans*, or *E. coli* (IMP-mutation).

From soft coral

An unidentified fungus LL-37H248 was isolated from an orange soft coral collected at Vancouver Island in Canada and spiroxins A (**27**)–E were obtained from the culture extract (McDonald et al., 1999, absolute stereochemistry: Wang et al., 2001). These compounds showed antibacterial activity against Gram-positive bacteria and cytotoxicity against cancer cell lines.

From jellyfish and mollusc

Two strains of *Emericella unguis* (M87-2 and M90B-10) gave two cyclic heptapeptides, unguisins A (**28**) and B and a chlorinated depside, guisinol (**29**) (Malmstrøm, 1999; Nielsen et al., 1999). The strain M87-2 was isolated from a solution of the medusa *Stomolopus meliagris* mixed with water and M90B-10 was from the soft part of an unidentified mollusc at Paria Bay in Venezuela. *E. unguis* M90A-2 isolated from scrape of the shell of an unidentified mollusc at the same site gave two new cyclic heptapeptides, unguisins C and D, together with unguisins A and B (Malmstrøm et al., 2002b). Unguisins A and B were reported to show moderate antibacterial activity against *S. aureus* (no quantitative data) (Malmstrøm, 1999). Guisinol exhibited moderate antibacterial activity against *S. aureus*.

Paraphaeosphaeria sp. N-119 isolated from the marine horse mussel *Modiolus auriculatus* in Okinawa afforded modiolides A (**30**) and B and modiolin (Tsuda et al., 2003). Modiolides A and B exhibited antibacterial activity against *M. luteus* (MIC = 16.7 μg/mL) and antifungal activity against *N. crassa* (MIC = 33.3 μg/mL).

Although these compounds were screened mostly for antimicrobial activities against medically and agrochemically significant bacteria, fungi, yeast, and viruses, they may be potentially toxic against marine species.

NEW CYTOTOXIC METABOLITES

Numerous culture broths of marine fungi have been tested for the cytotoxicity against cultured cancer cell lines for the search of new anticancer medicines and their leads. Cytotoxic metabolites may act as repellents, deterrents, and toxins against potential predators and fouling larvae and zoospores.

This section describes new cytotoxic metabolites produced by fungi isolated from marine organisms (Table 2). The structures of selected metabolites are shown in Figure 2.

Table 2. New cytotoxic metabolites of fungi isolated from marine organisms

Fungus	*Compound*	*Reference*
From alga		
Aspergillus insulicola	Insulicolide A	Rahbæk et al., 1997
Fusarium sp.	*N*-Methylsansalvamide	Cueto et al., 2000
Fusarium chlamydosporum	Fusaperazine A	Usami et al., 2002
Leptosphaeria sp.	Leptosins A–S	Takahashi et al., 1994a, b, 1995a, b; Yamada et al., 2002a, 2004
Penicillium sp.	Verticillins	Son et al., 1999
Penicillium sp.	Communesins A, B Penochalasins A–H Penostatins A–I	Iwamoto et al., 1998, 2001; Numata et al., 1993, 1996; Takahashi et al., 1996
Penicillium waksmanii	Pyrenocine E	Amagata et al., 1998b
Scytalidium sp.	Scytalidamides A, B	Tan et al., 2003
Trichoderma virens	Trichodermamide B	Garo et al., 2003
From seagrass		
Fusarium sp.	Sansalvamide A	Belofsky et al., 1999
Lignincola laevis	Phosphorohydrorazide thioate	Abraham et al., 1994

Contd.

Fungus	Compound	Reference
From sponge		
Aspergillus niger	Asperazine	Varoglu et al., 1997
Cladosporium herbarum	Herbarins A, B	Jadulco et al., 2002
Emericella variecolor	Evariquinone	Bringmann et al., 2003
Emericella variecolor	Varitriol	Malmstrøm et al., 2002a
Gymnascella dankaliensis	Dankasterone Gymnastatins A–E Gymnasterones A, B	Amagata et al., 1998c, d, 1999; Numata et al., 1997a
Myrothecium verrucaria	Trichothecenes	Amagata et al., 2003a
Penicillium sp.	Communesins C, D	Jadulco et al., 2004
Penicillium brocae	Brocaenols A, B	Bugni et al., 2003
Trichoderma harzianum	Harzialactone B Trichodenones A–C	Amagata et al., 1998a
From sea hare		
Periconia byssoides	Macrosphelides Pericosines A, B	Numata et al., 1997b; Yamada et al., 2001, 2002b
From other invertebrates		
Acremonium striatisporum	Virescenosides M–U	Afiyatullov et al., 2000, 2002, 2004
Hyphomycetes sp.	Kasarin	Suenaga et al., 2000
unidentified fungus	Spiroxins A–E	McDonald et al., 1999
From fish		
Aspergillus fumigatus	Fumiquinazolines A–G	Numata et al., 1992; Takahashi et al., 1995c
Penicillium fellutanum	Fellutamides A, B	Shigemori et al., 1991

From alga

Aspergillus insulicola isolated from various algae in the Bahamas (Rahbæk et al., 1997) and *Aspergillus versicolor* CNC 327 isolated from the green alga *Penicillus capitatus* also in the Bahamas (Belofsky et al., 1998) gave insulicolide A (**31**) and three non-cytotoxic compounds. Insulicolide A exhibited a mean IC_{50} of 1.1 μg/mL in the NCI's 60-cell line panel.

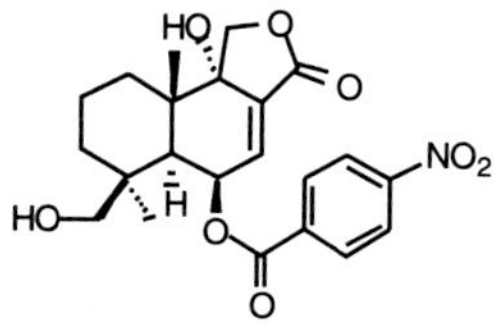

Insulicolide A (**31**)

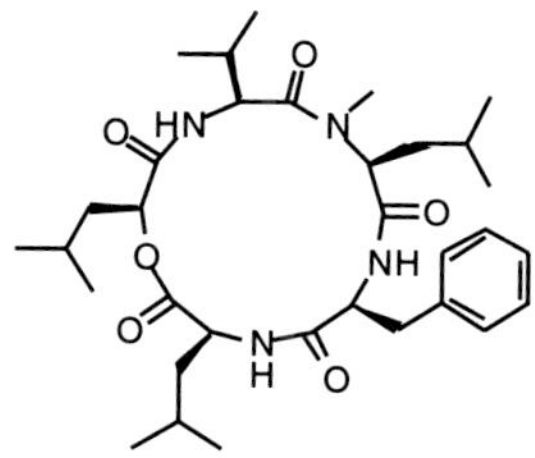

N-Methylsansalvamide (**32**)

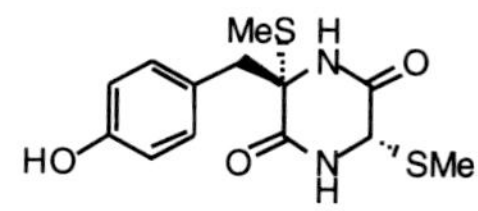

Fusaperazine A (**33**)

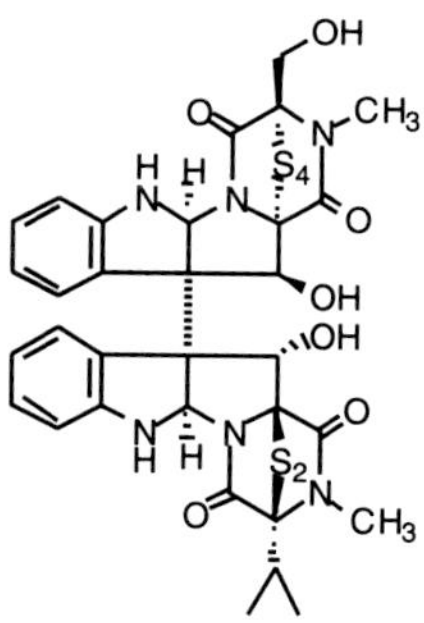

Leptosin A (**34**)

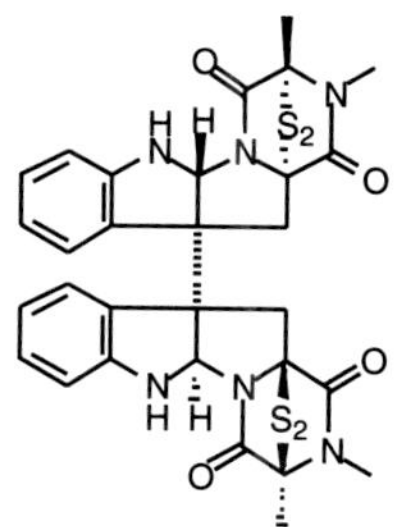

11,11'-Dideoxyverticillin A (**35**)

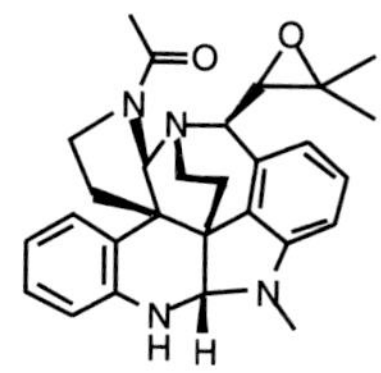

Communesin A (**36**)

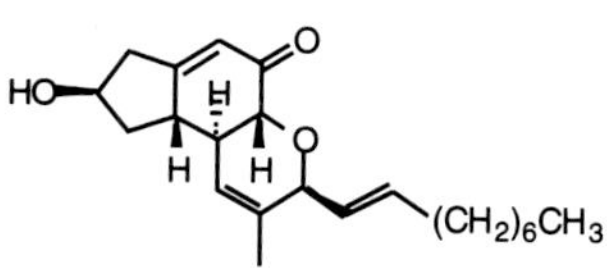

Penostatin A (**37**)

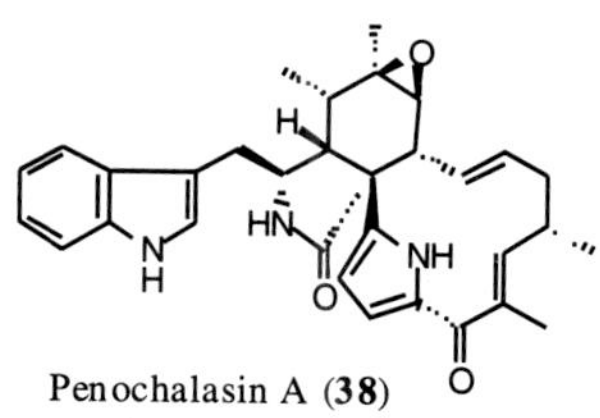

Penochalasin A (**38**)

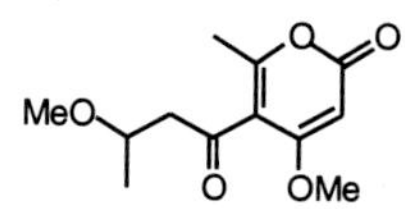

Pyrenocine E (**39**)

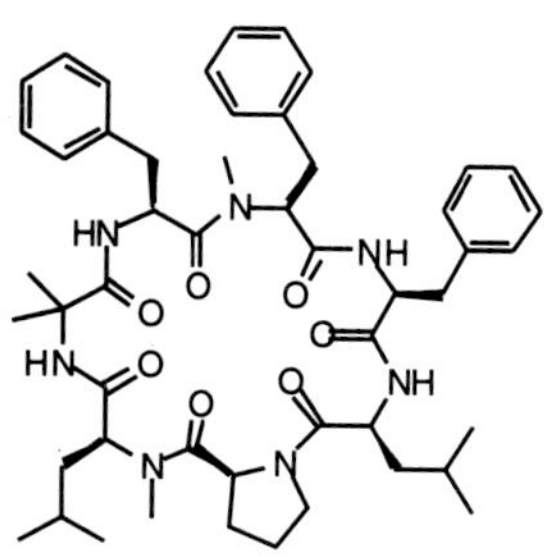

Scytalidamide A (**40**)

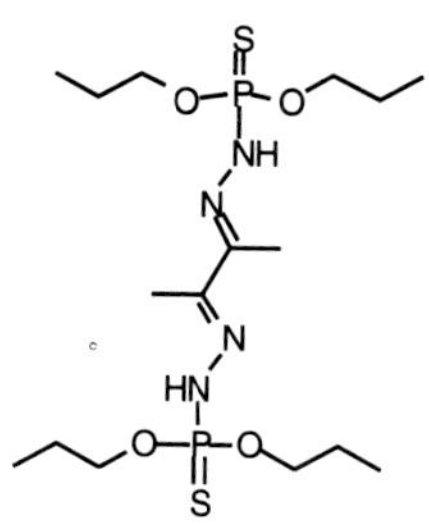

Phosphorohydrorazide Thioate (**41**)

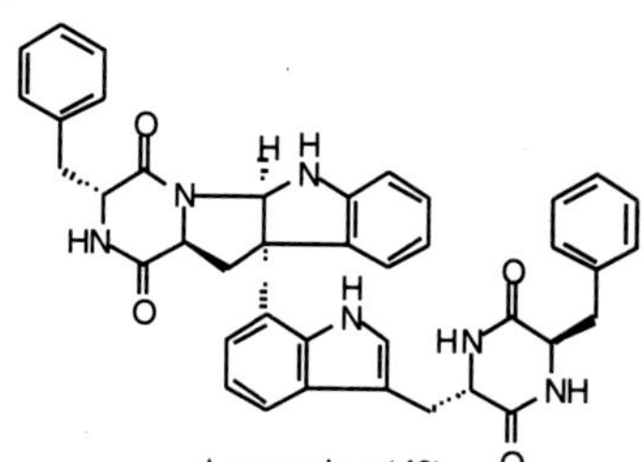

Asperazine (**42**)

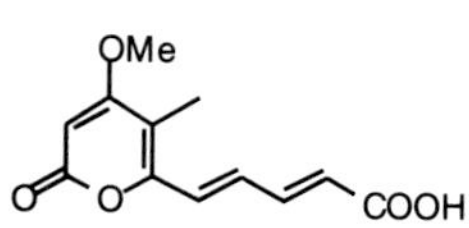

Herbarin A (**43**)

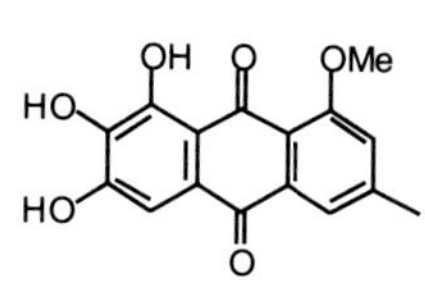

Evariquinone (**44**)

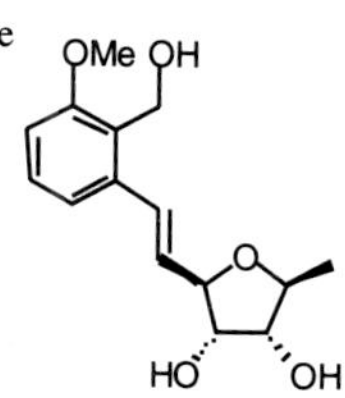

Varitriol (**45**)

Fig. 2. Structures of cytotoxic metabolites of fungi isolated from marine organisms

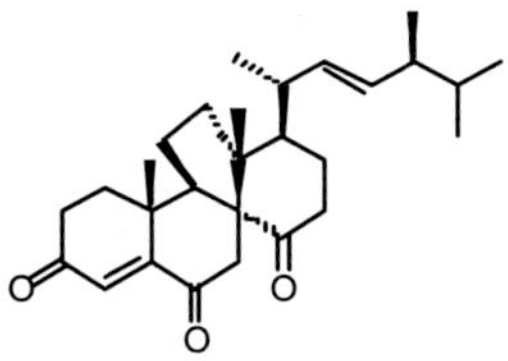

Dankasterone (**46**)

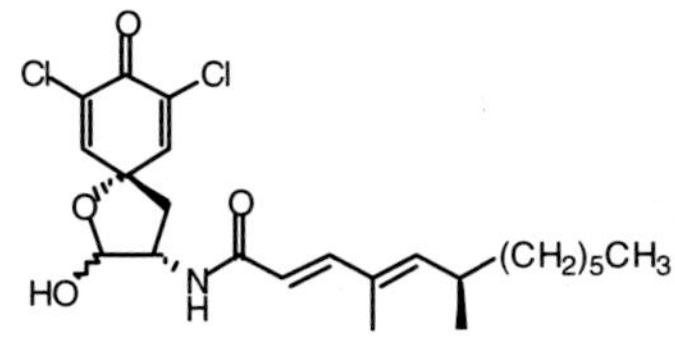

Gymnastatin A (**47**) α:β = 1:2

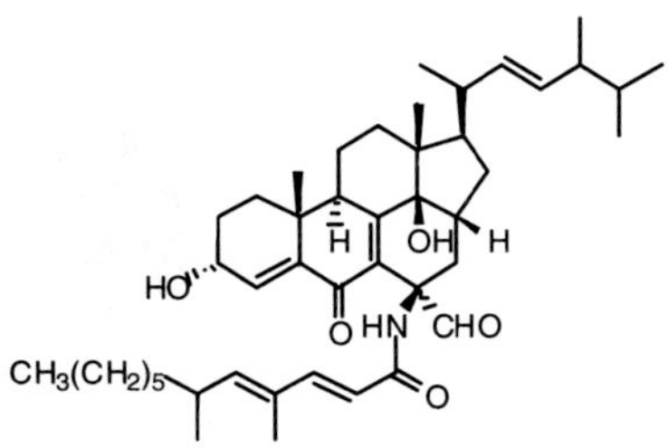

Gymnasterone A (**48**)

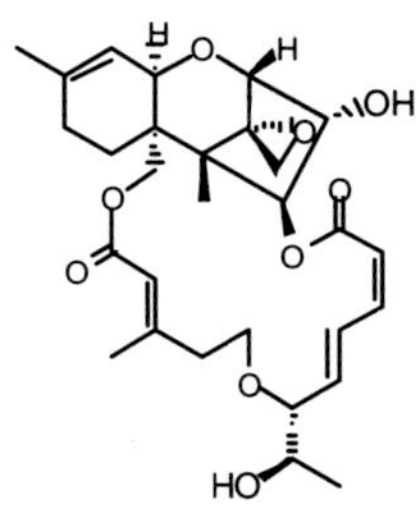

3-Hydroxyroridin E (**49**)

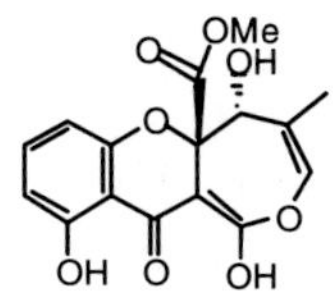

Brocaenol A (**50**)

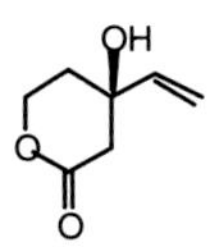

Harzialactone B (**51**)

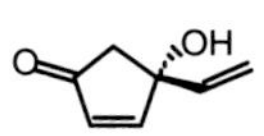

Trichodenone A (**52**)

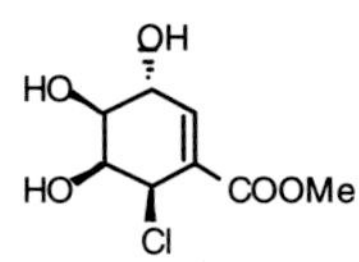

Pericosine A (**53**)

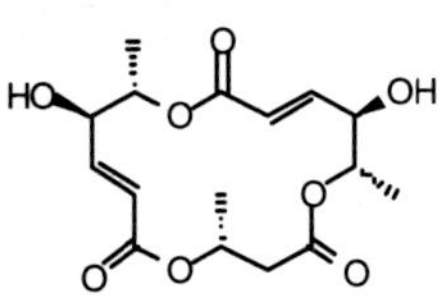

Macrosphelide E (**54**)

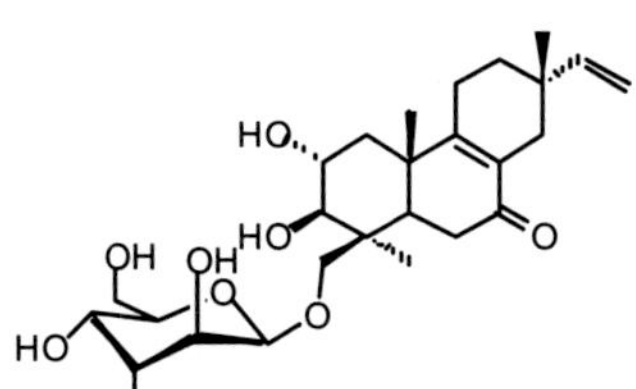

Virescenoside M (**55**)

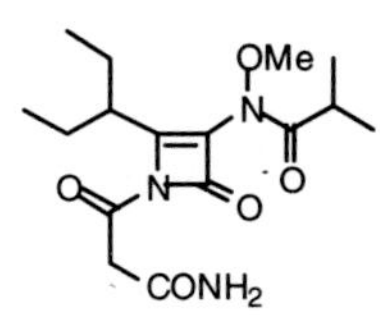

Kasarin (**56**)

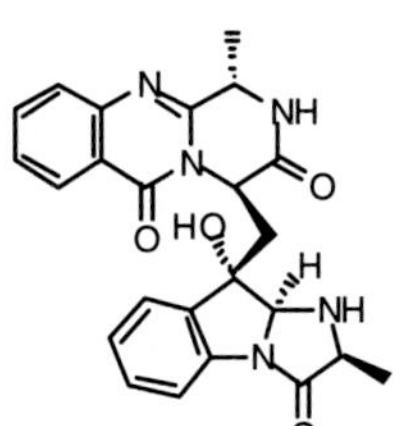

Fumiquinazoline A (**57**)

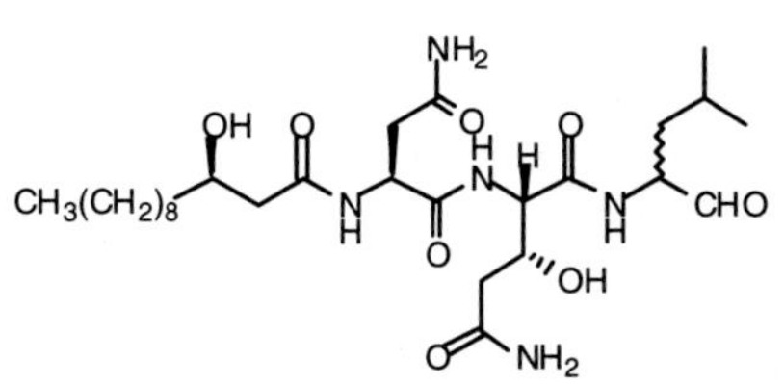

Fellutamide A (**58**)

Fig. 2 continued. Structures of cytotoxic metabolites of fungi isolated from marine organisms

Fusarium sp. CNL-619 was isolated from the green alga *Avrainvillea* sp. at US Virgin Islands and produced *N*-methylsansalvamide (**32**), which showed moderate cytotoxicity (GI_{50} = 8.3 μM) in the NCI's human tumor cell line screen (Cueto et al., 2000).

Fusarium chlamydosporum OUPS-N124 isolated from the red alga *Carpopeltis affinis* afforded fusaperazines A (**33**) and B (Usami et al., 2002). Fusaperazine A was modestly cytotoxic against the murine leukemia cell line P388 (ED_{50} = 22.8 μg/mL).

Leptosphaeria sp. OUPS-N80 (formerly OUPS-4) isolated from the brown alga *Sargassum tortile* is a prolific fungus and 24 new compounds, leptosins A (**34**)–S, have been obtained from this strain (Takahashi et al., 1994a, 1994b, 1995a, 1995b, Yamada et al., 2002a, 2004). Leptosins showed strong cytotoxicity against P388 with ED_{50} values in the ng order.

Penicillium sp. CNC-350 isolated from the surface of the green alga *Avrainvillea longicaulis* in the Bahamas gave two similar dimeric diketopiperazines, 11,11'-dideoxyverticillin A (**35**) and 11'-deoxyverticillin A (Son et al., 1999). These compounds had potent cytotoxicity against the human colon carcinoma cell line HCT-116 (IC_{50} = 30 ng/mL).

Penicillium sp. OUPS-N79 (formerly OUPS-79) isolated from the green alga *Enteromorpha intestinalis* at Tanabe Bay is also a prolific fungus, and 19 new compounds, communesins A (**36**) and B, penostatins A (**37**)–I, and penochalasins A (**38**)–H have been obtained (Numata et al., 1993, 1996, Takahashi et al., 1996, Iwamoto et al., 1998, 1999, 2001). These compounds exhibited potent cytotoxicity against P388. *Penicillium* sp. isolated from the Mediterranean sponge *Axinella verrucosa* gave cytotoxic communesins C and D together with B (Jadulco et al., 2004). Communesins A–D possess a novel carbon skeleton, and only one other compound with a similar skeleton, perophoramidine, has been isolated from the ascidian *Perophora nameii* (Verbitski et al., 2002). Therefore, perophoramidine may be of a microbial origin.

Penicillium waksmanii OUPS-N127 was isolated from the brown alga *Sargassum ringgoldianum* and produced pyrenocines D and E (**39**) (Amagata et al., 1998b). Pyrenocine E (**70**) showed cytotoxicity against P388 (ED_{50} = 1.30 μg/mL).

Scytalidium sp. CNC-310 isolated from the surface of the green alga *Halimeda* sp. collected in the Bahamas yielded scytalidamides A (**40**) and B (Tan et al., 2003). Scytalidamides A and B were cytotoxic against HCT-116 (IC_{50} = 2.7 and 11.0 μM, respectively).

Trichodermamide B (**5**, Figure 1), an antimicrobial metabolite of *Trichoderma virens* strains CNK266 and CNL910 (Table 1), exhibited significant cytotoxicity against HCT-116 (IC_{50} = 0.32 μg/mL) (Garo et al., 2003).

From seagrass

Sansalvamide A (**9**, Figure 1), an antiviral metabolite afforded from *Fusarium* sp. CNL 292 (Table 1), was cytotoxic with a mean IC_{50} value of 27.4 µg/mL in the NCI's 60-cell line panel (Belofsky et al., 1999).

An obligate marine fungus *Lignincola laevis* isolated from a marsh grass produced phosphorohydrorazide thioate **(41)** (Abraham et al., 1994). Compound **41** had a unique structure and exhibited cytotoxicity against the murine leukemia cell line L1210 at 0.25 µg/mL level.

From marine sponge

Aspergillus niger was isolated from the sponge *Hyrtios proteus* collected at Florida and gave asperazine (**42**), which showed leukemia selective cytotoxicity in the Corbett-Valeriote soft agar disc diffusion assay (Varoglu et al., 1997).

Cladosporium herbarum isolated from the Mediterranean sponge *Aplysina aerophoba* afforded herbarins A (**43**) and B (Jadulco et al., 2002). These compounds caused a brine shrimp (*Artemia salina*) mortality (75 and 65% at 50 µg, respectively).

Emericella variecolor E-00-6/3 isolated from the interior of the Mediterranean sponge *Haliclona valliculata* produced evariquinone (**44**) and isoemericellin (Bringmann et al., 2003). Evariquinone showed antiproliferative activity toward the human cervix carcinoma KB and non-small cell lung cancer NCI-H460 cells (60% and 69% inhibition, respectively, at 3.16 µg/mL).

The Caribbean sponge-derived *Emericella variecolor* M75-2 produced varitriol (**45**), varixanthone (**17**, Figure 1) (Table 1), dihydroterrein, and varioxirane (Malmstrøm et al., 2002a). Varitriol displayed more than 100-fold increased potency against selected renal, CNS, and breast cancer cell lines in the NCI's 60-cell panel.

Gymnascella dankaliensis OUPS-N134 isolated from the sponge *Halichondria japonica* collected at Osaka Bay, Japan afforded eight new cytotoxic compounds, dankasterone (**46**), gymnastatins A (**47**)–E, and gymnasterones A (**48**) and B (Numata et al., 1997a; Amagata et al., 1998c, 1998d, 1999). Gymnastatins A–C exhibited potent cytotoxicity against P388 (ED_{50} = 0.018, 0.108, and 0.106 µg/mL, respectively). Dankasterone and gymnasterone B showed moderate cytotoxicity against P388 (ED_{50} = 2.2 and 1.6 µg/mL, respectively). Gymnasterone A and gymnastatins D and E were less cytotoxic to P388 (ED_{50} = 10.1, 10.5, and 10.8 µg/mL, respectively).

Myrothecium verrucaria 973023 was isolated from the sponge *Spongia* sp. at Hawaii, and three cytotoxic trichothecenes, 3-hydroxyroridin E (**49**),

13'-acetyltrichoverrin B, and miophytocen C were obtained from the culture broth (Amagata et al., 2003a). These compounds showed significant inhibition against murine and human tumor cell lines.

Penicillium brocae F97S76 isolated from a tissue homogenate of the sponge *Zyzzya* sp. at Fiji gave brocaenols A (**50**), B, and C (Bugni et al., 2003). Brocaenols A and B showed very weak cytotoxicity against HCT-116 (IC_{50} = 20 and 50 µg/mL, respectively).

Trichoderma harzianum OUPS-N115 isolated from the sponge *Halichondria okadai* at Tanabe Bay, Japan afforded harzialactones A and B (**51**) and trichodenones A (**52**)–C (Amagata et al. 1998a). Trichodenones A–C exhibited cytotoxicity against P388 (ED_{50} = 0.21, 1.21, and 1.45 µg/mL, respectively) and harzialactone B had very week cytotoxicity (ED_{50} = 60 µg/mL).

From sea hare

Periconia byssoides OUPS-N133 isolated from the gastrointestinal tract of the sea hare *Aplysia kurodai* collected at Osaka Bay produced nine new compounds, pericosines A (**53**) and B, macrosphelides E (**54**)–I and L, and *seco*-macrosphelide E (Numata et al., 1997b, Yamada et al., 2001, 2002b, Nakamura et al., 2002). Pericosine A showed significant cytotoxicity against P388 (ED_{50} = 0.12 µg/mL), while B was less active (ED_{50} = 4.0 µg/mL). Macrosphelides inhibited the adhesion of the human leukemia HL-60 cells to human umbilical vein endothelial cells.

From other invertebrates

Acremonium striatisporum KMM 4401 isolated from superficial mycobiota of the holothurian *Eupentacta fraudatrix* collected at the Sea of Japan, and the Russian Federation gave nine compounds, virescenosides M (**55**)–U (Afiyatullov et al., 2000, 2002, 2004). Virescenosides were cytotoxic against Ehrlich carcinoma cells (IC_{50} = 10–100 µM). Virescenosides M, N, and P showed toxic effect to developing eggs of the sea urchin *Strongylocentrotus intermedius* (MIC_{50} = 2.7–20 µM).

Hyphomycetes sp. isolated from the zoanthid *Zoanthus* sp. at Amami Island, Japan afforded kasarin (**56**), which had very weak cytotoxicity against P388 (IC_{50} = 34 µg/mL) (Suenaga et al., 2000).

Antibacterial spiroxins A (**30**)–E obtained from the culture extract of an unidentified fungus LL-37H248 isolated from an orange soft coral collected at Vancouver Island in Canada (Table 1) exhibited cytotoxicity against a panel of 25 diverse cell lines (mean IC_{50} of 0.09 µg/mL) (McDonald et al., 1999).

From fish

Aspergillus fumigatus isolated from the gastrointestinal tract of the marine fish *Pseudolabrus japonicus* collected at Tanabe Bay, Japan produced fumiquinazolines A (**57**)–G (Numata et al., 1992, Takahashi et al., 1995c). These compounds showed cytotoxicity against P388 (ED_{50} = 6.1–52.0 μg/mL).

Penicillium fellutanum isolated from the gastrointestinal tract of the Japanese fish *Apogon endekataenia* gave two linear peptides, fellutamides A (**58**) and B (Shigemori et al., 1991). Fellutamide A was cytotoxic against KB, P388, and L1210 (IC_{50} = 0.5, 0.2, and 0.8 μg/mL, respectively).

Cytotoxic bryostatins, obtained from the bryozoan *Bugula neritina* (Pettit et al., 1982, 1996), have recently been revealed to be produced by the symbiotic bacterium, *Endobugula sertula* (Davidson et al., 2001), and prevent the larvae from predation by fish (Lopanik et al., 2004a, 2004b). Therefore, it is curious if the above cytotoxic compounds are used in a protection mechanism of the hosts or if they are produced as toxins against the hosts. Interestingly, cytotoxic compounds against cancer cells also show a lethal toxicity against the brine shrimp (*Artemia salina*) (Jadulco et al., 2002).

POSSIBLE ROLE OF FUNGAL METABOLITES IN THE CHEMICAL DEFENSE SYSTEMS OF MARINE SPONGES

Recently work on the possibility of fungi associated with marine sponges as a participant in the indirect protection mechanism of the hosts was conducted by us, and the results obtained up to date are described below.

Chemical defense systems of marine sponges

Marine sponges (Porifera) are sessile filter feeders that attach to solid substrates and feed on bacteria and other microscopic nutrients. Therefore, marine sponges are always exposed to the threats of infection, fouling, and adhesion by microorganisms and of potential predation.

Three chemical defense systems are considered for the protection of marine sponges (Figure 3). The first mechanism is called the indirect protection system that uses the chemical substances produced by surface-associated microorganisms. The second mechanism uses the chemical substances produced by sponge cells and endosymbiotic microorganisms. This is called the direct protection system. Rather smaller organic molecules are involved in these two systems. The third system is the selection mechanism of 'self' and 'non-self' microorganisms. This mechanism is similar to the immune system, and proteinaceous molecules are utilized here. These molecules are produced in response to the attack by 'non-self'

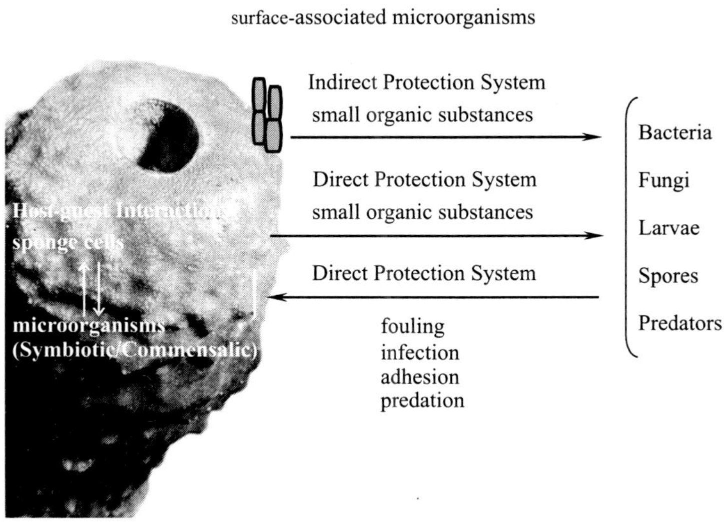

Fig. 3. Chemical defense systems of marine sponges

microorganisms. This mechanism is also the direct protection system.

Microbial associations on surfaces, in tissues, and internal spaces of marine sponges are well known (Hentschel et al., 2003). Bacteria including cyanobacteria are the main associates and large numbers are contained in some sponges. The amount of bacteria sometimes reach around a half of the tissue volumes. These bacteria play important roles in the chemical defense systems of sponges (Schröder et al., 2003). The bacterial metabolites contained in the culture broth of *Alteromonas* sp. isolated from the surface biofilms of the sponge *Halichondria okadai* inhibited the settlement of barnacle (Kon-ya et al., 1995). Recently, secondary metabolites of epibiotic bacteria on the surface of the sponge *Suberites domuncula* were proved to inhibit the growth of fouling bacteria (Thakur et al., 2003). Although the identification of symbiotic microorganisms that are responsible for the production of defensive metabolites obtained from tissue extracts is still very difficult, recent studies have provided strong evidences for the production of the metabolites. Piel et al. (2004) demonstrated the production of polyketide metabolites by an uncultivated endosymbiotic bacterium of the sponge *Theonella swinhoei*.

The association of marine ascomycetes of the genus *Koralionastes* with crustaceous sponges has been reported (Kohlmeyer and Volkmann-Kohlmeyer, 1990). Therefore, fungi may also have ecological roles in associating with marine sponges. The results are described in the next section that fungi isolated from marine sponges are possibly involved in the indirect chemical protection system of marine sponges.

Probable participation of fungi in the indirect protection system of marine sponges

This section describes the results from the study on antibacterial metabolites produced by fungi isolated from tissues of marine sponges. This study has been conducted to examine if these sponge-derived fungi participate in the indirect defense system of the hosts in utilizing their antibacterial metabolites.

Surface tissues of marine sponges were cut and sealed in sterilized plastic bags in water and stored in a cooler box or refrigerator. The tissue was cut into small pieces and smashed with sterilized seawater in a mortar with a pestle. The liquid portion was applied on an agar plate. Three pieces of the tissue remaining in the mortar were sucked to remove water, dried if necessary, and applied on an agar plate. Fungal mycelia grown from the tissues or in the agar plates of liquid portions were isolated and inoculated into slants. Each isolate was cultured in a plastic plate with 15 mL of 1/2 potato dextrose medium (50% seawater) for about three weeks at 20°C (steady culture). Methanol (5 mL) was added to the culture broth and the mixture was stored at –30°C until bioassays. The culture broths were tested against the marine bacterium *R. atlantica* TUF-D, which was isolated from slide grasses submerged in the coastal water at Kanagawa, Japan for one day (Namikoshi et al., 2003). *R. atlantica* is one of the common bacteria in the marine environment attaching on solid substrates and potentially fouling on marine sessile organisms such as sponges. Four terrestrial microorganisms, *S. aureus* IAM 12544T (Gram-positive bacterium), *E. coli* IAM 12119T (Gram-negative bacterium), *S. cerevisiae* IAM 1438T (yeast), and *Mucor hiemalis* IAM 6088 (filamentous fungus), were also tested for the growth inhibition.

Table 3 shows the results from isolation of fungi at Manado, North Sulawesi, Indonesia. One hundred and four sponges (89.7%) out of 116 species gave 226 fungal strains from tissue samples and 175 fungi were isolated from 77 liquid portions (66.4%). Thirty sponges afforded fungi only from tissue samples and three sponges gave fungi only from liquid portions. Interestingly, strains isolated from tissue samples were mostly different species from those obtained from liquid portions of the same sponges. The extracts of 116 sponges were also tested against *R. atlantica* and *M. hiemalis*. The antibacterial activity was detected for 12 extracts (10.3%), and two out of 12 sponges gave antibacterial fungal strains against *R. atlantica*. Fourteen sponge extracts (12.1%) exhibited antifungal activity, and 11 sponges (78.6%) out of 14 afforded fungi from their tissues. Therefore, these sponges show the selective antifungal property. The antibacterial activity against *R. atlantica* was detected for 15 out of 226 fungal strains (6.64%) isolated from tissue samples. The ratio of active strains was

Table 3. Isolation of fungi from marine sponges collected at Manado, North Sulawesi, Indonesia

No. of sponges collected	*No. of sponges yielded fungi (no. of fungi)*		
	From tissue	*From liquid*	*Total*
116	104 (226) (89.7%)	77 (175) (66.4%)	107 (401) (92.2%)

about three times higher than that of the strains isolated from liquid portions (4 out of 175 strains, 2.29%). Similar results were observed for other tropical north Pacific sponges.

Bioassay-guided isolation afforded three new (**60–62**) and five known (**59** and **63–66**) antibacterial metabolites against *R. atlantica* from 10 fungal strains (Figure 4). Penicillic acid (**59**) was obtained from seven strains of *Aspergillus* spp. collected at Manado in Indonesia, Palau, Pohnpei in the Federated States of Micronesia, and Chiba in Japan as an antibacterial metabolite against *R. atlantica*. Five out of seven strains exclusively produced penicillic acid as the sole antibacterial metabolite. Penicillic acid is well known as a mycotoxin and was first reported in 1936. Biological properties of penicillic acid have been reported to show toxicity to animals (Murnaghan, 1946, Hayes, 1977), inhibition of RNase and urease activity (Reiss, 1979; Tashiro et al., 1979), and phytotoxicity against plant roots and germination of seeds (Sassa et al., 1971, Keromnes and Thouvenot, 1985). However, these mycotoxic and phycotoxic activities were weak or marginal and almost no antibacterial activity was detected. In the experiments conducted, penicillic acid did not inhibit the growth of terrestrial bacteria at 100 μg/disc and exhibited strong antibacterial activity against *R. atlantica* (Table 4). From the property of biological activities and a wide distribution in the tropical and temperate North Pacific Ocean, it can be speculated that penicillic acid may be a common antibiotic against fouling bacteria produced by fungi associated with marine sponges.

A. ostianus TUF 01F313 isolated at Pohnpei produced penicillic acid (**59**) and three new antibacterial metabolites, 8-Chloro-9-hydroxy-8,9-deoxyasperlactone (**60**), 9-Chloro-8-hydroxy-8,9-deoxyasperlactone (**61**), and 9-Chloro-8-hydroxy-8,9-deoxyaspyrone (**62**) (Namikoshi et al., 2003). When cultured in the freshwater medium, this strain gave penicillic acid, asperlactone (**63**), and aspyrone (**64**) (Figure 5). Asperlactone and aspyrone were obtained together with penicillic acid from the seawater culture broth of *Aspergillus* sp. TUF 01F328 isolated at Pohnpei. Therefore, strain TUF 01F313 has the mechanism to incorporate Cl atoms in the metabolites to give **60–62** probably as a drain of the chloride ion (Namikoshi, 2006). This mechanism did not work with the Br ion that strain TUF 01F313 produced

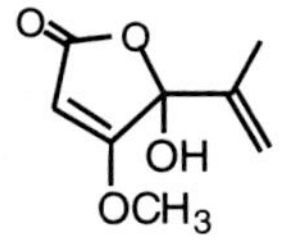

Penicillic Acid (**59**)

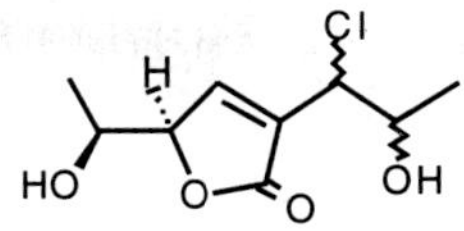

(**60**)

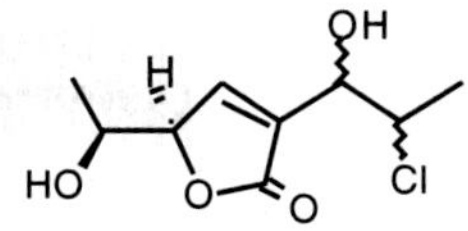

(**61**)

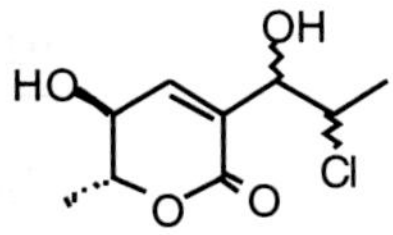

(**62**)

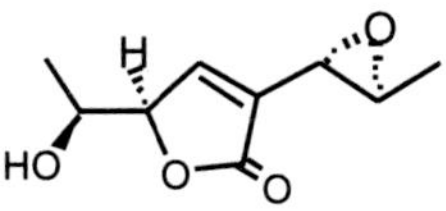

Asperlactone (**63**)

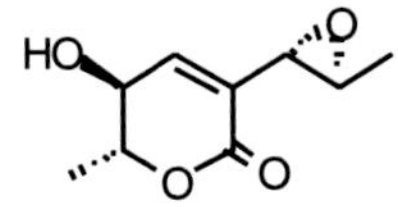

Aspyrone (**64**)

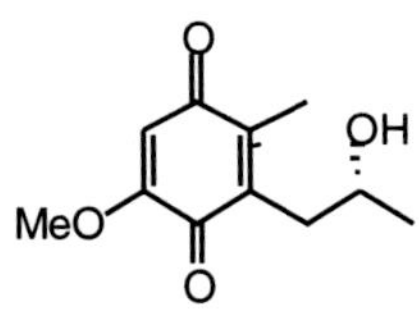

Anserinone B (**65**)

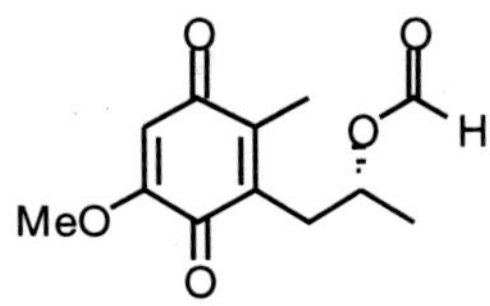

(+)-Formylanserinone B (**66**)

Fig. 4. Structures of antibacterial metabolites against the marine bacterium *Ruegeria atlantica* produced by fungi isolated from marine sponges

Table 4. Antibacterial activity of penicillic acid (**59**), compounds **60–62**, asperlactone (**63**), aspyrone (**64**), anserinone B (**65**), and (+)-formylanserinon B (**66**)

Compound	*R. atlantica*				*E. coli*		*S. aureus*			
	50[a]	25	10	5	50	25	50	25	10	5
59	–	–	19.8[b]	16.2	n. a.	n. a.	n. a.	n. a.	n. a.	n. a.
60	29.2	24.9	17	12.7	11.6	n. a.	13.2	10.2	n. a.	n. a.
61	14.1	10.1	n. a.	n. a.	n. a.	n. a.	9.9	n. a.	n. a.	n. a.
62	17.8	10.5	n. a.	n. a.	n. a.	n. a.	9.7	n. a.	n. a.	n. a.
63	35.2	25.4	20.9	13.4	n. a.	n. a.	n. a.	n. a.	n. a.	n. a.
64	33.5	30.9	20.3	15.8	n. a.	n. a.	n. a.	n. a.	n. a.	n. a.
65	15.3	n. a.	n. a.	n. a.	n. a.	n. a.	16.2	13.5	12.1	10.5
66	18.2	13.5	n. a.	n. a.	n. a.	n. a.	15.9	13.6	12.4	11.8

Test organisms: *Ruegeria atlantica* TUF-D, *Escherichia coli* IAM 12119T, *Staphylococcus aureus* IAM 12544T.

[a]Concentration (μg/disc).

[b]Inhibition zone (diameter, mm).

n. a.: Not active. (–): Not tested.

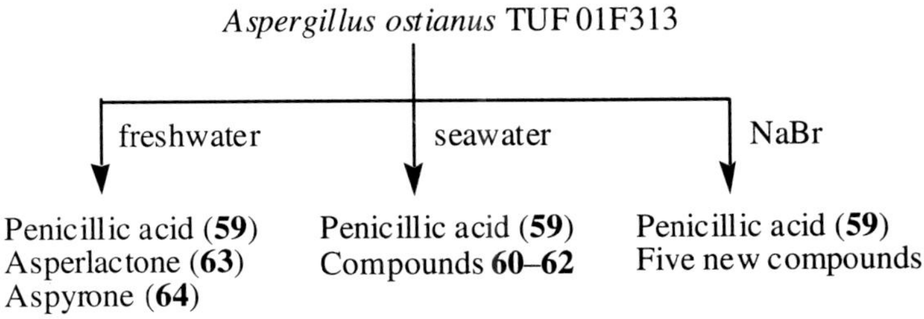

Fig. 5. Culture experiments of *Aspergillus ostianus* TUF 01F313 in different salt conditions

penicillic acid and five new compounds in the Br based medium (altered NaCl to NaBr in an artificial seawater) (Figure 5). Five new compounds are considered to be precursors or their variants in the biosynthesis of asperlactone and aspyrone.

Asperlactone and aspyrone were first reported in 1980 and 1967, respectively (Bereton et al., 1980, Mills and Turner, 1967) and showed weak antifungal (Torres et al., 1998), modest nematicidal (Kimura et al., 1996), and insect growth regulatory activities (Balcells et al., 1995). These compounds had similar antibacterial spectrum to that of penicillic acid (Table 4). Therefore, these compounds may also be produced as antibiotics against marine bacteria.

Anserinone B (**65**) and (+)-formylanserinone B (**66**) were isolated from three strains of *Aspergillus* spp. collected at Manado, Indonesia. These compounds have been obtained from *Penicillium corylophilum* a004181 and b004181 (mixture of two strains) isolated from a South Pacific deep water (–1335 m) sediment together with five related compounds, anserinone A, (–)-epoxyserinone A, (+)-epoxyserinone B, deoxyanserinone B, and hydroxymethylanserinone B (Gautschi et al., 2004). Interestingly, anserinones A and B were first isolated from the coprophilous fungus *Podospora anserina* as antifungal and antibacterial components (Wang et al., 1997). (+)-Formylanserinone B was reported to show significant cytotoxicity (Gautschi et al., 2004). Anserinone B and (+)-formylanserinone B exhibited strong antibacterial activity against *S. aureus* and to a lesser extent to *R. atlantica* (Table 4).

The above results suggested that fungi might participate in the indirect chemical defense system of marine sponges with their ability to produce strong antibiotics against marine bacteria. However, further investigations are necessary if these fungi are active on the surface of sponges and if removal of the fungi causes an infection or adhesion of bacteria.

It may be speculated that some fungi have acquired the ability to produce new secondary metabolites in adopting the respective marine habitats (Höller et al., 2000) and that some of them have re-colonized in terrestrial habitats, because some terrestrial fungi show the tolerance to the presence of NaCl in their culture broths (Tresner and Hayes, 1971).

CONCLUSION AND FUTURE PERSPECTIVES

Marine fungi (including 'marine-derived fungi') are rich sources of biologically active secondary metabolites and interesting research objects for chemical ecological studies. A possible role of antimicrobial metabolites produced by fungi isolated from marine organisms is suggested to be a negative cue against fouling, settlement, colonization, adhesion, and infection by potentially harmful microorganisms. Cytotoxic metabolites of fungi isolated from marine organisms may play as deterrents against larvae and potential predators.

It has been demonstrated that the possibility of fungi associated with marine sponges may participate in the indirect chemical defense system of the hosts in providing antibacterial metabolites. Penicillic acid (**59**) may be a ubiquitous antibiotic of marine fungi because this antibiotic strongly inhibited the growth of the marine bacterium *R. atlantica*, a common fouling species, with high selectivity from terrestrial bacteria and the producing fungi have been isolated from a large area of the North Pacific Ocean.

Chemical ecological studies on marine fungi and their association with the hosts provide a better understanding of the importance of fungi in the marine environment. It has become clear that tests of antimicrobial properties of marine fungal metabolites are not adequate to measure the roles of fungi in the chemical defense systems of the hosts. Assays that observe behavioral effects on potentially noxious microorganisms and planktons are also important in evaluating the secondary metabolites as agents of the chemical defense systems. Future studies on the deterrence of colonization, adhesion, and infection and modulation of morphology will also be as important as the effects on the growth of microorganisms to investigate the association of fungi with other organisms. Chemical relationships between fungi and their host organisms are very interesting future studies not only in understanding the marine fungal ecology but also in searching new bioprospects. Chemical ecological studies will provide novel secondary metabolites, which can be used for screening of new drug candidates and their lead compounds. Chemical cues for the prevention of fouling may be utilized for new less toxic marine antifoulants. Therefore, close collaborations have to be established among marine ecologists for their field-based observations, marine microbiologists to isolate and culture the microorganisms, pharmacologists to identify target receptors, molecular biologists to elucidate the genes, and chemists to determine the biologically active metabolites.

References

Abbanat, D., Leighton, M., Maiese, W., Jones, E.B., Pearce, C. and Greenstein, M. 1998. Cell wall active antifungal compounds produced by the marine fungus *Hypoxylon oceanicum* LL-15G256. I. Taxonomy and fermentation. Journal of Antibiotics, 51: 296-302.

Abdel-Wahab, M.A. and El-Sharouny, H.M. 2002. Ecology of subtropical mangrove fungi with emphasis on Kandelia candel mycota, pp. 247-265. In: Fungi in Marine Environments, K.D. Hyde (ed.), Fungal Diversity Press, Hong Kong.

Abraham, S.P., Hoang, T.D., Alam, M. and Jones, E.B.G. 1994. Chemistry of the cytotoxic principles of the marine fungus *Lignincola laevis*. Pure and Applied Chemistry, 66: 2391-2394.

Abrell, L.M., Borgeson, B. and Crews, P. 1996. A new polyketide, secocurvularin, from the salt water culture of a sponge derived fungus. Tetrahedron Letters, 37: 8983-8984.

Afiyatullov, S.S., Kuznetsova, T.A., Isakov, V.V., Pivkin, M.V., Prokof'eva, N.G. and Elyakov, G.B. 2000. New diterpenic altrosides of the fungus *Acremonium striatisporum* isolated from a sea cucumber. Journal of Natural Products, 63: 848-850.

Afiyatullov, S.S., Kalinocsky, A.I., Kuznetsova, T.A., Isakov, V.V., Pivkin, M.V., Dmitrenok, P.S. and Elyakov, G.B. 2002. New diterpene glycosides of the fungus *Acremonium striatisporum* isolated from a sea cucumber. Journal of Natural Products, 65: 641-644.

Afiyatullov, S.S., Kalinocsky, A.I., Kuznetsova, T.A., Pivkin, M.V., Prokofeva, N.G., Dmitrenok, P.S. and Elyakov, G.B. 2004. New glycosides of the fungus *Acremonium striatisporum* isolated from a sea cucumber. Journal of Natural Products, 67: 1047-1051.

Alva, P., McKenzie, E.H.C., Pointing, S.B., Pena-Muralla, R. and Hyde, K.D. 2002. Do sea grasses harbor endophytes? pp. 167-178. In: Fungi in Marine Environments. K.D. Hyde (ed.) Fungal Diversity Press, Hong Kong.

Amagata, T., Usami, Y., Minoura, K., Ito, T. and Numata, A. 1998a. Cytotoxic substances produced by a fungal strain from a sponge: physico-chemical properties and structures. Journal of Antibiotics, 51: 33-40.

Amagata, T., Minoura, K. and Numata, A. 1998b. Cytotoxic metabolites produced by a fungal strain from a *Sargassum* alga. Journal of Antibiotics, 51: 432-434.

Amagata, T., Minoura, K. and Numata, A. 1998c. Gymnasterones, novel cytotoxic metabolites produced by a fungal strain from a sponge. Tetrahedron Letters, 39: 3773-3774.

Amagata, T., Doi, M., Ohta, T., Minoura, K. and Numata, A. 1998d. Absolute stereostructures of novel cytotoxic metabolites, gymnastatins A-E, from a *Gymnascella* species separated from a *Halichondria* Sponge. Journal of the Chemical Society, Perkin Transaction, 1 3585-3599.

Amagata, T., Doi, M., Tohgo, M., Minoura, K. and Numata, A. 1999. Dankasterone, a new class of cytotoxic steroid produced by a *Gymnascella* species from a marine sponge. Chemical Communications, 1321-1322.

Amagata, T., Rath, C., Rigot, J.F., Tarlov, N., Tenney, K., Valeriote, F.A. and Crews,

P. 2003a. Structures and cytotoxic properties of trichoverroids and their macrolide analogues produced by saltwater culture of *Myrothecium verrucaria*. Journal of Medicinal Chemistry, 46: 4342-4350.

Amagata, T., Amagata, A., Tenney, K., Valeriote, F.A., Lobkovsky, E., Clardy, J. and Crews, P. 2003b. Unusual C25 steroids produced by a sponge-derived *Penicillium citrinum*. Organic Letters, 5: 4393-4396.

Armstrong, E., Yan, L.M., Boyd, K.G., Wright, P.C. and Burgess, J.G. 2001. The symbiotic role of marine microbes on living surfaces. Hydrobiologia, 461: 37-40.

Balcells, M., Canela, R., Coll, J., Sanchis, V. and Torres, M. 1995. Effect of fungal metabolites and some derivatives against *Tribolium castaneum* (Herbst) and *Nezara viridula* (L.). Pesticide Science, 45: 319-323.

Belofsky, G.N., Jensen, P.R., Renner, M.K. and Fenical, W. 1998. New cytotoxic sesquiterpenoid nitrobenzoyl esters from a marine isolate of the fungus *Aspergillus versicolor*. Tetrahedron, 54: 1715-1724.

Belofsky, G.N., Jensen, P.R. and Fenical, W. 1999. Sansalvamide: a new cytotoxic cyclic depsipeptide produced by a marine fungus of the genus *Fusarium*. Tetrahedron Letters, 40: 2913-2916.

Belofsky, G.N., Anguera, M., Jensen, P.R., Fenical, W. and Kock, M. 2000. Oxepinamides A-C and fumiquinazolines H-I: bioactive metabolites from a marine isolate of a fungus of the genus *Acremonium*. Chemistry—A European Journal, 6: 1355-1360.

Bereton, R.G., Garson, M.J. and Staunton, J. 1980 The use of fourier transform resolution techniques in a study of the biosynthesis of asperlactone from [1,2-$^{13}C_2$]acetate. Journal of the Chemical Society, Chemical Communications, 1165-1167.

Bernan, V.S. 2001. Metabolites of free-living, commensal and symbiotic benthic marine microorganisms, pp. 567-592. In: Marine Chemical Ecology. J.B. McClintock and B.J. Baker (eds.) CRC Press, Boca Raton, Fl., USA.

Bernan, V.S., Greenstein, M. and Maiese, W.M. 1997. Marine Microorganisms as a Source of New Natural Products. Advances in Applied Microbiology, 43: 57-90.

Biabani, M.A.F. and Laatsch, H. 1998. Advances in chemical studies on low-molecular-weight metabolites of marine fungi. Journal für Praktische Chemie Chemiker-Zeitung, 340: 589-607.

Bringmann, G., Lang, G., Steffens, S., Gunther, E. and Schaumann, K. 2003. Evariquinone, isoemericellin, and stromemycin from a sponge derived strain of the fungus *Emericella variecolor*. Phytochemistry, 63: 437-443.

Bugni, T.S. and Ireland, C.M. 2004. Marine-derived fungi: a chemically and biologically diverse group of microorganisms. Natural Product Reports, 21: 143-163.

Bugni, T.S., Abbanat, D., Bernan, V.S., Maiese, W.M., Greenstein, M., van Wagoner, R.M. and Ireland, C.M. 2000. Yanuthones: novel metabolites from a marine isolate of *Aspergillus niger*. Journal of Organic Chemistry, 65: 7195-7200.

Bugni, T.S., Bernan, V.S., Greenstein, M., Janso, J.E., Maiese, W.M., Mayne, C.L. and Ireland, C.M. 2003. Brocaenols A-C: novel polyketides from a marine derived *Penicillium brocae*. Journal of Organic Chemistry, 68: 2014-2017.

Byun, H.-G., Zhang, H., Mochizuki, M., Adachi, K., Shizuri, Y., Lee, W.-J. and Kim, S.-K. 2003. Novel antifungal diketopiperazine from marine fungus. Journal of Antibiotics, 56: 102-106.

Chen, C., Imamura, N., Nishijima, M., Adachi, K., Sakai, M. and Sano, H. 1996. Halymecins, new antimicroalgal substances produced by fungi isolated from marine algae. Journal of Antibiotics, 49: 998-1005.

Correa, J.A. 1997. Infectious diseases of marine algae: current knowledge and approaches. Progress in Phycological Research, 12: 149-180.

Cueto, M., Jensen, P.R. and Fenical, W. 2000. *N*-Methylsansalvamide, a cytotoxic cyclic depsipeptide from a marine fungus of the genus *Fusarium*. Phytochemistry, 55: 223-226.

Cueto, M., Jensen, P.R., Kauffman, C., Fenical, W., Lobkovsky, E. and Clardy, J. 2001. Pestalone, a new antibiotic produced by a marine fungus in response to bacterial challenge. Journal of Natural Products, 64: 1444-1446.

Cuomo, V., Palomba, I., Perretti, A., Guerriero, A., D'Ambrosio, M. and Pietra, F. 1995. Antimicrobial activities from marine fungi. Journal of Marine Biotechnology, 2: 199-204.

Davidson, B.S. 1995. New dimensions in natural products research: cultured marine microorganisms. Current Opinion in Biotechnology, 6: 284-291.

Davidson, S.K., Allen, S.W., Lim, G.E., Anderson, C.M. and Haygood, M.G. 2001. Evidence for the biosynthesis of bryostatins by the bacterial symbiont "*Candidatus* Endobugula sertula" of the bryozoan *Bugula neritina*. Applied and Environmental Microbiology, 67: 4531-4537.

Doshida, J., Hasegawa, H., Onuki, H. and Shimidzu, N. 1996. Exophilin A, a new antibiotic from a marine microorganism *Exophiala pisciphila*. Journal of Antibiotics, 49: 1105-1109.

Edrada, R.A., Heubes, M., Brauers, G., Wray, V., Berg, A., Gräfe, U., Wohlfarth, M., Mühlbacher, J., Schaumann, K., Sudarsono, Bringmann, G. and Proksch, P. 2002. Online analysis of xestodecalactones A–C, novel bioactive metabolites from the fungus *Penicillium* cf. *montanense* and their subsequent isolation from the sponge *Xestospongia exigua*. Journal of Natural Products, 65: 1598-1604.

Garo, E., Starks, C.M., Jensen, P.R., Fenical, W., Lobkovsky, E. and Clardy, J. 2003. Trichodermamides A and B, cytotoxic modified dipeptides from the marine-derived fungus *Trichoderma virens*. Journal of Natural Products, 66: 423-426.

Gautschi, J.T., Amagata, T., Amagata, A., Valeriote, F.A., Mooberry, S.L. and Crews, P. 2004. Expanding the strategies in natural product studies of marine-derived fungi: a chemical investigation of *Penicillium* obtained from deep water sediment. Journal of Natural Products 67: 362-367, (correction: Journal of Natural Products 2004 67: 1638).

Geiser, D.M., Taylor, J.W., Ritchie, K.B. and Smith, G.W. 1998. Cause of sea fan death in the West Indies. Nature, 394: 137-138.

Gil-Turnes, S. and Fenical, W. 1992. Embryos of *Homarus americanus* are protected by epibiotic bacteria. Biological Bulletin, 182: 105-108.

Gil-Turnes, S., Hay, M.E. and Fenical, W. 1989. Symbiotic marine bacteria chemically defend crustacean embryos from a pathogenic fungus. Science, 246: 116-118.

Harper, M.K., Bugni, T.S., Copp, B.R., James, R.D., Lindsay, B.S., Richardson, A.D., Schnabel, P.C., Tasdemir, D., Van Wagoner, R.M., Verbitski, S.M. and Ireland, C.M. 2001. Introduction to the chemical ecology of marine natural products, pp. 3-69. In: Marine Chemical Ecology. J.B. McClintock, and B.J. Baker (eds.), CRC Press, Boca Raton, Fl., USA.

Hayes, A.W. 1977. Effect of the mycotoxin penicillic acid on *Tetrahymena pyriformis*. Toxicon, 15: 497-504.

Hentschel, U., Fieseler, L., Wehrl, M., Gernert, C., Steinert, M., Hacker, J. and Horn, M. 2003. Microbial diversity of marine sponges, pp. 59-88. In: Sponges (Porifera). W.E.G. Müller (ed.), Springer, Berlin, Germany.

Hildebrand, M., Waggoner, L.E., Liu, H., Sudek, S., Allen, S., Anderson, C., Sherman, D.H. and Haygood, M. 2004. *bryA*: An unusual modular polyketide synthase gene from the uncultivated bacterial symbiont of the marine bryozoan *Bugula neritina*. Chemistry & Biology, 11: 1543-1552.

Höller, U., König, G.M. and Wright, A.D. 1999a. A new tyrosine kinase inhibitor from a marine isolate of *Ulocladium botrytis* and new metabolites from the marine fungi *Asteromyces cruciatus* and *Varicosporina ramulosa*. European Journal of Organic Chemistry, 2949-2955.

Höller, U., König, G.M. and Wright, A.D. 1999b. Three new metabolites from marine-derived fungi of the genera *Coniothyrium* and *Microsphaeropsis*. Journal of Natural Products, 62: 114-118.

Höller, U., Wright, A.D., Matthée, G.F., Konig, G.M., Draeger, S., Aust, H.-J. and Schulz, B. 2000. Fungi from marine sponges: diversity, biological activity and secondary metabilotes. Mycological Research, 104: 1354-1365.

Hwang, Y., Rowley, D., Rhodes, D., Gertch, J., Fenical, W. and Bushman, F. 1999. Mechanism of inhibition of a poxvirus topoisomerase by the marine natural product sansalvamide A. Molecular Pharmacology, 55: 1049-1053.

Hyde, K.D, Jones, E.B.G., Leaño, E., Pointing, S.B., Poonyth, A.D. and Vrijmoed, L.L.P. 1998. Role of fungi in marine ecosystems. Biodiversity and Conservation, 7: 1147-1161.

Iwamoto, C., Minoura, K., Hagishita, S., Nomoto, K. and Numata, A. 1998. Penostatins F-I, novel cytotoxic metabolites from a *Penicillium* species separated from an *Enteromorpha* marine alga. Journal of the Chemical Society, Perkin Transaction, 1 449-456.

Iwamoto, C., Minoura, K., Oka, T., Ohta, T., Hagashita, S. and Numata, A. 1999. Absolute stereostructures of novel cytotoxic metabolites, penostatins A-E, from a *Penicillium* species separated from an *Enteromorpha* alga. Tetrahedron, 55: 14353-14368.

Iwamoto, C., Yamada, T., Ito, Y., Minoura, K. and Numata, A. 2001. Cytotoxic cytochalasans from a *Penicillium* species separated from a marine alga. Tetrahedron, 57: 2997-3004.

Jadulco, R., Proksch, P., Wray, V., Sudarsono, Berg, A. and Gräfe, U. 2001. New macrolides and furan carboxylic acid derivative from the sponge-derived fungus *Cladosporium herbarum*. Journal of Natural Products, 64: 527-530.

Jadulco, R., Brauers, G., Edrada, R.A., Ebel, R., Wray, V. Sudarsono and Proksch, P. 2002. New metabolites from sponge-derived fungi *Curvularia lunata* and *Cladosporium herbarum*. Journal of Natural Products, 65: 730-733.

Jadulco, R., Edrada, R.A., Ebel, R., Berg, A., Schaumann, K., Wray, V., Steube, K. and Proksch, P. 2004. New communesin derivatives from the fungus *Penicillium* sp. derived from the Mediterranean sponge *Axinella verrucosa*. Journal of Natural Products, 67: 78-81.

Jenkins, K.M., Toske, S.G., Jensen, P.R. and Fenical, W. 1998a. Solanapyrones E–G, antialgal metabolites produced by a marine fungus. Phytochemistry, 49: 2299-2304.

Jenkins, K.M., Renner, M.K., Jensen, P.R. and Fenical, W. 1998b. Exumolides A and B: antimicroalgal cyclic depsipeptides produced by a marine fungus of the genus *Scytalidium*. Tetrahedron Letters, 39: 2463-2466.

Jensen, P.R. and Fenical, W. 2000. Marine microorganisms and drug discovery: current status and future potential, pp. 6-29. In: Drugs from the Sea. N. Fusetani (ed.), Karger, Basel, Switzerland.

Jensen, P.R. and Fenical, W. 2002. Secondary metabolites from marine fungi, pp. 293-315. In: Fungi in Marine Environments. K.D. Hyde (ed.), Fungal Diversity Press, Hong Kong.

Kendrick, B., Risk, M.J., Michaelides, J. and Bergman, K. 1982. Amphibious microborers: bioeroding fungi isolated from live corals. Bulletin of Marine Science, 32: 862-867.

Keromnes, J. and Thouvenot, D. 1985. Role of penicillic acid in the phytotoxicity of *Penicillium cyclopium* and *Penicillium canescens* to the germination of corn seeds. Applied and Environmental Microbiology, 49: 660-663.

Kimura, Y., Nakahara, S. and Fujioka, S. 1996. Aspyrone, a nematicidal compound isolated from the fungus, *Aspergillus melleus* Bioscience, Biotechnology, and Biochemistry, 60: 1375-1376.

Kjelleberg, S. and Steinberg, P.D. 2002. Defenses against bacterial colonization of marine plants, pp. 152-172. In: Phyllosphere Microbiology. S. Lindow, E. Hecht-Poinbar. and V.Elliott (eds.), American Phytopathological Society Press, St. Paul, USA.

Kobayashi, J. and Ishibashi, M. 1993. Bioactive metabolites of symbiotic marine microorganisms. Chemical Reviews, 93: 1753–1769.

Kohlmeyer, J. and Kohlmeyer, E. 1979 Marine Mycology: The Higher Fungi, Academic Press, New York, USA.

Kohlmeyer, J. and Volkmann-Kohlmeyer, B. 1990. New species of *Koralionastes* (*Ascomycotina*) from the Caribbean and Australia. Canadian Journal of Botany, 68: 1554–1559.

Kohlmeyer, J. and Volkmann-Kohlmeyer, B. (2003). Fungi from coral reefs: a commentary. Mycological Research, 107: 386-387.

Komatsu, K., Shigemori, H. and Kobayashi, J. 2001. Dictyonamides A and B, new peptides from marine-derived fungus. Journal of Organic Chemistry, 66: 6189-6192.

Kon-ya, K., Shimidzu, N., Otaki, N., Yokoyama, A., Adachi, K. and Miki, W. 1995. Inhibitory effect of bacterial ubiquinones on the settling of barnacle, *Balanus amphitrite*. Experientia, 51: 153-155.

Kumaresan, V., Suryanarayanan, T.S. and Johnson, J.A. 2002. Ecology of mangrove endophytes, pp. 145-166. In: Fungi in Marine Environments. K.D. Hyde (ed.), Fungal Diversity Press, Hong Kong.

Le Campion-Alsumard, T., Golubic, S. and Priess, K. 1995. Fungi in corals: symbiosis or disease? Interaction between polyps and fungi causes pearl-like skeleton biomineralization. Marine Ecology Progress Series. 117: 137-147.

Leaño, E.M. 2002. Ecology of straminipiles from mangrove habitats, pp. 111-134. In: Fungi in Marine Environments. K.D. Hyde (ed.), Fungal Diversity Press, Hong Kong.

Liberra, K. and Lindequist, U. 1995. Marine fungi—a prolific resource of biologically active natural products? Pharmazie, 50: 583-588.

Lin, W., Brauers, G., Ebel, R., Wray, V., Berg, A. Sudarsono and Proksch, P. 2003. Novel chromone derivatives from the fungus *Aspergillus versicolor* isolated from the marine sponge *Xestospongia exigua*. Journal of Natural Products, 66: 57-61.

Lintott, W.H. and Lintott, E.A. 2002. Marine fungi from New Zealand, pp. 285-292. In: Fungi in Marine Environments. K.D. Hyde (ed.), Fungal Diversity Press, Hong Kong.

Lopanik, N., Lindquist, N. and Targett, N. 2004a. Potent cytotoxins produced by a microbial symbiont protect host larvae from predation. Oecologia, 139: 131-139.

Lopanik, N., Gustafson K.R. and Lindquist N. 2004b. Structure of bryostatin 20: a symbiont-produced chemical defense for larvae of the host bryozoan, *Bugula neritina*. Journal of Natural Products, 67: 1412-1414.

Malmstrøm, J. 1999 Unguisins A and B: new cyclic peptides from the marine-derived fungus *Emericella unguis*. Journal of Natural Products, 62: 787-789.

Malmstrøm, J., Christophersen, C., Barrero, A.F., Oltra, J.E., Justicia, J. and Rosales, A. 2002a. Bioactive metabolites from a marine-derived strain of the fungus *Emericella variecolor*. Journal of Natural Products, 65: 364-367.

Malmstrøm, J., Ryager, A., Antoni, U. and Nielsen, P.H. 2002b. Unguisin C, a GABA-containing cyclic peptide from the fungus *Emericella unguis*. Phytochemistry, 60: 869-872.

McClintock, J.B. and Baker, B.J. 2001. Marine Chemical Ecology, CRC Press, Boca Raton. Fl., USA.

McDonald, L.A., Abbanat, D.R., Barbieri, L.R., Bernan, V.S., Discafani, C.M., Greenstein, M., Janota, K., Korshalla, J.D., Lassota, P., Tischler, M. and Carter, G.T. 1999. Spiroxins, DNA cleaving antitumor antibiotics from a marine-derived fungus. Tetrahedron Letters, 40: 2498–2492.

Mills, S.D. and Turner, W.B. 1967. A new metabolite of *Aspergillus melleus*. Journal of the Chemical Society (C), 2242-2244.

Murnaghan, M.F. 1946. The pharmacology of penicillic acid. Journal of Pharmacology and Experimental Therapeutics, 88: 119-132.

Nagai, K., Kamigiri, K., Matsumoto, H., Kawano, Y., Yamaoka, M., Shimoi, H., Watanabe, M. and Suzuki, K. 2002. YM-202204, a new antifungal antibiotic produced by marine fungus *Phoma* sp. Journal of Antibiotics, 55: 1036-1041.

Nakamura, H., Ono, M., Yamada, T., Numata, A. and Akita, H. 2002. Determination of the absolute stereostructure of *seco*-macrosphelide E produced by a fungal strain from a sea hare. Chemical and Pharmaceutical Bulletin, 50: 303-306.

Namikoshi, M. 2006. Biologically Active Natural Products from Marine Fungi pp. 285-359. In: Biomaterials from Aquatic and Terrestrial Organisms. M. Fingerman, and R. Nagabhushanam (eds.), Science Publishers, Inc., Enfield, New Hampshire, USA.

Namikoshi, M., Kobayashi, H., Yoshimoto, T. Meguro, S. and Akano, K. 2000. Isolation and characterization of bioactive metabolites from marine-derived filamentous fungi collected from tropical and sub-tropical coral reefs. Chemical and Pharmaceutical Bulletin, 48: 1452-1457.

Namikoshi, M., Negishi, R., Nagai, H., Dmitrenok, A. and Kobayashi, H. 2003. Three new chlorine containing antibiotics from a marine-derived fungus *Aspergillus ostianus* collected in Pohnpei. Journal of Antibiotics, 56: 755-761.

Nielsen, J., Nielsen, P.H. and Frisvad, J.C. 1999. Fungal depside, guisinol, from a marine derived strain of *Emericella unguis*. Phytochemistry, 50: 263-265.

Numata, A., Takahashi, C., Matsushita, T., Miyamoto, T., Kawai, K., Usami, Y., Matsumura, E., Inoue, M., Ohishi, H. and Shingu, T. 1992. Fumiquinazolines, novel metabolites of a fungus isolated from a saltfish. Tetrahedron Letters, 23: 1621-1624.

Numata, A., Takahashi, C., Ito, Y., Takada, T., Kawai, K., Usami, Y., Matsumura, E., Imachi, M., Ito, T. and Hasegawa, T. 1993. Communesins, cytotoxic metabolites of a fungus isolated from a marine alga. Tetrahedron Letters, 34: 2355-2358.

Numata, A., Takahashi, C., Ito, Y., Minoura, K., Yamada, T., Matsuda, C. and Nomoto, K. 1996. Penochalasins, a novel class of cytotoxic cytochalasans from a *Penicillium* species separated from a marine alga: structure determination and solution conformation. Journal of the Chemical Society, Perkin Transaction, 1 239-245.

Numata, A., Amagata, T., Minoura, K. and Ito, T. 1997a. Gymnastatins, Novel cytotoxic metabolites produced by a fungal strain from a sponge. Tetrahedron Letters, 38: 5675-5678.

Numata, A., Iritani, M., Yamada, T., Minoura, K., Matsumura, E., Yamori, T. and Tsuruo, T. 1997b. Novel antitumor metabolites produced by a fungal strain from a sea hare. Tetrahedron Letters, 38: 8215-8218.

Pallenberg, A.J. and White, J.D. 1986. The synthesis and absolute configuration of (+)-leptosphaerin. Tetrahedron Letters, 27: 5591-5594.

Pettit, G.R., Herald, C.L., Doubek, D.L. and Herald, D.L. 1982. Isolation and structure of bryostatin 1. Journal of the American Chemical Society, 104: 6846-6847.

Pettit, G.R., Gao, F., Blimberg, P.M., Herald, C.L., Kamano, Y., Lewin, N.E., Schmidt, J.M. and Chapuis, J.-C. 1996. Antineoplastic agents. 340. Isolation and structural elucidation of bryostatins 16-18. Journal of Natural Products, 59: 286-289.

Piel, J., Hui, D., Wen, G., Butzke, D., Platzer, M. Fusetani, N. and Matsunaga, S. 2004. Antitumor polyketide biosynthesis by an uncultivated bacterial symbiont of the marine sponge *Theonella* swinhoei. Proceedings of the National Academy of Science, U.S.A. 101: 16222-16227.

Pietra, F. 1997. Secondary metabolites from marine microorganisms: bacteria, protozoa, algae and fungi. Achievements and prospects. Natural Product Reports, 14: 453-464.

Pointing, S.B., Fryar, S.C., Hyde, K.D. and Mouzouras, R. 2000. Assessment of wood decay, fungal interactions and colonization, pp. 113-136. In: Marine Mycology: A Practical Approach. K.D. Hyde, and S.B. Pointing (eds.), Fungal Diversity Press, Hong Kong.

Proksch, P., Edrada, R.A. and Ebel, R. 2002. Drugs from the sea—current status and microbiological implications. Applied Microbiology and Biotechnology, 59: 125-134.

Rahbæk, L., Christophersen, C., Frisvad, J., Bengaard, H.S., Larsen, S. and Rassing, B.R. 1997. Insulicolide A: a new nitrobenzoyloxy-substituted sesquiterpene from the marine fungus *Aspergillus insulicola.* Journal of Natural Products, 60: 811-813.

Rand, T.G. 2000. Diseases of marine animals, pp. 21-48. In: Marine Mycology: A Practical Approach. K.D. Hyde, and S.B. Pointing (eds.) Fungal Diversity Press, Hong Kong.

Reiss, J. 1979. Inhibitory action of the mycotoxins patulin and penicillic acid on urease. Food and Cosmetics Toxicology, 17: 145-146.

Rollin, P. 1987. A Short Biomimetic Synthesis of Leptosphaerin. Tetrahedron Letters 28: 3813-3814.

Rowley, D.C., Kelly, S., Kauffman, C.A., Jensen, P.R. and Fenical, W. 2003. Halovirs A–E, new antiviral agents from a marine-derived fungus of the genus *Scytalidium*. Bioorganic and Medicinal Chemistry 11: 4263-4274.

Sassa, T., Hayakari, S., Ikeda, M. and Miura, Y. 1971. Plant growth inhibitors produced by fungi. I. Isolation and identification of penicillic acid and dihydropenicillic acid. Agricultural and Biological Chemistry, 35: 2130-2131.

Schiehser, G.A., White, J.D., Matsumoto, G.K., Pezzanite, J.O. and Clardy, J. 1986. The structure of leptosphaerin. Tetrahedron Letters, 27: 5587-5590.

Schmidt, E.W., Nelson, J.T., Rasko, D.A., Sudek, S., Eisen, J.A., Haygood, M.G. and Ravel, J. 2005. Patellamide A and C biosynthesis by a microcin-like pathway in *Prochloron didemni*, the cyanobacterial symbiont of *Lissoclinum patella*. Proceedings of the National Academy of Science, U.S.A. 102: 7315–7320.

Schlingmann, G., Milne, L., Williams, D.R. and Carter, G.T. 1998. Cell wall active antifungal compounds produced by the marine fungus *Hypoxylon oceanicum* LL-15G256. II. Isolation and structure determination. Journal of Antibiotics, 51: 303-316.

Schlingmann, G., Milne, L. and Carter, G.T. 2002. Isolation and identification of antifungal polyesters from the marine fungus *Hypoxylon oceanicum* LL-15G256. Tetrahedron, 58: 6825-6835.

Schröder, H.C., Brümmer, F., Fattorusso, E., Aiello, A., Menna, M., de Rosa, S., Batel, R. and Müller, W.E.G. 2003. Sustainable production of bioactive compounds from sponges: primmorphs as bioreactors, pp. 163-197. In: Sponges (Porifera). W.E.G. Müller (ed.), Springer, Berlin, Germany.

Shigemori, H., Wakuri, S., Yazawa, K., Nakamura, T., Sasaki, T. and Kobayashi, J. 1991. Fellutamides A and B, cytotoxic peptides from a marine fish-possessing fungus *Penicillium fellutanum*. Tetrahedron, 47: 8529-8534.

Shigemori, H., Komatsu, K., Mikami, Y. and Kobayashi, J. 1999. Seragakinone A, a new pentacyclic metabolite from a marine-derived fungus. Tetrahedron, 55: 14925-14930.

Son, B.W., Jensen, P.R., Kauffman, C.A. and Fenical, W. 1999. New cytotoxic epidithiodioxopiperazines related to verticillin A from a marine isolate of the fungus *Penicillium*. Natural Product Letters, 13: 213-222.

Steinberg, P.D., de Nys, R. and Kjelleberg, S. 2001. Chemical mediation of surface colonization, pp. 355-387. In: Marine Chemical Ecology. J.B. McClintock, and B.J. Baker (eds.) CRC Press, Boca Raton, Fl., USA.

Steinberg, P.D., de Nys, R. and Kjelleberg, S. 2002. Chemical cues for surface colonization. Journal of Chemical Ecology, 28: 1935-1951.

Suenaga, K., Aoyama, S., Xi, W., Arimoto, H., Yamaguchi, K., Yamada, K., Tsuji, T., Yamada, A. and Uemura, D. 2000. Isolation and structure of kasarin, a novel azetinone compound, isolated from a marine microorganism. Heterocycles, 52: 1033-1036.

Takahashi, C., Numata, A., Ito, Y., Matsumura, E., Araki, H., Iwaki, H. and Kushida, K. 1994a. Leptosins, antitumor metabolites of a fungus isolated from a marine alga. Journal of the Chemical Society, Perkin Transaction 1, 1859-1864.

Takahashi, C., Numata, A., Matsumura, E., Minoura, K., Eto, H., Shingu, T., Ito, T. and Hasegawa, T. 1994b. Leptosins I and J, cytotoxic substances produced by a *Leptosphaeria* sp. physico-chemical properties and structures. Journal of Antibiotics, 47: 1242-1249.

Takahashi, C., Takai, Y., Kimura, Y., Numata, A., Shigematsu, N. and Tanaka, H. 1995a. Cytotoxic metabolites from a fungal adherent of a marine alga. Phytochemistry, 38: 155-158.

Takahashi, C., Minoura, K., Yamada, T., Numata, A., Kushida, K., Shingu, T., Hagishita, S., Nakai, H., Sato, T. and Harada, H. 1995b. Potent cytotoxic metabolites from a *Leptosphaeria* species. Structure determination and conformational analysis. Tetrahedron, 51: 3483-3498.

Takahashi, C., Matsushita, T., Doi, M., Minoura, K., Shingu, T., Kumeda, Y. and Numata, A. 1995c. Fumiquinazolines A–G, novel metabolites of a fungus separated from a *Pseudolabrus* marine fish. Journal of the Chemical Society, Perkin Transaction 1, 2345-2353.

Takahashi, C., Numata, A., Yamada, T., Minoura, K., Enomoto, S., Konishi, K., Nakai, M., Matsuda, C. and Nomoto, K. 1996. Penostatins, novel cytotoxic metabolites from a *Penicillium* species separated from a green alga. Tetrahedron Letters, 37: 655-658.

Tan, L.T., Cheng, X.C., Jensen, P.R. and Fenical, W. 2003. Scytalidamides A and B, new cytotoxic cyclic heptapeptides from a marine fungus of the genus *Scytalidium*. Journal of Organic Chemistry, 68: 8767-8773.

Tashiro, F., Hirai, K. and Ueno, Y. 1979. Inhibitory effects of carcinogenic mycotoxins on deoxyribonucleic acid-dependent ribonucleic acid polymerase and ribonuclease H. Applied and Environmental Microbiology, 38: 191-196.

Thakur, N.L., Hentschel, U., Krosko, A., Pabel, C.T., Anil, A.C. and Müller, W.E.G. 2003. Antibacterial activity of the sponge *Suberites domuncula* and its primmorphs: potential basis for the defense. Aquatic Microbial Ecology, 31: 77-83.

Torres, M., Balcells, M., Sala, N., Sanchis, V. and Canela, R. 1998. Bactericidal and fungicidal activity of *Aspergillus ochraceus* metabolites and some derivatives. Pesticide Science, 53: 9-14.

Tresner, H.D. and Hayes, J.A. 1971. Sodium chloride tolerance of terrestrial fungi. Applied Microbiology, 22: 210-213.

Tsuda, M., Mugishima, T., Komatsu, K., Sone, T., Tanaka, M., Mikami, Y. and Kobayashi, J. 2003. Modiolides A and B, two new 10-membered macrolides from a marine-derived fungus. Journal of Natural Products, 66: 412-415.

Usami, Y., Aoki S., Hara T. and Numata A. 2002. New dioxopiperazine metabolites from a *Fusarium* species separated from a marine alga. Journal of Antibiotics, 55: 655-659.

Varoglu, M., Corbett, T.H., Valeriote, F.A. and Crews, P. 1997. Asperazine, a selective cytotoxic alkaloid from a sponge-derived culture of *Aspergillus niger*. Journal of Organic Chemistry, 62: 7078-7079.

Verbist, J.-F., Sallenave, C. and Pouchus, Y.-F. 2000. Marine fungal substances, pp. 979-1092. In: Studies in Natural Products Chemistry, Vol. 24. A. Rahman (ed.), Elsevier Science, New York, USA.

Verbitski, S.M., Mayne, C.L., Davis, R.A., Concepcion, G.P. and Ireland, C.M. 2002. Isolation, structure determination, and biological activity of a novel alkaloid, perophoramidine, from the Philippine ascidian *Perophora namei*. Journal of Organic Chemistry, 67: 7124-7126.

Wang, G.-Y.-S., Abrell, L.M., Avelar, A., Borgeson, B.M. and Crews, P. 1998. New hirsutane based sesquiterpenes from salt water cultures of a marine sponge-derived fungus and the terrestrial fungus *Coriolus consors*. Tetrahedron, 54: 7335-7342.

Wang, H.-J., Gloer, K.B. and Gloer, J.B. 1997. Anserinones A and B: new antifungal and antibacterial benzoquinones from the coprophilous fungus *Podospora anserina*. Journal of Natural Products 60: 629-631, (correction: Journal of Natural Products 2004 67: 1637).

Wang, T., Shirota, O., Nakanishi, K., Berova, N., McDonald, L.A., Barbieri, L.R. and Carter, G.T. 2001. Absolute stereochemistry of the spiroxins. Canadian Journal of Chemistry, 79: 1786-1791.

White, J.D., Badger, R.A., Kezar, H.S., III, Pallenberg, A.J. and Schiehser, G.A. 1989. Structure, synthesis and absolute configuration of leptosphaerin, a metabolite of the marine ascomycete *Leptosphaeria oraemaris*. Tetrahedron, 45: 6631-6644.

Williams, D.H., Stone, M.J., Hauck, P.R. and Rahman, S.K. 1989. Why are secondary metabolites (natural products) biosynthesized? Journal of Natural Products, 52: 1189-1208.

Yamada, T., Iritani, M., Doi, M., Minoura, K., Ito, T. and Numata, A. 2001. Absolute stereostructures of cell-adhesion inhibitors, macrosphelides C, E–G and I, produced by a *Periconia* species separated from an *Aplysia* sea hare. Journal of the Chemical Society, Perkin Transaction 1, 3046-3053.

Yamada, T., Iwamoto, C., Yamagaki, N., Yamanouchi, T., Minoura, K., Yamori, T., Uehara, Y., Andoh, T., Umemura, K. and Numata, A. 2002a. Leptosins

M–N_1, cytotoxic metabolites from a *Leptosphaeria* species separated from a marine alga. Structure determination and biological activities. Tetrahedron, 58: 479-487.

Yamada, T., Iritani, M., Minoura, K., Numata, A., Kobayashi, Y. and Wang, Y.-G. 2002b. Absolute stereostructures of cell adhesion inhibitors, macrosphelides H and L, from *Periconia byssoides* OUPS-N133. Journal of Antibiotics, 55: 147-154.

Yamada, T., Iwamoto, C., Yamagaki, N., Yamanouchi, T., Minoura, K., Hagishita, S. and Numata, A. 2004. Leptosins O–S, cytotoxic metabolites of a strain of *Leptosphaeria* sp. isolated from a marine alga. Heterocycles, 63: 641-653.

Zhang, H., Tomoda, H., Tabata, N., Miura, H., Namikoshi, M., Yamaguchi, Y., Masuma, R. and Omura, S. 2001. Cladospolide D, a new 12-membered macrolide antibiotic produced by *Cladosporium* sp. FT-0012. Journal of Antibiotics, 54: 635-641.

5

The Genus *Achlya* from Alkaline and Sewage Polluted Aquatic Environment

J.K. Misra and Anshul Pant

Mycological Research Unit, Department of Botany,
Sri Jai Narain Postgraduate College, Lucknow 226001, India
E-mail: jitrachravi@airtelmail.in

Abstract

Out of 44 validly known species of *Achlya* from all over the world, 36 are described from various water bodies in India. This chapter details a total of 9 species of *Achlya* that could be recovered from alkaline (9 species) and polluted waters (7 species)—the two extreme water systems. A comparison has also been made with those types that are described from other aquatic habitats to strengthen the fact that water molds including species of *Achlya* have enough potential to adapt to any environmental stresses and thus are plastic in nature.

INTRODUCTION

We know today 44 undoubtful species of *Achlya* from all over the world (Johnson et al., 2002). Of these 36 species (Table 1) are reported from India that include even doubtful ones (Khulbe, 2001). These Oomycota members are now not considered true fungi because they do not share monophyletic origin (Patterson and Sogin, 1992). But still mycologists study these interesting organisms and hopefully will also study them in the future.

Species of *Achlya* from different alkaline and polluted sites were isolated. The sites are freshwater environments, but extreme in nature. Extreme in the sense that—one group of sites are alkaline throughout (pH ranging between 7.3-8.8) and the others are polluted ones that receive water and waste material through rain runoff, regular domestic drainage with detergents and other chemicals used by washermen while cleaning clothes. Such sites are at the bank of a river—Gomati that flows through middle of the city of Lucknow and one permanent lake—Yamuna Jheel. This chapter describes the nine species of *Achlya* recovered from the aforesaid habitats. A comparison has also been made from those that are described from varying aquatic habitats by other Indian workers.

THE GENUS *ACHLYA* NEES

The genus *Achlya* was established by C.G. Nees von Esenbeck to include the achlorophyllous, filamentous organisms described by Carus (1823) as *Hydronema*. *Achlya prolifera* is a type species.

Methodology

Water samples were collected in sterilized bottles from different water bodies, both alkaline and polluted ones situated in and around Lucknow, India. The samples collected were brought back to the laboratory within a few hours after collection and each was baited separately with boiled hemp seed halves, wheat and maize grains, snake skins, ants and houseflies. The baits were periodically examined for the appearance of mycelium and those showing any infestation were thoroughly washed with distilled water and transferred to sterile distilled water in Petri plates having a corresponding bait. When the sporangia were formed the slides were prepared and the colony was left to form oogonia and antheridia. After the formation of sex organs, the identification was made using the available monographs and keys as provided by Coker (1923), Johnson (1956) and Johnson et al. (2002).

Terminology used to identify the genus

The terminology described below is based on the monograph of Johnson (1956) and pertains to the origin of antheridial branches and types of oospores. These terms have been used by the authors while identifying and describing their isolates.

1. *Androgynous:* Antheridial branch arises from the oogonial stalk but the antheridial cell is not a part of it (Figs. 1-3).
2. *Hypogynous:* No antheridial branch, antheridial cell originates from the oogonial stalk (Fig. 4).
3. *Exigynous:* Antheridial branch originating from the oogonial cell above the basal septum (Fig. 5).

4. *Monoclinous:* Antheridial branch originating from the same hypha from which oogonium originated (Figs. 5-6).
5. *Diclinous:* Both oogonium and antheridium are originating from different hyphae (Figs. 7-8).
6. *Centric:* With one or two peripheral layers of small oil droplets completely surrounding the central ooplasm (Fig. 9).
7. *Eccentric:* With one large oil globule disposed on one side of the oospore and not entirely enclosed by ooplasm (Fig. 10).
8. *Subcentric, Type I:* With one layer of small oil droplets on one side of the ooplasm and two or three layers on the opposing side (Fig. 11).

   Subcentric, Type II: As in type one, but with the single layer of oil droplets not formed, thus giving a lunate grouping of droplets partially surrounding the ooplasm (Fig. 12).

   Subcentric, Type III: With a single, circular layer of small oil droplets located eccentrically to the oospore wall (Fig. 13).

Description of different species isolated from alkaline and polluted water.

1. *Achlya americana* Humphrey

Mycelial growth extensive on hempseed halves, diffuse, hyphae stout, sparingly branched. Gemmae present. Sporangia fusiform, straight, 250-600 × 20-45 μm, renewed sympodially. Zoospore discharge achlyoid, spore cluster persistent at the mouth of the sporangium. Encysted spores 9-12 μm in diam. Ooogonia abundant, lateral, spherical, 40-80 μm in diam., oogonial wall smooth, oogonial stalks short and straight. Antheridial branches diclinous, antheridial cells tubular or clavate, attached by projections or laterally appressed. Oospores eccentric, spherical, 4-26 in number, 14-35 μm in diam.

Isolated from pond water (pH 8.0, temp. ranging between 17-37°C), collected from Canal-side pond, Telibagh, Lucknow. Recently this species was isolated from a site—Yamuna Jheel (pH 7.5, temp. 20°C) by Anshul Pant in the month of November, 2007.

This species has also been isolated by Chaudhuri and Kocchar (1935), Saksena and Rajgopalan (1958), Dayal and Tandon (1962), Srivastava (1967), Thakurji (1970), Hasija and Batra (1978), Khulbe (1980, 2001), Mer et al. (1980), Misra (1980, 1983b), Mer and Khulbe (1984), Khulbe (1985), Gupta and Mehrotra (1989) Mishra et al. (1990)

The isolate by Misra (1980) differs from that described by Chaudhuri and Kocchar (1935) in having thicker sporangia, bigger encysted spores,

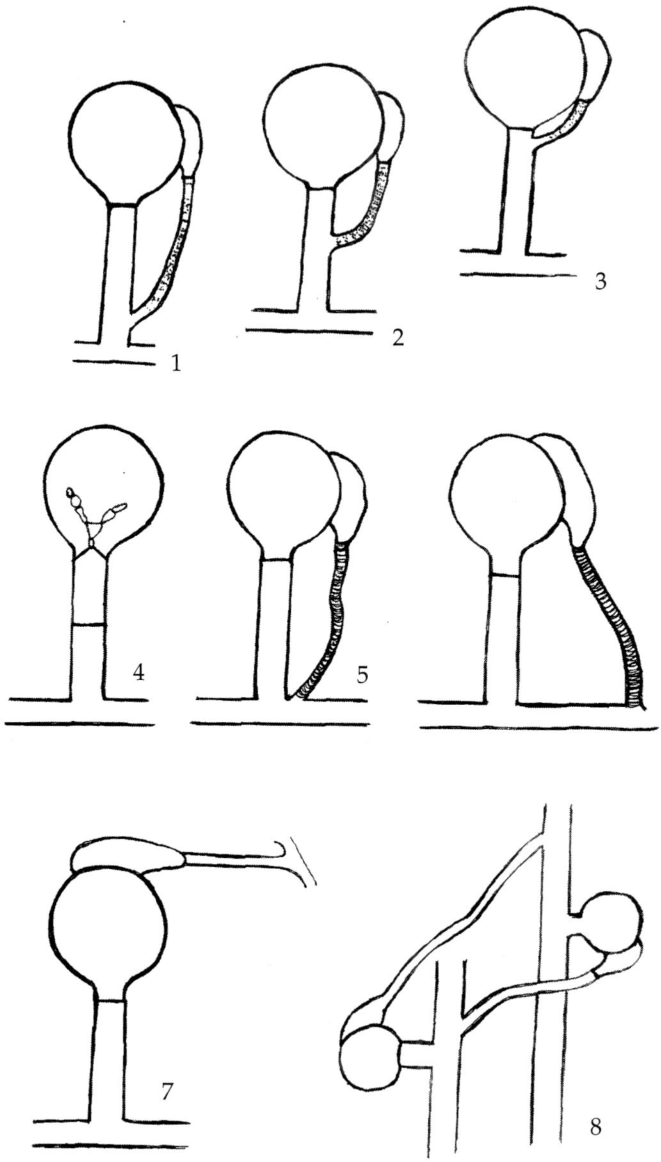

Fig. 1-3. Androgynous antheridia. **Fig. 4.** Hypogynous antheridium—antheridial cell originating from the oogonial stalk. **Figs. 5-6.** Monoclinous antheridia—antheridial branch originating from the same hypha from which oogonium originated. **Figs. 7-8.** Diclinous antheridia—both oogonium and antheridium are originating from different hyphae.

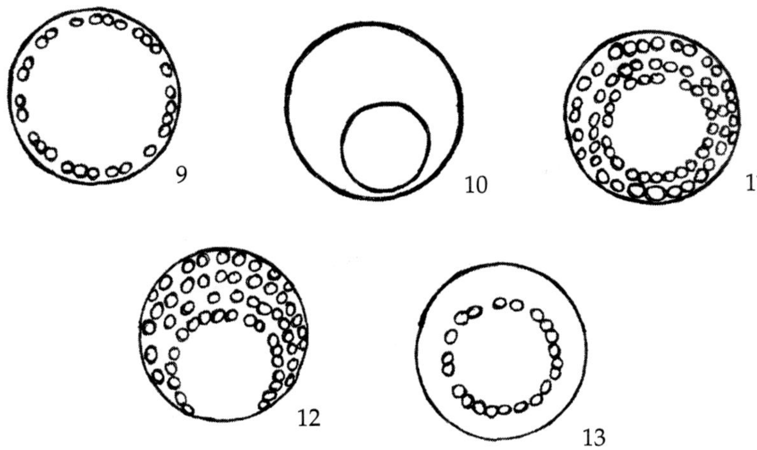

Fig. 9. Centric oospore—with one or two peripheral layers of small oil droplets completely surrounding the central ooplasm. **Fig. 10.** Eccentric oospore—with one large oil globule disposed on one side of the oospore. **Fig. 11.** Subcentric oospore type I—with one layer of small oil droplets on one side of the ooplasm and two or three layers on the opposing side. **Fig. 12.** Subcentric oospore type II—as in type one, layer of oil droplets giving a lunate grouping of droplets partially surrounding the ooplasm. **Fig. 13.** Subcentric oospore type III—with one circular layer of small oil droplets located eccentrically to the oospore wall

oogonia and oospores which are more in number and eccentric in nature. Further, only diclinous antheridia in the present isolate are seen which are reported to be occasional in their isolates.

The recent isolate of Pant's from the polluted site 4—Yamuna Jheel differs from that of Khulbe (1985) in having shorter sporangia, lesser number of subcentric oospores and having only diclinous antheridia. He has also isolated this species from a non-polluted site—6. The isolates had eccentric oospores. The percent frequency of this form was 25% considering only the polluted sites surveyed for a year (Pant unpublished).

2. ***Achlya diffusa*** Harvey ex Johnson

Mycelium extensive and diffuse. Hyphae slender and branched. Gemmae abundant, rod shaped or clavate, formed in chain by the segmentation of the hyphae. Sporangia abundant, measuring 333-804 × 44-67 µm in diam, renewed sympodially. Encysted spores 8-11 µm in diam. Oogonia abundant, laterally borne, spherical, 44-85 µm in diam, oogonial wall smooth, unpitted, oogonia like hyphal swelling seen. Antheridial branches diclinous, one to many per oogonium. Oospores 1-11 in number, usually

2-7, not filling the oogonium, 19-31 µm in diameter, usually 22-25 µm and eccentric.

Misra (1980) isolated this from pond water (pH 7.5) collected from the village Murawan Khera (Bengali ka pond) situated in the south east of Lucknow city on the Lucknow Rae -Bareli road. Pant isolated this species from three sites out of four polluted sites surveyed—Site 1 at the bank of river Gomati, pH 8.1, temp. 14°C and site 2 also at the other side of Gomati river, pH 7.9, temp. 15 C and site 4—Yamuna Jheel, pH 7.3, temp. 18°C in the months of Nov., 2007 and Jan. 2008. Its per cent occurrence was 75%. Furthermore, this species was also recovered from site 5 and 6 that are considered non-polluted in Pant's study (Pant unpublished).

This species has also been isolated by Srivastava (1967), Dayal and Thakurji (1968), Prabhuji and Srivastava (1977), Prabhuji (1984), Verma (1987), Khare (1992).

Our isolates differs from that of Srivastava (1967), Dayal and Thakurji (1968) in having smaller oogonia with smaller oospores which are fewer in number.

3. *A. dubia* Coker

Mycelium limited, sparsely branched. Gemmae sparsely formed. Sporangia abundant, filiform, measuring 186-600 × 80-100 µm in diam., renewed sympodially. Zoospore discharge thraustothecoid. Encysted spores 7-12 µm in diam. Oogonia abundant lateral, spherical, 20-60 µm in diam., predominantly 50 µm, oogonial wall smooth, pitted only under the point of attachment of antheridial cells. Antheridial branches diclinous, antheridial cells clavate or tubular, laterally appressed to the oogonium. Oospores spherical, usually not filling the oogonium, 1-10 in number, 15-30 µm in diam. and eccentric.

Misra (1980) also isolated from pond water (pH 7.9) which was collected from Lacchi Tara, Telibagh, Lucknow. Pant isolated this species from site—1, pH 8.1, temp. 20°C. Chaudhuri and Kocchar (1935) have also isolated this species.

The isolate of Misra (1980) differs from Chaudhuri and Kocchar (1935) isolate in having thicker hyphae, longer sporangia, smaller spores and oogonia. Moreso, the sporangia of this isolate are also longer than what Johnson Jr. (1956) and Srivastava (1967) have described. The oospores in the present isolate are also lesser in number. Pant's isolate of polluted site—1, (Gomati river bank behind Haathi Park, pH 8.1, temp. 20 isolated in the month of Nov., 2007) differs from that of Misra (1980) by having smaller sporangia with achlyoid discharge. Its per cent occurrence was 25% (Pant unpublished).

4. *A. hypogyna* Coker & Pemberton

Mycelium branched, moderately stout, tapering gradually towards the apex, measuring 36-189 µm at the base. Gemmae moderately abundant and variable in shape, often in chain and rod shaped formed by the segmentation of the hyphae. Sporangia cylindrical or clavate, renewed sympodially, 188-522 × 67-87 µm. Zoospore discharge achlyoid, spore cluster irregular and persistent. Encysted spores 7-8 µm in diam. Oogonia borne on short lateral branches, globular or very rarely oblong, 48-75 µm in diam., oogonial wall unpitted and smooth. Antheridia commonly hypogynous, diclinous, occasionally androgynous, usually 2-3 antheridia per oogonium, antheridial cells tubular or clavate, laterally appressed to the oogonium. Oospores spherical, 1-10 in number but usually 2-5, centric, 19-27 µm in diam, mostly 24 µm.

Isolated from pond water (pH 8.5), collected from village Devi Khera, situated in the vicinity of Lucknow Rae-Bareli road (Misra, 1980, Rai and Misra, 1976. Pant isolated this from a polluted site 4—Jamuna Jheel, pH 7.3, temp. 17°C in Dec. 2007. It was also recovered from a unpolluted site 5 in the same month. Pant's isolate differs from that of Misra (1980) in having lesser number of oospores. Its per cent occurrence was 25% (Pant unpublished). Gupta and Mehrotra (1989) have also isolated this species from Kurukshetra.

5. *A. oblongata* de Bary

Mycelium moderately stout and branched. Gemmae present, formed by the segmentation of the hyphae in chain. Sporangia abundant, fusiform, measuring 500- 667 × 55-78 µm, renewed sympodially. Zoospore discharge achlyoid, spore cluster persistent. Encysted spores 6-11 µm in diam. Oogonia abundant, borne laterally on main branches, oblong to subspherical, measuring 50-94 µm in diam. Antheridia diclinous, one to many per oogonium, laterally appressed, antheridial cells branched and bulbous. Oospores 4-12 in number, subcentric measuring 19-39 µm in diam.

Misra (1980, 1983b) isolated this species from pond water (pH 7.6), collected from Lacchi Tara, situated in the Telibagh area on the Lucknow Rae-Bareli road, Lucknow. Hasija and Khan (1983) have also isolated this form. Recently, Pant isolated this species from a polluted site—1 situated at Gomati River bank behind Haathi Park side, pH 8.0, temp. 17°C in the month of Dec. 2007 and from site 4—Yamuna Jheel, pH 7.2, temp. 14°C in the month of Jan. 2008. Its per cent occurrence was 50% (Pant unpublished).

Misra's isolate (Misra, 1980) differs from the one described earlier in having only achlyoid sporangia, smaller oogonia with lesser number of oospores which are bigger in size. Pant's isolate, however, differs from Dayal and Thakurji (1968), Misra (1980) in having shorter and thicker sporangia, smaller oogonia that are subspherical with lesser number of oospores (Pant unpublished).

6. *A. polyandra* Hildebrand

Mycelial growth dense, hyphae stout. Gemmae present, formed by the segmentation of the hyphae. Sporangia abundant, renewed by cymose branching, measuring 221-503 × 20-54 µm, commonly curved. Encysted spores 8-11 µm in diam. Oogonia abundant, borne laterally on the main branches, measuring 50-109 µm in diam, oogonial wall smooth, unpitted, oogonial stalks little curved in some cases. Antheridia both diclinous and androgynous, one to many per oogonium, short and clavate. Oospores 2-26 in number, commonly 4, measuring 17-29 µm in diam., centric and subcentric.

Misra (1980, 1983b) isolated from pond water (pH 8.0), collected from a pond situated near the District Jail, Lucknow. Pant isolated from site 1—situated at Gomati river bank behind the Haathi Park in the month of Dec. 2007, pH 8.1, temp. 19°C and from site 3—bank of Gomati river opposite Shani temple, pH 7.8, temp 15°C in the month of Jan. 2008. Its per cent occurrence was 50%. Pant's isolate differs from Misra's (1980) in having smaller oogonia, and having only subcentric oospores and diclinous antheridia (Pant unpublished).

7. *A. racemosa* Hildebrand

Mycelial growth moderate, hyphae thick. Gemmae not formed even in old cultures. Sporangia cylindrical or clavate, measuring 222-444 µm in diam, renewed by cymose branching. Encysted spores 6-8 µm in diam. Oogonia abundant, borne laterally in racemose manner on the main hyphae, measuring 33-50 µm in diam and spherical, oogonial wall unpitted except where the antheridia touch. Antheridia monoclinous and diclinous, one to many per oogonium, attached laterally or apically. In some cases antheridia are attached to the oogonium by projections. Oospores 1-6 in number, commonly 3, measuring 17-22 µm in diam and centric.

Isolated from soil (pH 8.0), collected from Govt. fisheries farm, Utrethiya, Lucknow. Pant isolated this species form site—1, pH 8.1, temp. 19°C in the month of Nov. 2007. Its per cent occurrence was 25% (Pant unpublished).

Misra's (1980) isolate differs from others in having all reproductive structures smaller in size. Pant's isolate differs from that of Misra (1980) in having smaller sporangia, oogonia and absence of monoclinous antheridia that are 3-4 in number (Pant unpublished).

8. A. *rodrigueziana* Wlf.

Mycelium extensive, dense, sparingly branched. Gemmae abundant, filiform. Sporangia abundant, filiform or fusiform, measuring 150-670 × 20-87 µm, renewed sympodially. Zoospore discharge achlyoid, spore cluster persistent at the mouth. Encysted spores 9-10 µm in diam. Oogonia abundant, lateral, terminal, spherical 25-31 µm in diam, oogonial wall unpitted. Antheridia present, diclinous, antheridial cells branched, fertilization tubes present, persistent. Oospores eccentric, spherical, not completely filling the oogonium, one in number, measuring 22-28 µm in diam.

Isolated from pond water (pH 8.1), collected from Major's pond, Telibagh, Lucknow by Misra (1980, 1983b). This species has not been isolated from any of the polluted sites surveyed.

9. A. *stellata* de Bary var. ***multispora*** Rai & Misra

Mycelium moderately extensive, diffuse, hyphae slender and branched. Gemmae absent. Sporangia moderately abundant, fusiform or clavate, 156-256 × 22-44 µm, renewed sympodially. Zoospore discharge achlyoid. Encysted spores 6-8 µm in diam. Oogonia abundant on lateral branches, globose, measuring 44-89 µm in diam inclusive of ornamentation, oogonial wall unpitted. Antheridia generally present, monoclinous, laterally appressed to the oogonium. Oospores commonly 3-5 in an oogonium, spherical, filling the oogonium, measuring 19-28 µm in diam and sub-centric.

Isolated from alkaline muddy soil (pH 8.0), collected from pond of fisheries farm, Utrethiya, Lucknow by Rai and Misra (1977a). This has not been isolated from polluted sites surveyed by Pant.

All 9 Achlyas described here were isolated from alkaline habitat while only seven were recovered from the polluted sites surveyed. Isolates from polluted sites have also shown variability in the sizes of their structures—sporangia, oogonia, etc. It, therefore, can be assumed that species of *Achlya* have preferences to the habitat and its water quality (Misra, 1981, 1982, 1983b). Isolation of seven forms from domestic and chemically polluted sites strengthen the idea of *Achlya* being plastic and adaptive in nature (Te- Strake, 1958, 1959; Johnson and Sparrow, 1961; Rai and Misra, 1977b; Misra, 1981, 1982, 1983a).

Table 1. Species of *Achlya* known from Indian water

1.	*Achlya americana* Humphrey
2.	*A. androcomposita* Hamid
3.	*A. apiculata* de Bary
4.	*A. aplanes* Maurizio
5.	*A. aplanes* Maurizio var. *indica* Saksena and Dayal
6.	*A. aquatica* Dayal and Thakurji
7.	*A. bisexualis* Coker & Couch
8.	*A. caroliniana* Coker
9.	*A. conspicua* Coker
10.	*A. colorata* Pringsheim
11.	*A. cornuta* Archer-
12.	*A. crenulata* Ziegler
13.	*A. de Baryana* Humphrey
14.	*A. diffusa* Harvey ex Johnson
15.	*A. dubia* Coker
16.	*A. dubia* var. *pigmenta* Chaudhuri and Kochhar
17.	*A. flagellata* Coker
18.	*A. hypogyna* Coker and Pemberton
19.	*A. imperfecta* Coker
20.	*A. kashyapia* Chaudhuri and Kochhar
21.	*A. klebsiana* Pieters
22.	*A. klebsiana* Pieters var. Kashyapia Chaudhuri
23.	*A. megasperma* Humphrey
24.	*A. oblongata* de Bary
25.	*A. oligacantha* de Bary
26.	*A. orion* Coker & Couch
27.	*A. oryzae* Ito et Nagai
28.	*A. papillosa* Humphrey
29.	*A. polyandra* Hildebrand
30.	*A. prolifera* Nees von Esenbeck
31.	*A. proliferoides* Coker
32.	*A. recemosa* Hildebrand
33.	*A. recurva* Cornu
34.	*A. rodrigueziana* Wolf
35.	*A. stellata* de Bary
36.	*A. stellata* de Bary var. *multispora* Rai and Misra

References

Carus, C.G. 1823. Beitrag zur Geschichte der unter Waser an verwesenden Thierkörpern sich erzengenden Schimmel-oder Algen-Gattemgen. Nova Acta Phys.-Medi Acad. Caes. Eop.-Corol. Cur, 11: 493-507.

Chaudhuri, H. and Kocchar, P.L. 1935. Indian water moulds-I. Proc. Indian Acad. Sci, 2B: 137-154.

Coker, W.C. 1923. The Saprolegniaceae With Notes on Other Water Molds. University of North Carolina Press, Chapel Hill, North Carolina, USA.

Dayal, R. and Tandon, R.N. 1962. Ecological studies of some Aquatic Phycomycetes I. Hydrobiologia, 20: 121-127.

Dayal, R. and Thakurji. 1968. Studies in aquatic fungi of Varanasi-V. A taxonomic study. Proc. Natn. Acad. Sci. India, 38B: 32-38.

Gupta, A.K. and Mehrotra, R.S. 1989. Occurrence, distribution and seasonal variation of watermolds as affected by chemical factors in tanks of Kurukshetra, India. Hydrobiologia, 173: 219-229.

Hasija, S.K. and Batra, S. 1978. The distribution of *Achlya americana* (Saprolegniaceae) in different aquatic habitats at Jabalpur, India. Hydrobiologia, 61: 277-279.

Hasija, S.K. and Khan, M.A. 1983. Some new records of aquatic fungi from Jabalpur, India. Proc. Natn. Acad. Sci. India, 53 (B): 71-72.

Johnson, T.W. Jr. and Sparrow, F.K. 1961. Fungi in Ocean and Estuaries. J. Cramer, Weinheim, Germany.

Johnson, T.W. Jr., 1956. The Genus Achlya: Morphology and Taxonomy. University of Michigan Press, Ann Arbor, USA.

Johnson, T.W. Jr., Seymour, R.L. and Padgett, D.E. 2002. Biology and Systematics of the Saprolegniaceae. Electronically Published in a CD by the authors.

Khare, A.K. 1992. Taxo-Ecological Studies on Watermolds of Bareilly. Ph.D. Thesis, Ruhelkhand University, Bareilly, India.

Khulbe, R.D. 1980. Occurrence of watermolds in some lakes of Nainital, India. Hydrobiologia, 74: 77-80.

Khulbe, R.D. 1985. Studies in Aquatic Phycomycetes of Nainital in Relation to Taxonomy, Physiology, Ecology and Pathology. D. Sc. Thesis, Kumaun University, Nainital, India.

Khulbe, R.D. 2001. A Manual of Aquatic Fungi (Chytridiomycetes & Oomycetes). Daya Publishing House, Delhi-110 035, India.

Mer, G.S. and Khulbe, R.D. 1984. Watermolds of Sat Tal (Nainital), India. Sydowia, Annales Mycologici Ser. II. 37: 195-207.

Mer, G.S., Sati, S.C. and Khulbe, R.D. 1980. Occurrence, distribution and seasonal periodicity of some aquatic fungi of Sat Tal (Nainital) India. Hydrobiologia, 76: 200-205.

Misra, J.K. 1980. Taxonomic and Ecological Studies on the Aquaitc Mycoflora of Alkaline Soils and Ponds of India. Ph.D. thesis. University of Lucknow, India.

Misra, J.K. 1981. Occurrence, distribution and seasonal periodicity of aquatic fungi as affected by hydrogen ion concentration of water of certain alkaline ponds of India. Trans. mycol. Soc. Japan, 22: 397-407.

Misra, J.K. 1982. Occurrence, distribution and seasonal periodicity of aquatic fungi as affected by chemical factors of water of certain alkaline ponds of India. Hydrobiologia, 97: 185-191.

Misra, J.K. 1983a. Occurrence, distribution and seasonal periodicity of aquatic fungi as affected by temperature of water of certain alkaline ponds of India. Indian Jour. Pl. Pathol, 1: 133-140.

Misra, J.K. 1983b. Aquatic mycoflora of alkaline ponds and soils. Bibliotheca Mycologica, 91: 425-455.

Mishra, R.P., Hasija, S.K. and Agarwal, G.P. 1990. Aquatic fungi from Ganga Sagar Lake, Jabalpur. Proc. Natn. Acad. Sci. India, 59B (III): 351-355.

Patterson, D.J. and Sogin, M.L. 1992. Eukaryote origins and Protistan diversity, pp. 13-46. In: The Origin and Evolution of the Cell. H. Hartman and K. Matsuno (eds.) World Scientific, Singapore.

Prabhuji, S.K. 1984. Studies on some water moulds occurring in certain soils of Gorakhpur. J. India Bot. Soc, 63: 387-396.

Prahbuji, S.K. and Srivastava, G.C. 1977. Some members of Saprolegniaceae occurring in the soils of Gorakhpur. Geobios, 4: 258-259.

Rai, J.N. and Misra, J.K. 1976. *Achlya hypogyna*—notable addition to Indian aquatic fungi. Current Science, 45: 543.

Rai, J.N. and Misra, J.K. 1977a. A new variety of *Achlya stellata* De Bary. Current Science, 46: 28.

Raj, J.N., and Misra, J.K. 1977b. Aquatic fungi from alkaline ponds and soils. Kavaka, 5: 73-78.

Saksena, S.B. and Rajgopalan, C. 1958. Studies on aquatic fungi of Sagar. J. Uni. Sagar, 7: 7-20.

Srivastava, G.C. 1967. Some species of Saprolegniaceae collected from Gorakhpur, India. Hydrobiologia, 30: 281-292.

Te- Strake, D. 1958. Estuarine Distribution and Saline Tolerance of Some Saprolegniaceae. Master's thesis, Duke University, Durham, UK.

Te- Strake, D. 1959. Estuarine distribution and saline tolerance of some Saprolegniaceae. Phyton (Buenos Aires) 12: 147-152.

Thakurji. 1970. Studies on aquatic fungi of Varanasi X- additional new interesting records of aquatic Phycomycetes. Hydrobiologia, 36: 179-186.

Verma, B.L. 1987. Zoosporic fungi in wheat fields of Kumaun Himalaya. Madras Agric. Journal, 74: 1-5.

6

Keratinolytic and Keratinophilic Fungi in Sewage Sludge: Factors Influencing their Occurrence

Krzysztof Ulfig

West Pomeranian University of Technology, Polymer Institute,
Dep. Biomaterials & Microbiological Technologies,
al. Piastów 17, 70-310 Szczecin, Poland
E-mail: k_ulfig@ps.pl

Abstract

The present study was conducted to determine the qualitative and quantitative composition of keratinolytic and keratinophilic fungi in sewage sludge. In particular, the aim was to determine the influence of sewage and sludge treatment technologies, as well as physico-chemical and microbiological factors on fungal composition in the sludge environment. Twenty-one sewage sludges from 16 wastewater treatment plants in Upper Silesia, Poland were examined for mycological, bacteriological, parasitological and physico-chemical properties. The wastewater treatment plants used different sewage and sludge treatment technologies. Each sludge was the 'final' product of sewage and sludge treatment processes at a given wastewater treatment plant. Keratinolytic and keratinophilic fungi were examined in sludges using the hair baiting method, with four incubation temperatures (23, 29, 33, and 37°C). Actidione-resistant fungal strains were identified. Keratinophilic fungi occurred abundantly in the sludges examined. The fungal composition reflected both sewage and sludge treatment technologies and the influence of 'combinations' of physico-chemical and microbiological factors characteristic for a given urban agglomeration, wastewater treatment plant or group of wastewater treatment plants. The most important factors affecting the fungal composition in sewage sludge were as follows: temperature, pH, ammonium nitrogen, proteolytic activity, organic carbon and total nitrogen, C:N ratio, total sulfur, C:S ratio, available phosphorus and particle size distribution. The study confirmed the division of the fungi examined into keratinolytic and keratinophilic. Keratinolytic fungi were able to decompose keratin, while keratinophilic fungi utilized simple and easy degradable components of keratinous remnants and the products of keratin decomposition. Keratinolytic fungi could eliminate keratinophilic fungi from hair. Keratinolytic fungi preferred neutral and alkaline sludges, while keratinophilic fungi occurred more frequently in acidic sludges. Quantitiative relationships between *Trichophyton terrestre*, with its teleomorph *Arthroderma quadrifidum* and *Chrysosporium keratinophilum* reflected ammonium nitrogen concentration and pH, C:N ratio, as well as colloidal loam and water contents in sewage sludge. The composition of keratinolytic fungi can be an indicator of the sludge organic matter stabilization process. This conclusion addresses both sewage and sludge treatment technologies and microbiological organic matter transformations. Due to the correlation with fecal coliform quantities, the composition of keratinolytic fungi can also be useful as a rough indicator of the sludge hygienization process. The relationships can be used in both wastewater treatment plants and soil reclamation practice. From the mycological point of view, opportunistic fungi, especially *Microsporum gypseum* and *Pseudallescheria boydii*, pose the major health risk in sewage sludge. In the light of the available data, pathogenic fungi should be regarded as an important element of public health risk posed by sewage sludge, especially when applied to land. The need for future studies are described below.

INTRODUCTION

The occurrence of fungi in sewage and sewage treatment facilities, e.g., activated sludge reactors and trickling filters has been known for long, since abundant fungal growth causes bacteria elimination and activated sludge swelling. However, till Cooke's research (1959, 1963, and others) little had been known on the occurrence of fungi in sewage and their role in sewage treatment processes. Cooke (1957) and Cooke and Pipes (1970) published the list of fungal species occurring in sewage and polluted waters. The list was then completed by Diener et al. (1976). Many keratinophilic fungal species, but not keratinolytic ones, were included in the list. Tomlinson and Williams (1975) made further considerable progress in the elucidation of the fungal role in sewage treatment processes. On the list of fungi occurring in trickling filters, the researchers included *Trichosporon cutaneum*, a species with keratinolytic activity and pathogenic properties to humans (de Hoog et al., 2000). Subsequently, de

Bertoldi (1981) put the dermatophytes, *Trichophyton* sp. and *Epidermophyton* sp., on the list of pathogenic fungi in sewage sludge. The European Commission (2001) also mentioned these dermatophytes, together with the genus *Trichosporon*.

Studies using the hair baiting method have shown that soil keratinolytic fungi occur in sewage in small quantities (Ulfig, 1986a). In contrast to sewage, these fungi inhabit sewage sludge in extreme abundance. Preliminary studies on the occurrence of keratinolytic fungi in sewage sludge from Upper Silesia, Poland, indicated that the fungal qualitative and quantitative composition was dependent on sewage and sludge treatment technologies and on the influence of physico-chemical and microbiological factors (Ulfig, 1991). In a later study (Ulfig et al., 1996), the hypothesis was suggested that the fungal composition could be a useful tool in evaluation of sludge organic matter stabilization processes. The incidence of keratinolytic and keratinophilic fungi in sewage sludge was also studied in Egypt (Abdel-Hafez and El-Sharouny, 1990). Results, especially the prevalence of *Chrysosporium* species indicated the low degree of sludge organic matter stabilization. The fungal composition in Egyptian sludge also could reflect climatic conditions (high ambient temperatures).

Studies of fungi in sewage- or sludge-reclaimed soils have also been performed. Cooke (1971) published the first report on this subject, but it concerned the whole soil fungal population. In this population, only *Trichosporon cutaneum* can be regarded as keratinolytic. According to Lima et al. (1996), soil reclamation with sewage sludge considerably increases fungal quantities. A separate problem is the impact of sewage sludge pollutants, especially heavy metals on mycorrhizal fungi (Oudeh et al., 2002, and others). Abdel-Hafez and El Sharouny (1987) studied the seasonal changes of keratinolytic and keratinophilic fungi in sludge-reclaimed soils in Egypt. Subsequently, Ulfig and Korcz (1994) performed a field experiment, in which dewatering, mineralization and structuralization of soil mixture with sludge could explain the changes in fungal composition of sludge-reclaimed soil. These observations supported the hypothesis that fungi could be used as indicators of organic matter stabilization processes.

Special attention should be paid to the data obtained by Ali-Shtayeh et al. (1999) and Ali-Shtayeh and Jamous (2000). The researchers observed many similarities in qualitative and quantitative compositions of actidione-resistant fungi between sewage-irrigated and control soils. The highest number of these fungi was isolated from raw sewage, followed by sewage-irrigated soil, sewage-free soil, and soil highly contaminated with sewage. However, the number of keratinolytic fungi was much higher in sewage-irrigated soils. The composition of keratinolytic fungi was found to reflect the influence of sewage irrigation, temperature, organic matter content, and soil pH.

In general, the occurrence of pathogenic bacteria, viruses, and zooparasites (protozoa, cestodes and nematodes) in sewage sludge has been relatively well recognized (Bitton et al., 1980; Kowal, 1982; Dumontet et al., 1999; Environmental Protection Agency, 1999; European Commission, 2001). However, little is known on the incidence of pathogenic fungi in the sludge environment. The list of pathogenic fungi in sewage and sewage sludge was published by World Health Organization (1981). However, Straub et al. (1993) considered the fungi as posing a minimal health risk when sludge is applied to land. This opinion came as opportunistic fungi are ubiquitous saprophytic organisms, and even when the sludge is pasteurized, its recolonization by these fungi takes place. It seems that after this opinion later studies disregarded the problem of pathogenic fungi in sewage sludge (Ross et al., 1992; Harrison et al., 1999; Environmental Protection Agency, 1999; European Union, 2000). Some publications (European Commission, 2001; Harrison and Oakes, 2002) mentioned pathogenic fungal species in sewage sludge, but did not give any comment on the subject. Finally, Podgórski (1997) paid attention to the occurrence of toxigenous fungi in sewage sludge, which could have destructive influence on soil microflora.

However, all quoted studies have not considered a very important sewage sludge characteristic. Sludges contain exceptionally high quantities of keratinous substrata of human and animal origin, mainly hair observed even with the naked eye. Keratinolytic fungi are specialized in decomposition of keratin, being the main component of the substrata. Keratinophilic fungi accompany keratinolytic fungi, utilizing non-protein components of the substrata or the products of keratin decomposition (Majchrowicz and Dominik, 1969). Under favorable environmental conditions, the abundance of keratinous substrata implies the abundant growth of keratinolytic and keratinophilic fungi. This provides a unique opportunity for studying different ecological relationships, including the influence of sewage and sewage sludge treatment technologies on fungal composition.

Theoretically, all keratinolytic fungi possess potentially pathogenic properties to humans and animals (Filipello-Marchisio, 2000). Indeed, many mycoses caused by keratinolytic and also keratinophilic fungi have been reported (de Hoog et al., 2000). Some of the fungi also produce harmful secondary metabolites, including antibiotics and mycotoxins (Kaliciński et al., 1975; Deshmukh and Agrawal, 1998). Therefore, studies of the factors influencing these fungi in sewage sludge and sludge-amended soils are of ecological and epidemiological significance. Due to the extensive sludge land use, there are serious gaps in the knowledge on the ecology of keratinolytic and keratinophilic fungi, especially in sewage sludge and sludge-amended soils, and increasing number of opportunistic mycoses, thus the need for the present study.

The aim was to determine the qualitative and quantitative composition of keratinolytic and keratinophilic fungi in sewage sludge. In particular, the study determined the influence of sewage and sewage sludge treatment technologies, as well as physico-chemical and microbiological factors on fungal composition in the sludge environment. An evaluation was also performed whether the fungal composition could be used as an indicator of wastewater treatment technologies and sludge organic matter stabilization and hygienization. Results also contributed to the evaluation of the health risk posed by pathogenic fungi in sewage sludge. Actidione-resistant fungi were identified.

MATERIAL AND METHODS

Twenty-one sewage sludges from 16 wastewater treatment plants in Upper Silesia, Poland were examined. Characteristics of sewage and sewage sludge technologies are presented in Table 1. The sludges examined can be divided into two general groups. The first group included sludges from wastewater treatment plants using conventional technologies (primary, secondary, excess, and mixed sludges, after anaerobic stabilization, dewatered usually in sludge drying beds). The second group included sludges from wastewater treatment plants using non-conventional technologies (excess sludges after extended aeration, without primary settling tank, and integrated biological process for P, N, and C removal, stabilized or not stabilized anaerobically, dewatered mechanically and in sludge drying bed, lagooned or piled). It should be stressed that each sludge was the 'final' product of sewage and sewage sludge treatment processes at a given wastewater treatment plant.

Sludges were sampled using a metal spade disinfected with 60% ethyl alcohol. At each location, ca. 5 kg of sludge was collected in a clean plastic pail disinfected in the same way. Each sample was taken from five points of a sludge drying bed or lagoon (corners and the middle), cleaned from stones and larger particles, thoroughly crumbled, and mixed. These 'wet' sludge samples were delivered to the laboratory within 2-5 hours for microbiological and physico-chemical analyses. Sludges were sampled in the autumn-winter season.

Keratinolytic and keratinophilic fungi were examined in sludges using the hair baiting method (Vanbreuseghem, 1952) in own modification

Table 1. Sewage sludges and their treatment technologies

Wastewater treatment plant	*Symbol*	*Sludge treatment technologies*
Sosnowiec Radocha I	SOS1	Mixed sludge (primary + excess), stabilized in sludge anaerobic digestion chamber, condensed and dewatered mechanically, stored in sludge pile
Sosnowiec Porąbka	SOS2	Primary sludge, stabilized in Imhoff's tank and in sludge anaerobic digestion chamber, dewatered in sludge drying bed for over 12 months
Sosnowiec Kazimierz	SOS3	Mixed sludge (primary + excess), stabilized in Imhoff's tank, dewatered in sludge drying bed for 7 months
Sosnowiec Zagórze	SOS4	Mixed sludge (primary + excess), stabilized in a sludge anaerobic digestion chamber, dewatered in sludge drying bed for 4 months
Siemianowice Śląskie Centrum	SIE1	Excess sludge after extended aeration (without primary settling tank), after integrated biological process for P, N and C removal, stabilized in a sludge anaerobic digestion chamber, dewatered in sludge drying bed for 6 months (without coagulant addition)
Siemianowice Śląskie Centrum	SIE2	As above; after conditioning with a coagulant, dewatered in beltpress
Siemianowice Śląskie Centrum	SIE3	As above; without a coagulant, after pouring into a sludge drying bed
Siemianowice Śląskie Centrum	SIE4	As above; with a coagulant, dewatered in beltpress, and in sludge drying bed for 12 months
Siemianowice Śląskie Centrum	SIE5	As above; without a flocculant, dewatered in sludge drying bed for 18 months
Siemianowice Śląskie Centrum	SIE6	As above; without a flocculant, dewatered in sludge drying bed for 12 months
Bytom Śródmieście	B1	Primary sludge, stabilized in Dorr's tank and in sludge anaerobic digestion chamber, dewatered in sludge drying bed

Contd.

Table 1 continued

Wastewater treatment plant	*Symbol*	*Sludge treatment technologies*
Bytom Łagiewniki	B2	Mixed sludge (primary + secondary), after biological trickling filter, dewatered in sludge drying bed
Bytom Rozbark	B3	Mixed sludge (primary + secondary), after biological trickling filter, stabilized in Imhoff's tank, dewatered in sludge drying bed
Bytom Szombierki	B4	Primary sludge, stabilized in Imhoff's tank, dewatered in sludge drying bed for 18 months
Bytom Bobrek	B5	Mixed sludge (primary + secondary), after biological trickling filter, stabilized in Imhoff's tank, dewatered in sludge drying bed for 12 months
Bytom Miechowice	B6	Excess sludge after extended aeration (without primary settling tank), after integrated biological process for N and C removal in the Biomix system, dewatered by centrifuge, piled with plant residues for ca. 1-2 years
Bytom Radzionków	B7	Primary sludge, stabilized in a sludge anaerobic digestion chamber, dewatered in sludge drying bed for 12 months
Olkusz	OL	Mixed sludge (primary + excess), stabilized in a sludge anaerobic digestion chamber, conditioned with a flocculant, dewatered by centrifuge and piled
D[1]browa Górnicza Centrum	DA	Mixed sludge (primary + excess), stabilized in a sludge anaerobic digestion chamber, dewatered in beltpress, piled for 2 weeks
Katowice Panewniki	PA	Excess sludge after extended aeration (without primary settling tank), after integrated biological process for P, N and C removal, stabilized in a sludge anaerobic digestion chamber, lagooned for several days
Katowice Gigablok	GI	Excess sludge after extended aeration (without settling tank), condensed and dewatered in sludge drying bed

(Ulfig, 2003). Sterilized standard Petri dishes were filled with 40 g of sludge and covered with 0.4 g of detergent-defatted, fine cut, and autoclaved childrens' hair in each. The dishes were incubated in the dark at 23, 29, 33 and 37°C for four months. Ten Petri dishes responded to each temperature and sludge sample. During incubation, stable moisture conditions (ca. 40%) were maintained in the dishes. After 1, 2, 3 and 4 months of incubation, microscopic observations of hair and inoculations of hair attacked by fungi on Sabouraud 1:10 + mineral salts (TK medium; Takashio, 1973), supplemented with chloramphenicol (100 mg/l) and actidione (500 mg/l), were performed. The inoculated Petri dishes were incubated at 23 and 37°C for 10 days. The rule was accepted that the growth of a given species on hair, confirmed by its growth on TK medium with antibiotics meant the appearance of the species in a given Petri dish. The fungal growth indices were as follows: number of appearances; isolation frequency (number of Petri dishes positive for fungal growth*100/ total number of Petri dishes set up); and number of species.

Pure fungal strains were identified to species or genera using selected taxonomic monographs (Padhye and Carmichael, 1971; Sigler and Carmichael, 1976; Raper and Fennell, 1977; van Oorschot, 1980; Domsch et al., 1980; von Arx, 1986, 1987; Currah, 1985; Cano and Guarro, 1990; Guarro et al., 1999; and others). Fungal abilities for hair degradation were examined using the *in vitro* test by Ulfig et al. (1998a). Strains with strong and moderate keratinolytic activity, forming penetrating bodies, pockets or radial hyphae in hair were recognized as keratinolytic. Fungi with no or weak keratinolytic properties, colonizing hair superficially, were ranked as keratinophilic. Not all (there were too many), but only groups of strains representative for each species, were tested for their keratinolytic properties. On the basis of the test results, fungal species were included in keratinolytic and keratinophilic groups.

Physico-chemical parameters measured in sludge samples were as follows: particle size distribution, moisture, pH in H_2O, pH in 1M KCl, conductivity, total nitrogen, organic carbon, C:N ratio, total phosphorus, available phosphorus, available potassium, total sulfur, C:S ratio, sulfate sulfur, exchange acidity H_w, sorption capacity T, nitrate nitrogen, nitrite nitrogen, ammonium nitrogen, phosphates, heavy metals (Fe, Mn, Zn, Cd, Pb, Cu, Cr, Ni, Hg, and As), TPH (aliphatic hydrocarbons; non-polar compounds), TPOC (aliphatic hydrocarbons; polar + non-polar compounds), and PAHs (naphthalene, acenaphthene, fluorine, phenanthrene, anthracene, fluoranthene, pyrene, benzo(a)anthracene, chrysene, benzo(b)-fluoranthene, benzo(k)fluoranthene, benzo(a)pyrene, benzo(g,h,i)perylene, dibenzo(a,h)anthracene, and indeno(1,2,3-cd)pyrene). Physico-chemical analyses were performed according to ISO, EPA and Polish standards.

Microbiological parameters measured in sludge samples were as follows: Most Probable Number (MPN) of total and fecal coliforms, MPN of fecal streptococci; total bacterial number on SMA at 23°C; number of mesophilic and thermophilic bacteria on SMA at 37 and 45°C; *Salmonella* (qualitative identification), total fungal number on MEA supplemented with chloramphenicol (100 mg/l) at 23°C; number of mesophilic and thermophilic fungi on YpSs supplemented with chloramphenicol (100 mg/l) at 37 and 45°C; and the number of actidione-resistant fungi on Wiegand medium supplemented with chloramphenicol (100 mg/l) and actidione (500 mg/l) at 23 and 37°C (Prochacki, 1975; Geldreich, 1975; Countryside Hygiene Institute, 1985). Number of parasite ova (*Ascaris* sp., *Trichuris* sp. and *Toxocara* sp.) and proteolytic and dehydrogenase activities (Russel, 1972) were also examined in sludge samples. The terms 'mesophilic' and 'thermophilic' were used in the sense of incubation temperatures.

Results (total and for each incubation temperature separately) were analyzed statistically. Parametric methods, i.e., one-way ANOVA, simple linear correlations (Pearson method), cluster analysis (Ward/1-r Pearson method), and factor analysis (principal components) were used (Matthews and Farewell, 1996; Dobosz, 2001). Before statistical analysis, the data were transformed using the equation $y = \ln(x + 2)$. Statistical calculations were performed at $P \leq 0.05$. Species with frequencies >1% were taken for statistical analyses.

RESULTS

The qualitative and quantitative composition of keratinolytic and keratinophilic fungi in sewage sludge is presented in Tables 2 and 3, respectively.

Altogether, 1107 appearances of keratinolytic fungi belonging to 16 species were observed. *Microsporum gypseum* with its teleomorph *Arthroderma gypseum*, *Chrysosporium indicum*, *Trichophyton ajelloi* with its teleomorph *Arthroderma uncinatum*, *Chrysosporium* anamorph of *Aphanoascus clathratus*, *Chrysosporium zonatum*, *Chrysosporium keratinophilum* with its teleomorph *Aphanoascus keratinophilus*, *Trichophyton terrestre* with its teleomorph *Arthroderma quadrifidum*, *Arthrographis kalrai*, *Chrysosporium* anamorphs of *Aphanoascus reticulisporus/fulvescens* with the teleomorph *Aphanoascus reticulisporus*, and *Chrysosporium* anamorph of *Arthroderma curreyi* were the species with frequencies >1%. The *Chrysosporium* anamorphs of *Aphanoascus reticulisporus* and *Aphanoascus fulvescens* are morphologically indistinguishable. Therefore, these anamorphs were grouped together. The highest numbers of appearances and species were found at 23°C. This temperature favored the species as follows: *Trichophyton ajelloi* with its teleomorph *Arthroderma uncinatum*, *Trichophyton*

Table 2. The qualitative and quantitative composition of keratinolytic fungi in sewage sludge

Species and indices of fungal growth	*Number of appearances and fungal indices at temperature*				*Total*	*Frequency*
	23°C	29°C	33°C	37°C		
Microsporum gypseum (Bodin) Guiart & Grigorakis	54	154	35	–	243	22
Teleomorph *Arthroderma gypseum* (Nann.) Weitzman *et al.*	4	–	–	–	4	0.4
Chrysosporium indicum (Randhawa & Sandhu) Garg	1	4	174	18	197	17.8
Trichophyton ajelloi (Vanbreuseghem) Ajello	116	19	–	–	135	12.2
Teleomorph *Arthroderma uncinatum* Dawson & Gentles	8	3	–	–	11	1
Chrysosporium anamorph of *Aphanoascus clathratus* Cano & Guarro	7	33	49	36	125	11.3
Chrysosporium zonatum Al.-Musallam & Tan	1	4	12	56	73	6.6
Chrysosporium keratinophilum D.Frey ex Carmichael	25	27	14	1	67	6.1
Teleomorph *Aphanoascus keratinophilus* Punsola & Cano	–	–	9	1	10	0.9
Trichophyton terrestre Durie & Frey	62	3	–	–	65	5.9
Teleomorph *Arthroderma quadrifidum* Dawson & Gentles	43	–	–	–	43	3.9
Arthrographis kalrai (Tewari & Macpher.) Sigler & Carmichael	34	1	–	3	38	3.4
Chrysosporium anamorph of *Aphanoascus reticulisporus/fulvescens*	8	9	16	1	34	3.1
Teleomorph *Aphanoascus reticulisporus* (Routien) Hubálek	7	–	6	–	13	1.2

Contd.

Table 2 continued

Species and indices of fungal growth	*Number of appearances and fungal indices at temperature*				*Total*	*Frequency*
	23°C	29°C	33°C	37°C		
Chrysosporium anamorph of *Arthroderma curreyi* Berkeley	27	–	–	–	27	2.4
Malbranchea fulva Sigler & Carmichael	1	–	5	1	7	0.6
Chrysosporium pannicola (Corda) van Oorschot & Stalpers	5	–	–	–	5	0.5
Amauroascus mutatus (Quelet) Rammeloo	4	–	–	–	4	0.4
Scopulariopsis brevicaulis (Sacc.) Bain.	–	–	2	2	4	0.4
Myceliophthora vellerea (Sacc. & Speg.) van Oorschot	1	–	–	–	1	0.1
Myceliophthora sp.	1	–	–	–	1	0.1
Number of appearances	409	257	322	119	1107	–
Frequency of isolation	90.9	94.3	95.2	37.1	79.4	–
Number of species	16	9	8	8	16	–

Table 3. The qualitative and quantitative composition of keratinophilic fungi in sewage sludge

Species and indices of fungal growth	*Number of appearances and fungal indices at temperature*				*Total*	*Frequency*
	23°C	29°C	33°C	37°C		
Pseudallescheria boydii (Shear) McGinnis *et al.*	40	73	78	143	334	34.7
Verticillium lecani (Zimm.) Viegas	189	56	2	-	247	25.7
Aspergillus fumigatus Fres.	2	54	75	98	229	23.8
Penicillium janthinellum Biourge	7	22	44	2	75	7.8
Aspergillus alutaceus Berk. & Curt.	-	-	12	11	23	2.4
Aspergillus terreus Thom	1	4	5	13	23	2.4
Paecilomyces lilacinus (Thom) Samson	4	4	-	-	8	0.8
Narasimhella marginospora (Kuehn & Orr) v. Arx	1	1	2	1	5	0.5
Mycelia sterilia (white)	-	1	3	-	4	0.4
Verticillium psalliotae Treschow	4	-	-	-	4	0.4
Syncephalastrum racemosum Cohn ex Schroeter	3	-	-	-	3	0.3
Fusarium oxysporum Schlecht.	2	-	-	-	2	0.2
Acremonium strictum W.Gams	1	-	-	-	1	0.1

Contd.

Table 3 continued

Species and indices of fungal growth	*Number of appearances and fungal indices at temperature*				*Total*	*Frequency*
	23°C	29°C	33°C	37°C		
Botryotrichum piluliferum Sacc. & March.	1	-	-	-	1	0.1
Neosartoria fischeri (Wehmer) Malloch & Cain	-	-	-	1	1	0.1
Paecilomyces marquandii (Massee) Hughes	1	-	-	-	1	0.1
Penicillium nigricans Bain. ex Thom	-	-	1	-	1	0.1
Number of appearances	256	215	222	269	962	-
Frequency of isolation	93.3	79	66.7	92.4	82.9	-
Number of species	13	8	9	7	17	-

terrestre with its teleomorph *Arthroderma quadrifidum*, *Arthrographis kalrai*, and *Chrysosporium* anamorph of *Arthroderma curreyi*. Subsequently, *Microsporum gypseum* predominated at 29°C, although this species was also relatively frequently isolated at 23 and 33°C. *Chrysosporium indicum*, *Chrysosporium* anamorph of *Aphanoascus clathratus* and *Chrysosporium* anamorphs of *Aphanoascus reticulisporus/fulvescens* prevailed at 33°C, although *Chrysosporium* anamorph of *Aphanoascus clathratus* and *Chrysosporium* anamorphs of *Aphanoascus reticulisporus/fulvescens* were also frequently isolated at 29/37°C and 23/29°C, respectively. The highest isolation frequencies were found at 29 and 37°C. *Chrysosporium zonatum* prevailed at 37°C.

Altogether, 962 appearances of keratinophilic fungi belonging to 17 species were observed. *Pseudallescheria boydii*, *Verticillium lecani*, *Aspergillus fumigatus*, *Penicillium janthinellum*, *Aspergillus alutaceus* and *Aspergillus terreus* were the species with frequencies >1%. The highest isolation frequency and the highest number of species were found at 23°C. The number of appearances was also high at this temperature. *Verticillium lecani* grew best at 23°C, although the species also occurred frequently at 29°C. The highest number of *Pseudallescheria boydii*, *Aspergillus fumigatus* and *Aspergillus terreus* appearances was observed at 37°C, although the fungi also relatively frequently occurred at 29 and 33°C. The highest number of *Penicillium janthinellum* appearances was noticed at 33°C, although the number of its appearances at 29°C was also high. *Aspergillus alutaceus* occurred at 33 and 37°C, with similar numbers of appearances. The highest number of appearances was found at 37°C.

The isolation frequency of keratinolytic fungi was negatively correlated with the number of appearances and isolation frequency of keratinophilic fungi ($r = -0.45$ and -0.51), especially *Pseudallescheria boydii* ($r = -0.57$).

Statistics for microbiological and physico-chemical characteristics of sewage sludge are presented in Tables 3 and 4, respectively. Among physico-chemical parameters, the highest variability coefficients (standard deviation/arithmetic mean) were found for nitrate nitrogen, TPOC, nitrite nitrogen, phosphates, and Hg content. The smallest variability coefficients were determined for pH in H_2O, pH in 1M KCl, moisture, 1-0,1 mm fraction and organic carbon contents. The variability among microbiological parameters was generally high. The highest variability coefficients were determined for MPN of fecal coliforms, number of thermophilic bacteria, and MPN of fecal streptococci. The lowest variability coefficients were found for a total number of bacteria and proteolytic and dehydrogenase activities. *Salmonella* (B and C groups) was isolated from five sludge samples. Parasite ova (*Ascaris* sp. and *Trichuris* sp.) in quantities

Table 4. Physico-chemical characteristics of sewage sludge

Parameter	*Unit*	*Mean*	*Stand. Dev.*	*Variability coefficient*	*Median*	*Lower quartile*	*Upper quartile*	*Mini-mum*	*Maxi-mum*
Fraction 1-0.1 mm	%	62	16	0.26	67	45	74	36	87
Fraction 0.1-0.05 mm	As above	5	6	1.1	3	1	7	1	22
Fraction 0.05-0.02 mm	As above	5	8	1.55	2	1	3	0	26
Fraction 0.02-0.005 mm	As above	3	4	1.33	1	0	2	0	13
Fraction 0.005-0.002 mm	As above	1	2	1.57	1	0	1	0	10
Fraction <0.002 mm	As above	23	13	0.57	21	15	28	5	59
Moisture	As above	60.6	12.6	0.21	62.3	49.4	69.3	40	81.4
pH in H_2O	–	6.9	0.8	0.11	6.9	6.6	7.4	4.6	7.9
pH in 1M KCl	–	7	0.6	0.09	7.1	6.5	7.4	5.9	7.9
Conductivity	mS/cm in 25°C	3286	3205	0.98	2375	1331	3378	811	13243
Total nitrogen	% d.w.	2.43	1.27	0.52	2.3	1.56	3.25	0.54	5.1
Organic carbon	As above	24.2	6.7	0.28	23.3	20.2	27.7	10.9	37.8
C:N	–	13.4	12.7	0.95	10.3	7.4	13.8	5.8	66.4
Total phosphorus	mg/kg d.w.	11415	17043	1.49	5673	3399	8117	1287	66424

Contd.

Table 4 continued

Parameter	*Unit*	*Mean*	*Stand. Dev.*	*Variability coefficient*	*Median*	*Lower quartile*	*Upper quartile*	*Mini-mum*	*Maxi-mum*
Available phosphorus	mg P_2O_5/100 g d.w.	882	953	1.08	925	148	1100	36	3670
Phosphates	mg/kg d.w.	299.3	499.4	1.67	123.3	18	280.8	0.4	2076.8
Available potassium	mg K_2O/100 g d.w.	100.2	150.4	1.5	35	26.3	101.6	8.3	661.9
Total sulfur	% d.w.	0.63	0.38	0.61	0.45	0.38	0.75	0.21	1.66
Sulfate sulfur	As above	0.32	0.3	0.94	0.23	0.13	0.34	0.04	1.34
Exchange acidity H_W	mval/100 g d.w.	6.03	6.46	1.07	3.8	2.16	5.96	0.04	26
Sorption capacity T	mval/100 g d.w.	31.98	20.16	0.63	34.7	15.49	42.9	1.78	71.02
Ammonium nitrogen	mg N-NH_4/kg d.w.	199	279.7	1.4	54	14.9	294	7.9	1006.4
Nitrite nitrogen	mg N-NO_2/kg d.w.	4.5	7.7	1.69	1.3	0.3	4.9	0.1	34.8
Nitrate nitrogen	mg N-NO_3/kg d.w.	1.8	4.8	2.6	0.5	0.3	1.3	0	22.3
Fe	mg/kg d.w.	16876	7316	0.43	17449	12084	21548	1326	32500
Mn	As above	361.5	216.8	0.6	287.5	240.9	419.6	132.4	1143
Zn	As above	2321	1270	0.55	2289	1479	2841	566	6718
Cd	As above	20.6	28.6	1.39	11	7.9	20.3	3.3	133.4

Contd.

Table 4 continued

Parameter	*Unit*	*Mean*	*Stand. Dev.*	*Variability coefficient*	*Median*	*Lower quartile*	*Upper quartile*	*Mini-mum*	*Maxi-mum*
Pb	As above	329.5	199	0.60	275.8	196.2	450	20.7	766.2
Cu	As above	185	100.6	0.54	158.4	123.5	231	44.5	474
Cr	As above	112	101.2	0.90	73.8	64.7	106.9	6	373.6
Ni	As above	48.6	35.3	0.73	34.9	28.4	46.3	14.1	167.3
Hg	As above	5.4	9	1.67	1.6	1.4	3.4	0.5	34.3
As	As above	11.2	7	0.63	7.9	6.5	17.3	1.4	24
TPH	g/kg d.w.	3.6	3.4	0.95	2.6	1.3	4.6	0.1	13
TPOC	g/kg d.w.	23.2	44.8	1.93	7.2	4.5	22.9	0.9	194.3
PAHs	mg/kg d.w.	18.58	13.07	0.70	13.73	8.59	30.99	3.04	47.94

d.w.—dry weight.

between 20-60/kg d.w. (dry weight) were observed in four sludge samples.

Many statistical relationships between keratinolytic/keratinophilic fungi data and physico-chemical/microbiological parameters were observed. The most important relationships are presented below. *Microsporum gypseum* preferred sludges with total sulfur content >1% d.w. (Figure 1). These sludges also contained sulfate concentrations >0.4% d.w. and had C:S ratio <30. *Chrysosporium indicum* occurred with significantly higher frequencies in sludges with available phosphorus content >200 mg P_2O_5/100 g d.w. (Figure 2). At 23°C, *Chrysosporium keratinophilum* preferred sludges with ammonium nitrogen content >400 mg N-NH_4/kg d.w. (Figure 3). Also, *Chrysosporium keratinophilum* occurred with significantly higher frequency in sludges with proteolytic activity >2 g N-NH_3/100 g d.w. In contrast, *Trichophyton terrestre,* with its teleomorph *Arthroderma quadrifidum,* preferred sludges with lower proteolytic activity (Figure 4). The number of *Chrysosporium keratinophilum* appearances at 23°C was positively correlated with fecal coliforms (Figure 5). The lower sludge C:N ratio, the higher number of *Chrysosporium keratinophilum* appearances. Geophilic dermatophytes, i.e., *Arthroderma quadrifidum* and *Arthroderma uncinatum* with their anamorphs, preferred sludges with C:N ratio between 10-15 (Figure 6). The higher colloidal loam fraction (<0.002 mm) content, the lower *Trichophyton terrestre/Arthroderma quadrifidum* frequencies (Figure 7).

The frequency of *Chrysosporium* anamorph of *Aphanoascus clathratus* at 33°C was associated with proteolytic activity, ammonium nitrogen content, and alkaline reaction.

Most of the prevailing keratinolytic species occurred in neutral and alkaline sludges. *Chrysosporium keratinophilum, Chrysosporium* anamorph of *Aphanoascus clathratus, Chrysosporium zonatum, Arthrographis kalrai* and *Chrysosporium* anamorphs of *Aphanoascus reticulisporus/fulvescens* were alkalophilic. Among the keratinolytic species, only *Chrysosporium* anamorph of *Arthroderma curreyi* prevailed in acidic sludges (pH in H_2O <7). In contrast, keratinophilic species such as *Verticillium lecani, Aspergillus fumigatus, Penicillium janthinellum* and *Aspergillus alutaceus* preferred acidic sludges. Among the prevailing keratinophilic species, only *Aspergillus terreus* occurred in alkaline sludges.

High Cd, Zn and Hg quantities (over the Polish sewage sludge standards) were determined in some sludge that were examined. However, the factor analysis and other statistical methods showed that heavy metals did not significantly affect the composition of keratinolytic and keratinophilic fungi in the sludge samples examined.

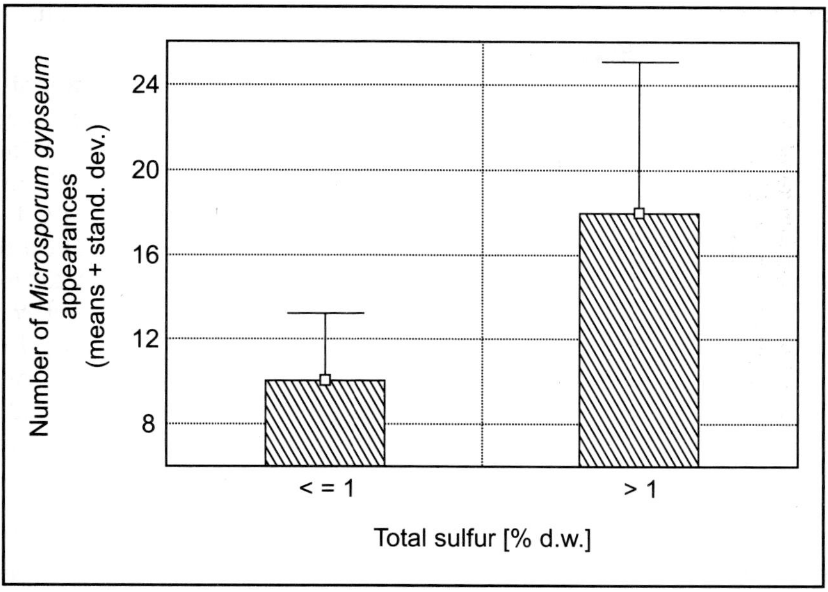

Fig. 1. Relationship between the number of *Microsporum gypseum* appearances and total sulfur content in sewage sludge

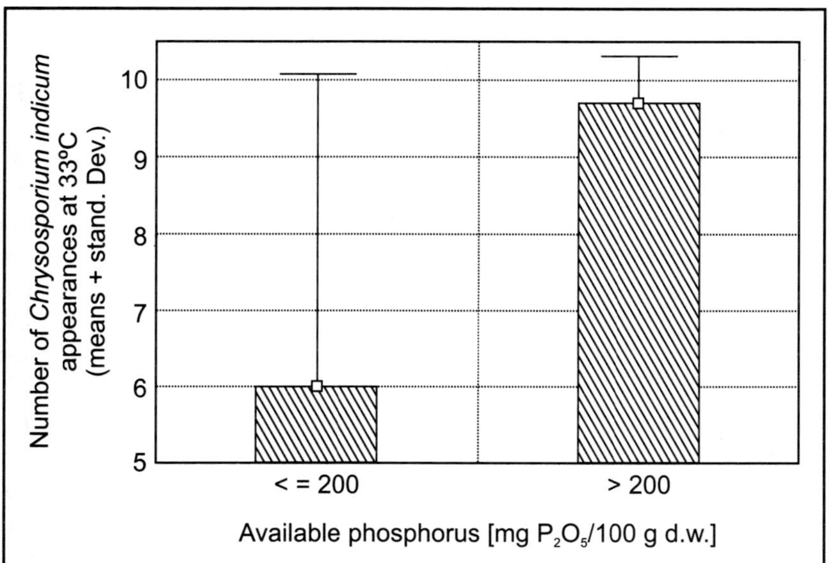

Fig. 2. Relationship between the number of *Chrysosporium indicum* appearances at 33°C and available phosphorus content in sewage sludge

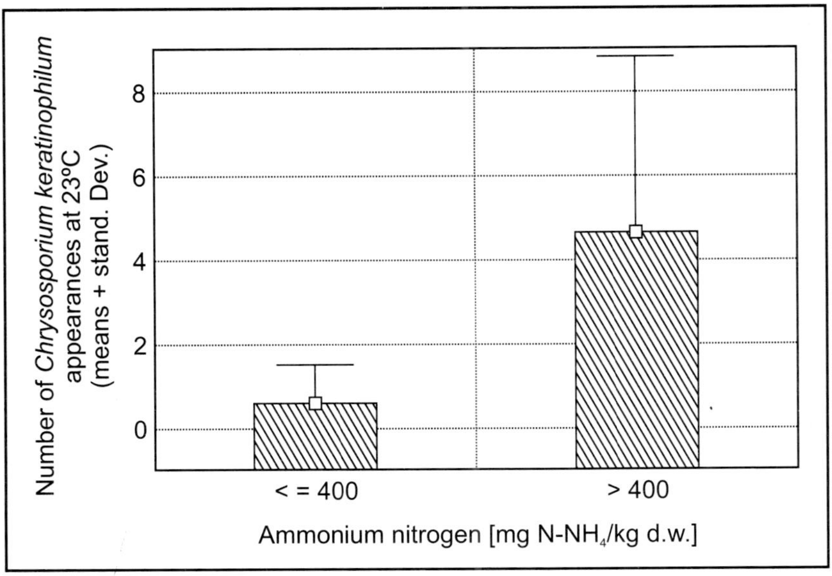

Fig. 3. Relationship between the number of *Chrysosporium keratinophilum* appearances at 23°C and ammonium nitrogen content in sewage sludge

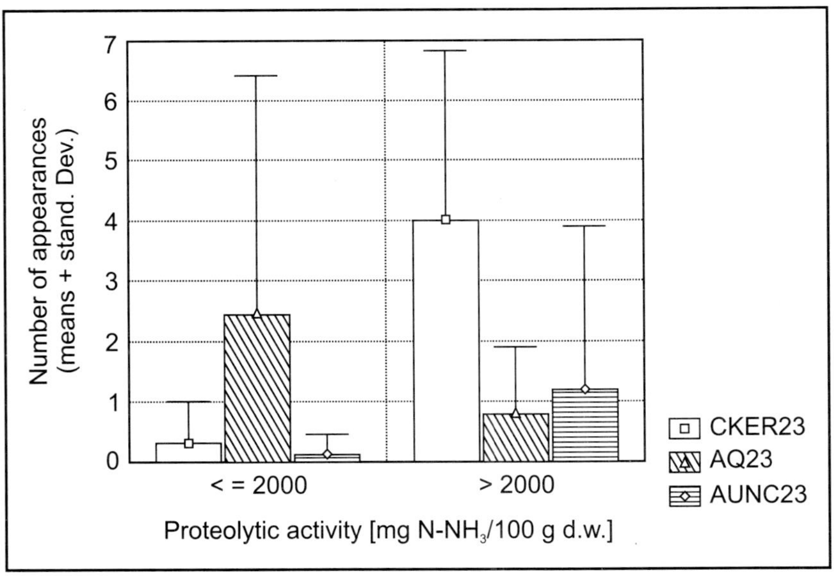

Fig. 4. Relationship between the number of *Chrysosporium keratinophilum* (CKER23), *Arthroderma quadrifidum* (AQ23) and *Arthroderma uncinatum* (AUNC23) appearances at 23°C with proteolytic activity in sewage sludge

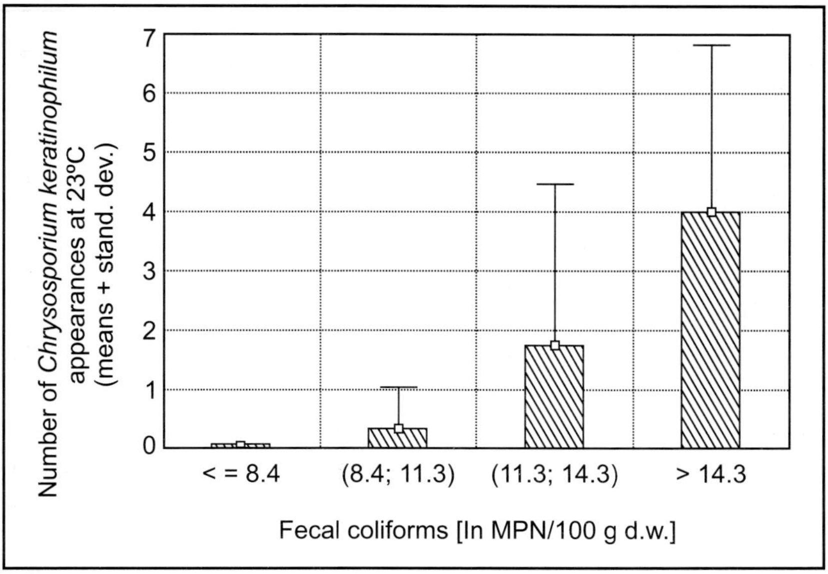

Fig. 5. Relationship between the number of *Chrysosporium keratinophilum* appearances at 23°C and fecal coliform quantities in sewage sludge

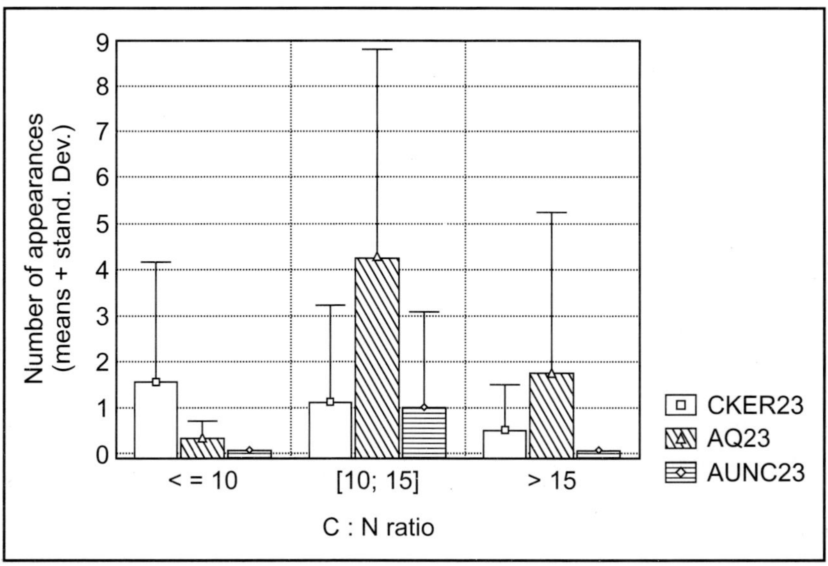

Fig. 6. Relationship between the number of *Chrysosporium keratinophilum* (CKER23), *Arthroderma quadrifidum* (AQ23) and *Arthroderma uncinatum* (AUNC23) appearances at 23°C with C:N ratio in sewage sludge

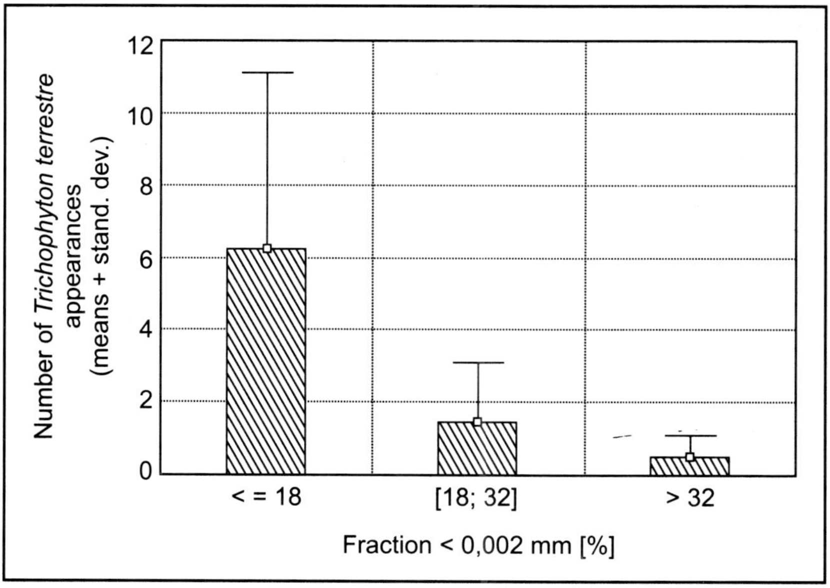

Fig. 7. Relationship between the number of *Trichophyton terrestre* appearances and colloidal loam (fraction < 0.002 mm) content in sewage sludge

Chrysosporium anamorphs of *Aphanoascus reticulisporus/fulvescens* occurred in sludges with TPOC >10 g/kg d.w. and total PAHs >20 mg/kg d.w.

No relationships were found between keratinolytic and keratinophilic fungi with the presence of *Salmonella* and parasite ova in sewage sludge.

The cluster analysis of the fungi data is illustrated in Figure 8. The cluster analysis divided the fungi into two groups. The first group included most of keratinolytic fungi and *Aspergillus terreus* with some other keratinophilic species (OTHERS 2). The second group included most of keratinophilic fungi and keratinolytic *Microsporum gypseum, Chrysosporium* anamorph of *Arthroderma curreyi* and the OTHERS fungi. *Microsporum gypseum* together with *Pseudallescheria boydii* formed one of the subgroups.

Figure 9 also illustrates the cluster analysis of the fungi data but in relation to sewage sludges (cases). The cluster analysis divided the sludges into two major groups and five subgroups. The first group included sludges from Siemianowice Śląskie-Centrum wastewater treatment plant (I subgroup) and sludges from wastewater treatment plants in Olkusz, Dgbrowa Górnicza and Katowice (Gigablok and Panewniki) (II subgroup). The second group included sludges from wastewater treatment plants in

Sosnowiec (III subgroup) and Bytom (IV and V subgroups). Sludges from wastewater treatment plants using non-conventional technologies (after extended aeration and the integrated biological process for P, N and C removal) were in I, II and III subgroups. The diagram was similar to the one obtained from the cluster analysis of physico-chemical data.

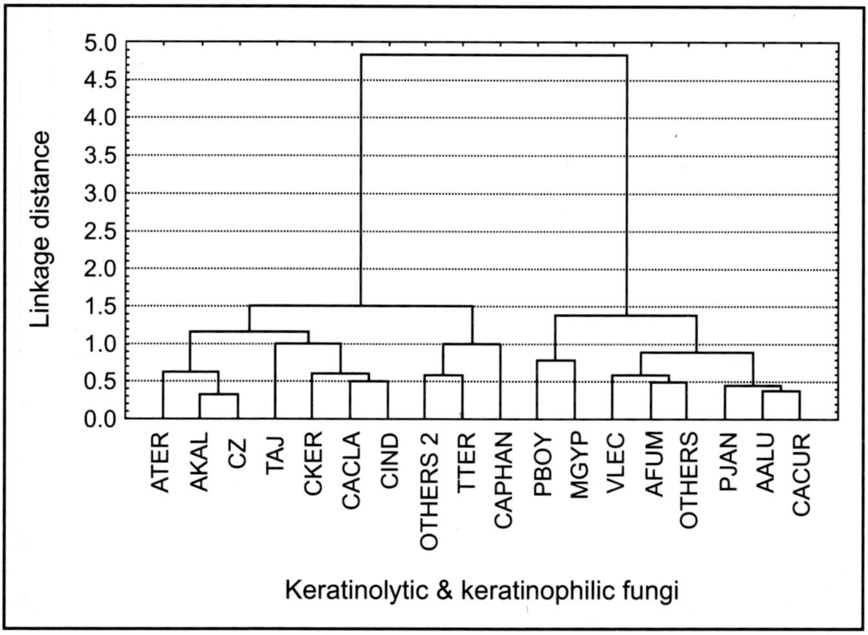

Fig. 8. Cluster analysis diagram for keratinolytic and keratinophilic fungi in sewage sludge

Abbreviations: ATER—*Aspergillus terreus;* AKAL—*Arthrographis kalrai;* CZ—*Chrysosporium zonatum;* TAJ—*Trichophyton ajelloi;* CKER—*Chrysosporium keratinophilum;* CACLA—*Chrysosporium* anamorph of *Aphanoascus clathratus;* CIND—*Chrysosporium indicum;* OTHERS 2—other keratinophilic fungi; TTER—*Trichophyton terrestre;* CAPHAN—*Chrysosporium* anamorphs of *Aphanoascus reticulisporus/fulvescens;* PBOY—*Pseudallescheria boydii;* MGYP—*Microsporum gypseum;* VLEC—*Verticillium lecani;* AFUM—*Aspergillus fumigatus;* OTHERS—other keratinolytic fungi; PJAN—*Penicillium janthinellum;* AALU—*Aspergillus alutaceus;* CACUR—*Chrysosporium* anamorph of *Arthroderma curreyi*

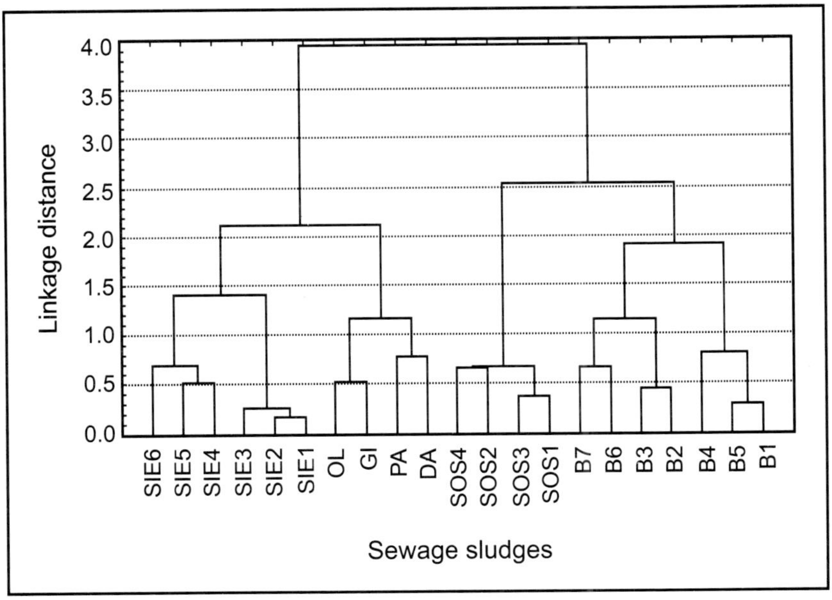

Fig. 9. Cluster analysis diagram for sewage sludge (based on the fungi data)

DISCUSSION

Ecological relationships

Sewage sludges from 16 wastewater treatment plants using different sewage and sewage sludge treatment technologies were examined. The statistical analysis showed many relationships between the fungi data and physico-chemical/microbiological parameters. Especially, the cluster analysis based on the fungi data divided the sludges into two groups and five subgroups. The three subgroups included sludges from one or several wastewater treatment plants, located in 1-3 towns, and using both conventional and non-conventional technologies. One of the three subgroups included sludges from the Siemianowice Śląskie-Centrum wastewater treatment plant, which used the original sedimentation method for the anaerobic phase of the biological phosphorus removal process (Kowal, 1998). The diagram based on the fungi data was close to that based on the physico-chemical data. It can be concluded, therefore, that not only sewage and sludge treatment technologies but, first of all, the influence of 'combinations' of physico-chemical and microbiological factors influenced the fungal composition in sewage sludge. However, this conclusion requires a detailed explanation.

The results from literature data (Filipello-Marchisio, 2000; Ali-Shtayeh and Jamous, 2000; and others) and own observations lead to the conclusion that fungal keratinolytic activity on hair can not be clearly distinguished. It is accepted, however, that keratinolytic activity is stronger and more frequent in some species than in others. These species should be considered specialized in keratin decomposition. In general, the present study confirmed the division of the fungi examined into keratinolytic and keratinophilic. Keratinophilic fungi were isolated from hair during the first month of incubation. If these fungi did not compete with keratinolytic fungi, they remained on the hair till the end of the experiments. If keratinophilic fungi had to compete with keratinolytic fungi, they could be even eliminated from hair by the latter. Basically, keratinophilic fungi were not able to initiate keratin decomposition, presumably utilized simple and easy available hair components or the products of keratin decomposition. In the cluster analysis, however, some keratinolytic species were grouped together with keratinophilic species. Subsequently, there were several keratinophilic species in the keratinolytic fungi group. This resulted from the influence of physico-chemical and microbiological factors on fungal composition. These observations concerned actidione-resistant strains.

It can also be concluded that the ascendancy of keratinolytic fungi over keratinophilic fungi on hair resulted not only from nutrition and room competition but also due to the influence of pH. It is known that geophilic dermatophytes and other keratinolytic fungi prefer neutral and alkaline environments (see literature below). Thus, the results obtained for sewage sludge coincide with literature data. It is also known, however, that keratinolytic fungi tolerate wide pH ranges (Garg et al., 1985; Filipello-Marchisio et al., 1991; Senapati, 1999; Simpanya, 2000; Ulfig, 2000). The results also confirmed this phenomenon.

Most keratinolytic and keratinophilic fungi are mesophilic in nature (Garg et al., 1985; Kushwaha, 2000). The mesophilic nature of the fungi was observed in this study. However, the results also confirmed the high variability in fungal temperature spectra and reflected the influence of environmental factors on fungal growth at different temperatures. For instance, in pure culture *Microsporum gypseum* and *Chrysosporium indicum* differ in biomass production but have almost identical temperature spectra. Their optimal growth occurs at 29-30°C, but the fungi still display good growth at 37°C (Dvořák and Hubálek, 1969; Senapati, 1999). On hair laid on sewage sludge, *Microsporum gypseum* and *Chrysosporium indicum* predominated at 29 and 33°C, respectively. No growth of the dermatophyte was observed at 37°C, and the number of *Chrysosporium indicum* appearances at this temperature was considerably lower than at 33°C.

These 'modifications' of the fungal temperature spectra could have resulted from the influence of environmental factors in sewage sludge.

Trichophyton terrestre has been frequently isolated from environments with low organic matter (carbon) content. Moreover, Korniłłowicz-Kowalska and Bohacz (2002) showed the negative correlation between the *Trichophyton terrestre* isolation frequency and total nitrogen content in arable soils. Subsequently, *Trichophyton ajelloi* is considered ubiquitous, and its incidence is not dependent on soil organic matter (Chmel et al. 1972; Ulfig, 2000; Korniłłowicz-Kowalska and Bohacz, 2002; and others). In a recent study (Ulfig et al., 2002), attention was paid to low nitrogen and sulfur contents as the factors favoring the incidence of this fungus in soil. However, carbon, nitrogen and sulfur contents in sewage sludge were much higher than in soils examined. Therefore, sewage sludges and conditions for fungal growth were additionally characterized with C:N and C:S ratios.

It is accepted that in ecologically stabilized soils of the moderate humid climate, the C:N ratio ranges between 10-12 (Buckman and Brady, 1971). The best *Trichophyton terrestre* and *Trichophyton ajelloi* growth, associated with *Arthroderma quadrifidum* and *Arthroderma uncinatum* ascomata production, took place in sludges with C:N 10-15 and pH from neutral to alkaline (pH 6.9-7.9). Although, dermatophyte anamorphs also occurred with high frequencies in sludges with C:N>15. It can be assumed, therefore, that sludges with C:N 10-15 are stabilized as regards carbon and nitrogen transformations, and the dermatophyte ascomata abundance should be considered the indicator of this stabilization. This problem requires further elucidation.

The hypothesis suggested that keratin substrata are more and more available to keratinolytic fungi when sludge organic matter stabilization, dewatering, and structuralization processes proceed and aeration conditions improve (Ulfig, 1991). In non-stabilized, moist, bad aerated sludges, fungi from the genus *Chrysosporium* prevail. With progressing organic matter stabilization and improvement of environmental conditions, keratin substrata are more and more intensively decomposed by geophilic dermatophytes, especially *Trichophyton terrestre* with its teleomorph *Arthroderma quadrifidum*. It was also supposed that the predominance of geophilic dermatophytes was associated with the decrease of fecal coliform quantities. Therefore, the composition of keratinolytic fungi could be used as a rough indicator of the sludge hygienization process. A study of keratinolytic fungi in sludge-reclaimed soil confirmed the above-mentioned hypothesis (Ulfig and Korcz, 1994). However, the above-quoted studies did not determine factors influencing fungal composition in sludge and soil. Owing to more physico-chemical and microbiological data available, in a later study (Ulfig et al., 1996) positive correlations between the incidence of *Chrysosporium keratinophilum* at 23°C with total nitrogen content, volatile compounds, moisture, and pH in H_2O, as well as a negative correlation between *Chrysosporium keratinophilum* and C:N ratio were found. In this study, the frequency of *Chrysosporium keratinophilum* at 23 and 29°C and the frequency of *Chrysosporium* anamorph of *Aphanoascus clathratus* at 33°C were associated with proteolytic activity, ammonium nitrogen content, and alkaline pH. Thus, the hypothesis can be evolved with additional elements. Certain sludges are characterized by high organic nitrogen content, accumulated chiefly in microbial biomass, and in consequence by low C:N ratio (<10). Proteolytic microorganisms decompose nitrogen organic compounds. This decomposition is often called nitrogen mineralization or ammonification. High ammonium release is typical for sewage sludge and sludge-amended soils (Ryan and Keeney, 1975; Tester et al., 1977; Terry et al., 1978; Parker and Sommers, 1983; Hallet et al., 1999). Ammonium causes medium alkalization. It can also inhibit further ammonification and maintain ammonium concentration at a stable level. It appears that keratin substrata play a minor role in the ammonification process, as their microbial decomposition is too slow. However, under high ammonium concentration the characteristic composition of keratinolytic fungi, with the predomination of *Chrysosporium keratinophilum* and some other fungi from this genus occurs. High colloidal loam content and high moisture additionally support the growth of these fungi. The hypothesis, based on statistical relationships, should be confirmed in model experiments.

It is likely that nitrogen transformations, especially ammonification, associated with pH changes also causes characteristic changes in fungal compositions of soils and bottom sediments rich in nitrogen and ammonium, e.g., in manure-fertilized soils or in pastures and farms (Kaul and Sumbali, 1999). In another study (Ulfig, 1991), in putrefying sludges, with high water content, the growth of keratinolytic fungi was weak and it was impossible to maintain hair on the sludge surface. This hair sinking into the sludge caused the need for using additional hair portions. Understandingly, *Chrysosporium keratinophilum* prevailed on the putrefying sludges at 23°C. Thus, the composition of keratinolytic and keratinophilic fungi is also associated with the sludge putrefaction process. Therefore, the fungal composition can be the indicator of the sludge organic matter stabilization process in both the technological sense and microbiological organic matter transformation aspect. It appears that the fungal composition can be of practical significance in wastewater treatment plants and soil reclamation practice.

A separate problem is the influence of sludge particle size distribution on fungal composition. The results indicate that the higher content of the colloidal loam fraction (<0.002 mm), the lower *Trichophyton terrestre* and its teleomorph *Arthroderma quadrifidum* frequency. However, explanation of environmental factors associated with this fraction requires elucidation in model experiments.

Phosphorus is the essential element in fungal metabolism (Alkiewicz, 1966; Jennings, 1995; and others). The sludges examined were characterized with high but variable total phosphorus, available phosphorus and phosphate contents. The sludge phosphorus content was chiefly dependent on the specified wastewater treatment technology. It is understandable that high available phosphorus and phosphate quantities were found in sludges after the biological phosphorus removal process, e.g., in sludges from the Siemianowice Śląskie-Centrum wastewater treatment plant. *Chrysosporium indicum* was the keratinolytic fungus frequently isolated from sludges, especially at 33°C. However, the growth of this fungus on hair laid on sludges with high available phosphorus content was especially abundant. Korniłłowicz-Kowalska and Bohacz (2002) found the positive correlation between phosphorus content and *Trichophyton georgiae* and *Chrysosporium tuberculatum* frequencies in arable soils. However, no data have been found to explain the abundant growth of *Chrysosporium indicum* in sludges with high available phosphorus content.

Ulfig et al. (1996) and Płaza et al. (1998) found that heavy metals could considerably impact keratinolytic and keratinophilic fungi in sewage sludge. In the present study, however, the metals were not the most important factors affecting fungal composition. A separate problem is the availability of heavy metals to fungi and the influence of these metals on microorganisms under changing pH in sewage sludge.

During a study of keratinolytic fungi in bottom sediments (Ulfig et al., 1997a, b), *Chrysosporium keratinophilum* was found to be resistant to high phenol and petroleum hydrocarbon contents. It was suggested in later studies (Ulfig et al., 1998b; Ulfig et al., 2003) that keratinolytic fungi could be used as indicators of leakage toxicity and bioremediation progress of petroleum-contaminated soil. Finally, many keratinolytic fungi strains with the ability for removing petroleum hydrocarbons from different media during peptone and/or keratin decomposition were isolated from soils, bottom sediments and sewage sludge (Ulfig et al., 2000; Przystaś et al., 2000). In the present study, positive correlations between the incidence of *Chrysosporium* anamorphs of *Aphanoascus reticulisporus/fulvescens* with TPH/TPOC and PAH contents were found. An ability of these strains for utilizing different petroleum hydrocarbons should be examined in separate experiments.

Health risk from fungal pathogens

A separate problem is to evaluate the health risk posed by fungal pathogens occurring in sewage sludge. The results of the present study along with historical data contribute to the above-mentioned problem. Pathogenic fungi isolated from sludges examined belong to BSL-1 (BioSafety Level-1) and BSL-2 categories (de Hoog, 1996; de Hoog et al., 2000). The BSL-2 fungi are of high epidemiological significance. These fungi are opportunists, having keratinolytic or keratinophilic properties, and belonging to dermatophytes and related species and to other fungal groups. Zoophilic dermatophytes, i.e., *Microsporum canis*, *Microsporum persicolor* and *Trichophyton mentagrophytes* var. *mentagrophytes* were isolated from bottom sediments (Ulfig and Ulfig, 1990; Ulfig et al., 1998a) and from air contaminated with sewage bioaerosol (Ulfig, 1994). In earlier studies (Ulfig and Korcz, 1983; Ulfig, 1986a; Ulfig, 1991; Ulfig and Korcz, 1994; Ulfig et al., 1996), both zoophilic and anthropophilic dermatophytes were not isolated from sewage and sewage sludge. In the present study, in which many sludge samples were examined and four incubation temperatures used, these dermatophytes were not isolated either. It was also proved that the survival of zoophilic dermatophytes in sewage sludge did not exceed one month. The survival of these dermatophytes was affected by physico-chemical and microbiological factors, especially by the strong antagonistic impact of sludge microflora (Ulfig, 1986b; Ulfig, 1988). The survival of anthropophilic dermatophytes in sewage sludge was not studied. It was observed earlier, however, that the survival time of these dermatophytes in soil did not exceed eight days (Grin and Ožegović, 1963). Thus, the available results testify that the sewage sludge environment does not favor the survival of zoophilic and anthropophilic dermatophytes. Therefore, the infection risk posed by these dermatophytes in contact with sewage sludge and sludge-amended soil should be considered minimal. However, the problem of the health risk posed by other fungi occurring in sewage sludge remains.

Among the dermatophytes and related fungi prevailing in sewage sludge, attention should be paid to *Microsporum gypseum*, *Chrysosporium keratinophilum*, *Trichophyton terrestre* and *Trichophyton ajelloi*. With the exception of *Microsporum gypseum*, the above-mentioned fungi heave been rarely reported as agents responsible for mycoses (Hubálek and Hornich, 1977; Zabawski and Baran, 1998; Filipello-Marchisio et al., 1995). However, *Microsporum gypseum* represents the intermediate phase in the evolution from saprophytism to parasitism. The number of mycoses caused by this dermatophyte is relatively high (Hayashi and Toshitani, 1983; Porro et al., 1997; Śpiewak, 1998). The fungus often attacks immunocompromised patients (de Hoog et al., 2000). It appears, therefore, that *Microsporum gypseum* should not be treated at the same epidemiological level as the other fungi are. It is known that one of the conditions for infection initiation is the inoculation of adequate spore quantity in the skin. (Hashimoto, 1991). Sludges contain high amounts of keratin remnants and provide favorable conditions for the growth of *Microsporum gypseum* (high total sulfur content, low C:S ratio). It can be supposed, therefore, that sludges also contain sufficient fungal inoculum for infection initiation.

The study also paid attention to the relatively high frequency of *Chrysosporium zonatum* in sewage sludge. According to Sigler et al. (1998) and Sigler (2002a), the fungus can colonize the lungs and cause deep mycoses in immunocompromised patients. Hayashi et al. (2002) also described a case of lung mycosis caused by the fungus in a patient with healthy immunological system. Although the number of mycoses caused by *Chrysosporium zonatum* is low, the appearance of this species on the list of pathogenic fungi reflects the general increase in the number of opportunistic mycoses. This increase may be associated with increasing contamination of the environment with organic waste, including keratinous remnants and xenobiotics, and with more and more favorable conditions for growth and expansion of opportunistic fungi.

Pseudallescheria boydii is another opportunistic species, earlier known as one of the *mycetoma* causative agents (Prochacki, 1975). The fungus causes mycoses in patients with injuries and in persons who have drowned after swallowing contaminated water. *Pseudallescheria boydii* shows special affinity to the nervous system. Untreated mycoses are usually fatal. Treatment is difficult due to the resistance of this fungus to antifungal agents. Mycoses of the respiratory system (lungs, nose) and eyes also are known (de Hoog et al., 2000; Zabawski and Baran, 1998; Sigler, 2002b). *Pseudallescheria boydii* has been found to occur in sewage and wastewater treatment facilities, natural soils, manure, pot soil, and in waters polluted with sewage (Cooke, 1957; Jeffery et al., 1997; de Hoog et al., 2000). However, the fungus has not been isolated from sewage sludge (Abdel-Hafez and El-Sharouny, 1990) and soils reclaimed with sewage or sludge (Cooke, 1971; Ali-Shtayeh et al., 1999). Since *Pseudallescheria boydii* was the most common keratinophilic fungus in the present study, these observations are difficult to explain.

Aspergillus fumigatus is another opportunistic fungus frequently occurring in sewage sludge (Stevens et al., 2000; Brookman and Denning, 2000). This thermotolerant fungus is isolated mainly from composts and environments, in which temperatures range between 40-60°C or reach even higher values (Rosenberg, 1975; Johri, 1980; Dumontet et al., 1999). It is understandable, therefore, that the higher the temperature, the higher the frequencies of the fungus on hair. At 45°C, no strongly keratinolytic fungi were observed on hair. At this temperature, *Aspergillus fumigatus* was the predominating species (Ulfig, 2003). The fungus produces cellulolytic and lignolytic enzymes (Tansey et al., 1977; Domsch et al. 1980). Its growth on hair, though without visible substrate degradation, results certainly from the production of proteolytic and lipolytic enzymes by the fungus (Al-Musallam and Radwan, 1990; Onifade et al., 1998). It may be concluded, therefore, that at high temperatures (maximal growth temperature is 70°C) the abundance of keratinous remnants favors the growth of *Aspergillus fumigatus* in sewage sludge. This may be of epidemiological importance.

CONCLUSIONS

- Rich qualitative and quantitative composition of keratinolytic and keratinophilic fungi characterizes sewage sludge.
- The composition of keratinolytic and keratinophilic fungi reflects both sewage and sludge treatment technologies and the influence of 'combinations' of physico-chemical and microbiological factors characteristic for a given urban agglomeration, wastewater treatment plant or group of wastewater treatment plants.
- The most important factors influencing the composition of keratinolytic and keratinophilic fungi in sewage sludge are as follows: temperature, pH, ammonium nitrogen, proteolytic activity, organic carbon and total nitrogen, C:N ratio, total sulfur, C:S ratio, available phosphorus and particle size distribution.
- The study confirmed the division of the fungi examined into keratinolytic and keratinophilic. Keratinolytic fungi are able to decompose keratin, while keratinophilic fungi utilize simple and easy degradable components of keratinous remnants and the products of keratin decomposition. Keratinolytic fungi can eliminate keratinophilic fungi from hair. Keratinolytic fungi prefer neutral and alkaline sludges, while keratinophilic fungi occur more frequently in acidic sludges.
- Quantitiative relationships between *Trichophyton terrestre* with its teleomorph *Arthroderma quadrifidum* and *Chrysosporium keratinophilum* reflect ammonium nitrogen concentration and pH, C:N ratio, as well as colloidal loam and water contents in sewage sludge.
- The composition of keratinolytic fungi can be the indicator of the sewage sludge organic matter stabilization process. This conclusion addresses both sewage and sludge treatment technologies and microbiological organic matter transformations. Due to the correla-

tion with fecal coliform quantities, the composition of keratinolytic fungi can also be the rough indicator of sludge hygienization. The relationships can be used in both wastewater treatment plants and soil reclamation practice.

- From the mycological point of view, opportunistic fungi, especially *Microsporum gypseum* and *Pseudallescheria boydii*, pose the major health risk in sewage sludge. Temperature in the range of 33-37°C favors the growth of *Pseudallescheria boydii*, whereas the temperature 29°C is optimal for the growth of *Microsporum gypseum* in sewage sludge. High total sulfur content (>1% d.w.) and low C:S ratio (<30:1) favor the growth of both species.
- In the light of available data, pathogenic fungi should be regarded as the important element of public health risk posed by sewage sludge, especially when applied to land.
- Future studies of keratinolytic and keratinophilic fungi in sewage sludge should focus on the following issues: (1) confirmation of the statistical relationships in model experiments; (2) elaboration of a method for determination of an absolute number of fungal propagules able to attack keratin; and (3) assessment of health and environmental risks posed by pathogenic and toxigenous fungi in sewage sludge and sludge-amended soils.

References

Abdel-Hafez, A.I.I. and El-Sharouny, H.M.M. 1987. Seasonal fluctuations of fungi in Egyptian soils receiving city sewage effluents. Cryptogamia, 8: 235-249.

Abdel-Hafez, A.I.I. and El-Sharouny, H.M.M. 1990. The occurrence of keratinophilic fungi in sewage sludge from Egypt. Journal of Basic Microbiology, 30: 73-79.

Al-Musallam, A.A. and Radwan, S.S. 1990. Wool-colonizing microorganisms capable of utilizing wool lipids and fatty acids as sole sources of carbon and energy. Journal of Applied Bacteriology, 69: 806-813.

Ali-Shtayeh, M.S., Jamous, R.M.F. and Abu-Ghdeib, S.I. 1999. Ecology of cycloheximide-resistant fungi in field soils receiving raw city wastewater or normal irrigation water. Mycopathologia, 144: 39-54.

Ali Shtayeh, M.S. and Jamous, R.M.F. 2000. Keratinophilic fungi and related dermatophytes in polluted soil and water habitats, pp. 51-59. In: Biology of Dermatophytes and other Keratinophilic Fungi. R.K.S. Kushwaha, and J. Guarro (eds.), Revista Iberoamericana de Micología.

Alkiewicz, J. 1966. Medical Mycology (in Polish), PZWL, Warszawa, Poland.

Bitton, G., Damron, B.L., Edds, G.T. and Davidson, J.M. (eds.) 1980. Sludge—Health Risk of Land Application. Ann Arbor Science, Ann Arbor, USA.

Brookman, J.L. and Denning, D.W. 2000. Molecular genetics in *Aspergillus fumigatus*. Current Opinion in Microbiology, 3: 468-474.

Buckman, H.C. and Brady, N.C. 1971. Soil and its Properties (in Polish). PWRiL, Warszawa, Poland.

Cano, J.F. and Guarro, J. 1990. The genus *Aphanoascus*. Mycological Research, 94: 355-377.

Chmel, L., Hasilíková, A., Hraško, J. and Vláčilíková, A. 1972. The influence of some ecological factors on keratinophilic fungi in the soil. Sabouraudia, 10: 26-34.

Cooke, W.B. 1957. Checklist of fungi isolated from polluted water and sludge. Annals of. Mycology, 2: 146-175.

Cooke, W.B. 1959. Trickling filter ecology. Ecology, 40: 173-291.

Cooke, W.B. 1963. A Laboratory Guide to Fungi in Polluted Waters, Sewage, and Sewage Treatment Systems, Public Health Services. Pub., 999-WP-1, USDHEW, Cincinnati, USA.

Cooke, W.B. 1971. The effect of sludge additives on soil fungus populations, Mycopathologia et Mycologia Applicata, 44: 205-219.

Cooke, W.B. and Pipes, W.O. 1970. The occurrence of fungi in activated sludge. Mycopathologia et Mycologia Applicata, 40: 249-270.

Countryside Hygiene Institute. 1985. Methods for sewage sludge sanitary control (in Polish). Lublin, Poland.

Currah, R.S. 1985. Taxonomy of the Onygenales: Arthrodermataceae, Gymnoascaceae, Myxotrichaceae and Onygenaceae. Mycotaxon, 24: 1 216.

de Bertoldi, M. 1981. Pathogenic Fungi Associated with Land Application of Sludge, World Health Organization, Regional Office for Europe.

de Hoog, G.S. 1996. Risk assessment of fungi reported from humans and animals. Mycoses, 39: 407-417.

de Hoog, G.S., Guarro, J., Gene, J. and Figueras, M.J. 2000. Atlas of Clinical Fungi, Centraalbureau voor Schimmelcultures and Universitat Rovira i Virgili, Utrecht, Reus, The Netherlands.

Deshmukh, S.K. and Agrawal, S.C. 1998. Biology of keratinophilic fungi and related dermatophytes, pp. 253-272. In: Microbes: For Health, Wealth and Sustainable Environment. A. Varma (ed.). Malhotra Publishing House, New Delhi, India.

Diener, U.L., Morgan-Jones, G., Hagler, W.M. and Davis, N.D. 1976. Mycoflora of activated sewage sludge. Mycopathologia, 58: 155-156.

Dobosz, M. 2001. Computer Statistical Analysis of Research Data (in Polish), Akademicka Oficyna Wydawnicza, Warszawa, Poland.

Domsch, K.H., Gams, W. and Traute-Heidi, A. 1980. Compendium of Soil Fungi. Academic Press, London-San Francisco.

Dumontet, S., Dinel, H. and Baloda, S.B. 1999. Pathogen reduction in sewage sludge by composting and other biological treatments: a review. Biological Agriculture and Horticulture, 16: 409-430.

Dvořák, J. and Hubálek, Z. 1969. The growth of dermatophytes at 4°C and 37°C; the relation of this character to others. Mycopathologia et Mycologia Applicata, 38: 305-308.

Environmental Protection Agency U.S. 1999. Environmental Regulations and

Technology, Control of Pathogens and Vector Attraction in Sewage Sludge, EPA/625/R-92/013, Washington, DC., USA.
European Commission 2001. Evaluation of sludge treatment for pathogen reduction, Office for Official Publications of the European Communities, Luxembourg.
European Union 2000. Working document on sludge, Brussels.Belguim. (http://europa.eu.int/comm/environment/sludge/sludge_en.pdf).
Filipello Marchisio, V. 2000. Keratinophilic fungi: their role in nature and degradation of keratinic substrates, pp. 86-92. In: Biology of Dermatophytes and other Keratinophilic Fungi. R.K.S. Kushwaha, and J. Guarro (eds.), Revista Iberoamericana de Micología.
Filipello-Marchisio, V., Curetti D., Cassinelli C. and Bordese C. 1991. Keratinolytic and keratinophilic fungi in the soils of Papua New Guinea. Mycopathologia, 115: 113-119.
Filipello-Marchisio, V., Preve, L. and Tullio, V. 1995. Fungi responsible for skin mycoses in Turin (Italy). Mycoses, 39: 141-150.
Garg, A.P., Gandotra, S., Mukerji, K.G. and Pugh, G.J.F. 1985. Ecology of keratinophilic fungi. Proceedings of Indian Academy of Sciences (Plant Sciences), 94: 149-163.
Geldreich, E.E. 1975. Handbook for Evaluating Water Bacteriological Laboratories, Municipal Environmental Research Laboratory, Office of Research and Development, U.S. Environmental Protection Agency, EPA-670/9-75-006, Cincinnati, USA.
Grin, E.I. and Ožegović, L. 1963. Influence of the soil on certain dermatophytes and their evolutionary trend. Mycopathologia et Mycologia Applicata, 21: 23-28.
Guarro J., Gené J. and Stchigel A.M. 1999. Developments in fungal taxonomy. Clinical Microbiology Review, 12: 454-500.
Hallet, R.A., Bowden, W.B. and Smith, C.T. 1999. Nitrogen dynamics in forest soils after municipal sludge additions. Water, Air and Soil Pollution, 112: 259-278.
Harrison, E.Z. and Oakes, S.R. 2002. Investigation of alleged health incidents associated with land application of sewage sludges. New solutions, 12: 387-408.
Harrison, E.Z., McBride, M.B. and Bouldin, D.R. 1999. Land application of sewage sludges: an appraisal of the US regulations. International Journal Environmental Pollution, 11: 1-36.
Hashimoto, T. 1991. Infectious propagules of dermatophytes, pp. 181-200. In: The Fungal Spore and Disease Initiation in Plants and Animals. G.T. Cole, and H.C. Hoch (eds.), Plenum Press, New York, London.
Hayashi, N. and Toshitani, S. 1983. Human infections with *Microsporum gypseum* in Japan. Mykosen, 26: 337-345.
Hayashi, S., Naitoh, K., Matsubara, S., Nakahara, Y., Nagasawa, Z., Tanabe, I., Kusaba, K., Tadano, J., Nishimura, K. and Sigler, L. 2002. Pulmonary colonization of *Chrysosporium zonatum* associated with allergic inflammation in an immunocompetent subject. Journal of Clinical Microbiology, 40: 1113-1115.
Hubálek, Z. and Hornich, M. 1977. Experimental infection of white mouse with

Chrysosporium and *Paecilomyces*.Mycopathologia 62: 173-178.

Jeffery, S., Foran, C.L., Williams, E.A., Ly, T., Muir, D.B. and Roeber, R.U. 1997. Potting mixes—reservoirs of human fungal pathogens, Proceedings of the international symposium on growing media and plant nutrition in agriculture. Freising, Acta Horticulture, 450: 527-533.

Jennings, D.H. 1995. The Physiology of Fungal Nutrition, Cambridge University Press, Cambridge, UK.

Johri, B.N. 1980. Biology of thermophilous fungi, pp. 265-278. In: Recent Advances in Biology of Micro-organisms Vol. 1. K.S. Bilgrami, and K.M. Vyas (eds.), B. Singh and M.P. Singh Publishers, Dehra Dun, India.

Kaliciński, J., Prochacki, H. and Engelhardt-Zasada, C. 1975. Aflatoxin-like compounds produced by dermatophytes. Mycopathologia, 55: 23-24.

Kaul, S. and Sumbali, G. 1999. Impact of some ecological factors on the occurrence of poultry soil-inhabiting keratinophiles. Mycopathologia, 143: 155-159.

Korniłłowicz-Kowalska, T. and Bohacz, J. 2002. Some correlations between the occurrence frequency of keratinophilic fungi and selected soil properties. Acta Mycologica, 37: 101-116.

Kowal, N.E. 1982. Health Aspects of Land Treatment, Microbiological. U.S. EPA, 600/1-82-007, Cincinnati, USA.

Kowal, T. 1998. The sedimentation method for conducting the anaerobic phase of the biological wastewater phosphorus removal (in Polish). Gaz, Woda i Technika Sanitarna, 3: 124-128.

Kushwaha, R.K.S. 2000. The genus *Chrysosporium*, its physiology and biotechnological potential, pp. 66-76. In: Biology of Dermatophytes and other Keratinophilic Fungi. R.K.S. Kushwaha and J. Guarro (eds.), Revista Iberoamericana de Micología.

Lima, J.A., Nahas, E. and Gomes, A.C. 1996. Microbial populations and activities in sewage sludge and phosphate fertilizer-amended soil. Applied Soil Ecology, 4: 75-82.

Majchrowicz, I. and Dominik, T. 1969. Further contribution to the knowledge of keratinolytic and keratinophilic soil fungi of the region of Szczecin—keratinolytic and keratinophilic fungi in the immediate surroundings of cattle. Ekologia Polska (Seria A) 17: 87-116.

Matthews, D.E. and Farewell, V.T. 1996. Using and Understanding Medical Statistics, Karger, Basel-Sydney.

Onifade, A.A., Al-Sane, N.A., Al.-Musallam, A.A. and Al-Zarban, S. 1998. A review: potentials for biotechnological applications of keratin-degrading microorganisms and their enzymes for nutritional improvement of feathers and other keratins as livestock feed resources. Bioresource Technology, 66: 1-11.

Oudeh, M., Khan, M. and Scullion, J. 2002. Plant accumulation of potentially toxic elements in sewage sludge as affected by soil organic matter and mycorrhizal fungi. Environmental Pollution, 116: 293-300.

Padhye, A.A. and Carmichael, J.W. 1971. The genus *Arthroderma* Berkeley. Canadian Journal of Botany, 49: 1525-1540.

Parker C.F. and Sommers L.E. 1983. Mineralization of nitrogen in sewage sludges, Journal of Environmental Quality, 12: 150-156.

Płaza, G., Łukasik, W. and Ulfig, K. 1998. Effect of cadmium on growth of potentially pathogenic fungi. Mycopathologia, 141: 93-100.

Podgórski, L. 1997 Sanitary aspects of sewage sludge control (in Polish), p. 137. In: Gospodarka osadami ściekowymi. Wojewódzki Inspektorat Ochrony Środowiska w Rzeszowie.

Porro, A.M., Yoshioka, M.C.N., Kaminski, S.K., Palmeira, M.C.A., Fischman, O. and Alchorne, M.M.A. 1997. Disseminated dermatophytosis caused by *Microsporum gypseum* in two patients with the acquired immunodeficiency syndrome. Mycopathologia, 137: 9-12.

Prochacki, H. 1975. Fundamentals of Medical Mycology (in Polish). PZWL, Warszawa, Poland.

Przystaś, W., Ulfig, K., Miksch, K. and Grabowska, I. 2000. A preliminary study on the petroleum hydrocarbon removal by keratinolytic fungi (in Polish). Zeszyty Naukowe Politechniki Śląskiej, Inżynieria Środowiska z. 45, 195-202.

Raper, K.B. and Fennel, K.D. 1977. The Genus *Aspergillus*. R.E. Krieger Publishing Company, Huntington, New York, USA.

Rosenberg, S.L. 1975. Temperature and pH optima for 21 species of thermophilic and thermotolerant fungi. Canadian Journal of Microbiology, 21: 1535-1540.

Ross, A.D., Lawrie, R.A., Keneally, J.P. and Whatmuff, M.S. 1992. Risk characterization and management of sewage sludge on agriculture land—implications for the environment and the food-chain. Australian Veterinary Journal, 69: 177-181.

Russel, S. 1972. Methods for Determination of Soil Enzymes (in Polish). PTG, Warszawa, Poland.

Ryan, J.A. and Keeney, D.R. 1975. Ammonia volatilization from surface-applied wastewater sludge. Journal of Water Pollution Control Federation, 47: 386-393.

Senapati, V. 1999. Studies on nutritional pattern and growth retardation of some gymnoascaceous fungi, Ph.D. thesis, Sambalpur University, Rourkela, India.

Sigler, L. 2002a. Ascomycetes. The Onygenaceae and other fungi from the order Onygenales, pp. 195-236. In: Pathogenic Fungi in Humans and Animals, 2nd edition D.H. Howard (ed.), Mercel Dekker, Inc., New York-Basel.

Sigler, L. 2002b. Miscellaneous opportunistic fungi. Microascaceae and other Ascomycetes, Hyphomycetes, Coelomycetes, and Basidiomycetes, pp. 637-676. In: Pathogenic Fungi in Humans and Animals, 2nd edition D.H. Howard (ed.), Mercel Dekker, Inc., New York-Basel.

Sigler, L. and Carmichael, J.W. 1976. Taxonomy of *Malbranchea* and some other Hyphomycetes with arthroconidia. Mycotaxon, 4: 349 488.

Sigler, L., Flis, A.L. and Carmichael, J.W. 1998. The genus *Uncinocarpus* (Onygenaceae) and its synonym *Brunneospora*: new concepts, combinations and connections to anamorphs in *Chrysosporium*, and further evidence of relationship with *Coccidioides immitis*. Canadian Journal of Botany, 76: 1624-1636.

Simpanya, M.F. 2000. Dermatophytes: their taxonomy, ecology and pathogenicity, pp. 1-12. In: Biology of Dermatophytes and other Keratinophilic Fungi. R.K.S Kushwaha, and J. Guarro.(eds.), Revista Iberoamericana de Micología.

Stevens, D.C., Kan, V.L., Judson, M.A., Morrison, V.A., Dummer, S., Denning, D.W., Bennett, J.E., Walsh, T.J., Patterson, T.F. and Pankey, G.A. 2000. Practice guidelines for diseases caused by *Aspergillus*. Clinical Infections Diseases, 30: 696-709.

Straub, T.M., Pepper, I.L. and Gerba, C.P. 1993. Hazards from pathogenic microorganisms in land-disposed sewage sludge. Reviews of Environmental Contamination and Toxicology, 132: 55-91.

Śpiewak, R. 1998. Zoophilic and geophilic fungi as a cause of skin disease in farmers. Annals of Agriculture and Environmental Medicine, 5: 97-102.

Takashio, M. 1973. Etudes des phénoménes de reproduction liés au vieillissement et au rajeunissement des cultures de champignons. Annales de la Société Belge de Medeciné Tropicale, 53: 427-580.

Tansey, M.R., Murrman, D.N., Behnke, B.K. and Behnke, E.R. 1977. Enrichment, isolation, and assay of growth of thermophilic and thermotolerant fungi in lignin-containing media. Mycologia, 64: 463-476.

Terry, R.E., Nelson, D.W., Sommers, L.E. and Meyer, G.J. 1978. Ammonia volatilization from wastewater sludge applied to land. Journal of Water Pollution Control Federation, 50: 2657-2665.

Tester, G.F., Sikora, L.J., Taylor, J.M. and Parr, J.F. 1977. Decomposition of sewage sludge compost in soil: I. Carbon and nitrogen transformation. Journal of Environmental Quality, 6: 459-463.

Tomlinson, T.G. and Williams, L.L. 1975. Fungi, Academic Press, London, UK.

Ulfig, K. 1986a. Keratinophilic fungi in sewage and waters (in Polish). Ochrona Środowiska (Wydawnictwo PZITS, Wrocław), 488/3 (29): 43.

Ulfig, K. 1986b. A preliminary study on the growth and survival of selected pathogenic dermatophytes on bottom sediments and sewage sludge (in Polish). Roczn PZH 37: 336-340.

Ulfig, K. 1988. The growth and survival of selected pathogenic dermatophytes in bottom sediments and sewage sludge (in Polish). Roczniki PZH 39: 151-155.

Ulfig, K. 1991. Keratinophilic fungi in sewage sludge (in Polish). Roczniki PZH 42: 309-315.

Ulfig, K. 1994. The occurrence of keratinolytic fungi in the polluted environment of the Gliwice–Łabędy district (in Polish). Roczniki PZH 45: 337-346.

Ulfig, K. 2000. The occurrence of keratinolytic fungi in waste and waste contaminated habitats, pp. 44-50. In: Biology of Dermatophytes and other Keratinophilic Fungi. R.K.S. Kushwaha and J. Guarro (eds.). Revista Iberoamericana de Micología.

Ulfig, K. 2003. Studies of keratinolytic and keratinophilic fungi in sewage sludge by means of a multi-temperature hair baiting method. Polish Journal of Environmental Studies, 12: 461-466.

Ulfig, K. and Korcz, M. 1983. Isolation of keratinophilic fungi from sewage sludge. Sabouraudia, 21: 247-250.

Ulfig, K. and Ulfig, A. 1990. Keratinophilic fungi in bottom sediments of surface waters. Journal of Medical and Veterinary Mycology, 28: 419-422.

Ulfig, K. and Korcz, M. 1994. Keratinolytic fungi in sewage sludge applied to

devastated urban soil. A preliminary experiment. International Journal of Environmental Health Research, 4: 244-253.
Ulfig, K., Terakowski, M., Płaza, G. and Kosarewicz, O. 1996. Keratinolytic fungi in sewage sludge. Mycopathologia, 136: 41-46.
Ulfig, K., Guarro, J., Cano, J., Gené, J., Vidal, P. and Figueras, M.J. 1997a. General assessment of the occurrence of keratinolytic fungi in river and marine beach sediments of Catalonian waters (Spain). Water, Air and Soil Pollution, 94: 275-287.
Ulfig, K., Guarro, J., Cano, J., Gené, J., Vidal, P. and Figueras, M.J. 1997b. The seasonal changes of keratinolytic fungi in sediments of Catalonian rivers (Spain). Water, Air and Soil Pollution, 96: 269-290.
Ulfig, K., Guarro, J., Cano, J., Gené, J., Vidal, P., Figueras, M.J. and Łukasik, W. 1998a. A preliminary study of the occurrence of actidione-resistant fungi in sediments of Catalonian river mouths (Spain). I. Keratinolytic fungi and related Onygenales, Mycopathologia, 141, 143-151.
Ulfig, K., Płaza, G., Łukasik, K., Krajewska, J., Mańko, T., Wypych, J., Dziewiêcka, B. and Worsztynowicz, A. 1998b. Selected filamentous fungi as the indicators of leachate toxicity and bioremediation progress (in Polish). Ogólnopolskie Sympozjum Naukowo-Techniczne "Bioremediacja Gruntów", Wisła-Bukowa, pp. 47-53.
Ulfig. K., Płaza. G., Wypych. J., Łukasik. K., Dziewięcka. B., Mańko. T., Krajewska. J., Terakowski. M. and Staszewski. T. 2000. A preliminary study on the petroleum hydrocarbon removal by the strain *Trichophyton ajelloi* R66 during keratin biodegradation (in Polish). Zeszyty Naukowe Politechniki Śląskiej, Inżynieria Środowiska, 45: 101-108.
Ulfig, K., Płaza, G., Terakowski, M. and Worsztynowicz, A. 2002. Keratinolytic fungi in an acidic waste lagoon at a petroleum refinery. Roczniki PZH 53: 267-276.
Ulfig, K., Płaza, G., Worsztynowicz, A., Mańko, T., Tien, A.J. and Brigmon, R.L. 2003. Keratinolytic fungi as indicators of hydrocarbon contamination and bioremediation progress in a petroleum refinery. Polish Journal of Environmental Studies, 12: 245-250.
van Oorschot, C.A.N. 1980. A revision of *Chrysosporium* and allied genera. Studies in Mycology, 20: 1-89.
Vanbreuseghem, R. 1952. Technique biologique pour l'isolment des dermatophytes du sol. Annales de la Société Belge de Medeciné Tropicale, 32: 173.
von Arx J.A. 1986. The ascomycete genus *Gymnoascus*. Persoonia, 13: 173-183.
von Arx J.A. 1987. A re-evaluation of the Eurotiales. Persoonia, 13: 273-300.
World Health Organization 1981. The risk to health of microbes in sewage sludge applied to land, Euro Repts and Studies No 54, Copenhagen, Denmark.
Zabawski, J. and Baran, E. 1998. The characteristics of more frequently occurring pathogenic and opportunistic fungi from subphyla: Zygomycotina, Ascomycotina and Deuteromycotina (in Polish), pp. 37-254. In: Zarys mikologii lekarskiej. E. Baran (ed.) Wrocław, Poland.

7

Fungi in Snow Environments: Psychrophilic Molds—A Group of Pathogens Affecting Plants under Snow

Naoyuki Matsumoto[1]* and Tamotsu Hoshino[2]

[1]National Institute for Agro-Environmental Sciences
3-1-3 Kan-non dai, Tsukuba 305-8604, Japan
E-mail: nowmat@affrc.go.jp

*Present address: National Agricultural Research Institute for Hokkaido Region, 1 Hitsujigaoka, Toyohira-ku, Sapporo 062-8555, Japan

[2]National Institute of Advanced Industrial Science and Technology
2-17-2-1 Tsukisamu-higasi, Toyohira-ku, Sapporo 062-8517, Japan
E-mail: tamotsu.hoshino@aist.go.jp

Abstract

Snow molds cause damage to plants under snow. While snow cover protects plants from freezing, it maintains darkness, humidity and low temperature. The habitat under snow characterizes the pathogens as opportunistic parasites. Plant photosynthesis is inhibited, and disease resistance decreases with time. Low temperature suppresses the activity of most microorganisms, and only low temperature-tolerant microorganisms can prevail. The most critical issue for snow molds is the duration of snow cover. They have evolved different survival strategies. Biological control is promising so far as low temperature-tolerant antagonists are introduced to the habitat under snow where diversity of active microorganisms is low and a vacant niche is available for the antagonists. Global warming certainly affects the incidence of snow molds but may not reduce snow mold damage since farmers choose to grow more productive but less winter hardy cultivars and crops. Snow molds are not only economically important but represent good materials for biological science due to their unique features.

INTRODUCTION

A group of fungi can tolerate low temperatures and attack plants under snow cover. They are referred to as snow molds. Snow molds are unique in that they prevail when their hosts lie dormant during the winter. However, not all of them prevail exclusively under snow. Matsumoto (1994) divided them into obligate and facultative snow mold fungi (Table 1). Facultative snow molds are relatively fast-growing and occur under moist environmental conditions when plants are not necessarily covered with snow or even in summer; e.g. *Sclerotinia trifoliorum* incites clover rot in damp winter (Lester and Large 1958), and *Microdochium nivale* causes leaf blotch of wheat in summer (Asuyama, 1940). Obligate snow molds prevail exclusively under snow.

Snow mold fungi are opportunistic parasites, and their incidence is largely affected by climatic factors before snow cover and its duration. Snow cover protects plants from freezing damage and represents the habitat essential for disease development with moisture and constant low temperature. The ambient temperature under snow does not vary significantly from place to place or with time, but the duration of snow cover differs greatly, producing contrasting survival strategies especially in most ubiquitous *Typhula* spp.

The recent tendency of global warming certainly affects the incidence of snow molds: *Sclerotinia borealis* has been replaced by *T. ishikariensis* in eastern Hokkaido, Japan where winter used to be much colder with severe soil freezing and less snow. Examples are illustrated along with the development of new crop cultivars in this chapter.

Environment under snow cover

Snow molds are essentially opportunistic parasites that attack hosts under snow cover where low temperatures limit species diversity. The resistance of plants is lowered by interrupted photosynthesis and the eventual exhaustion of reserve materials further predisposes plants to the attack of snow molds.

Snow cover and plant damage due to snow molds

As a rule, damage due to snow mold fungi increases with the duration of snow cover, but this relationship varies with individual snow mold fungi. For example, *Sclerotinia borealis* develops in regions characterized by less snow, lower temperatures and frozen soil, while *T. ishikariensis* is restricted to regions characterized by deep snow on unfrozen soil with temperatures near 0°C (Tomiyama, 1955). Grassland stands in localities where snow cover remains for more than 150 days suffer recurrent damage from *T. ishikariensis* (Årsvoll, 1973). Conversely, *T. incarnata* occurs in regions where snow cover lasts at least 90 days (Årsvoll, 1973) and *M. nivale* requires only 60-90 days of snow cover to cause serious injury to grass and winter cereals (Årsvoll, 1973).

The more persistent the snow cover, the more severe the damage caused by snow molds. The slow growth of *S. borealis* requires a long snow cover period to cause severe damage. Nissinen (1996) found a positive association between maximum depth of snow cover and sclerotinia snow mold incidence. In an eight year study, Ozaki (1979) correlated the occurrence of sclerotinia snow mold and its damage to orchardgrass. He experimentally demonstrated a 50% increase in disease severity when snow cover exceeding 40 cm deep was maintained for 72 days compared to disease levels in plots where snow cover of similar depth lasted for 42 days.

Poor species diversity

Approximately 40 fungi have been described as pathogens of wheat seedlings, but only five have been shown to be pathogenic under snow (Wiese, 1977). This figure is small considering that in some areas, wheat seedlings are covered with snow for 150 days or more. The Phytopathological Society of Japan (Anonymous 1980) also reported that on orchardgrass, five snow mold fungi out of a total of 29 fungal leaf parasites were found. Eighty-nine fungal species were discovered on winter wheat seedlings and other plants collected in winter and early spring (Bruehl et al 1966), and 33 species from forage grasses just after thawing of snow cover (Årsvoll, 1975), but most of them were mesophiles and probably dormant. Species diversity is thus limited and low ambient temperature restricts the activity of microflora under snow cover. However, a recent survey with DNA sequence-based methods substantially broadened our understanding of the diversity of fungi in tundra soils (Schadt et al., 2003); abundant, previously unknown fungi were active beneath the snow. These fungi are not likely pathogenic but could influence the incidence of snow molds.

At an ambient temperatures of around 0°C, the activity of most microorganisms is significantly reduced. However, snow mold fungi have evolved the ability to grow at low temperatures. This feature characterizes snow molds as opportunistic parasites in the practical absence of antagonists under snow. Their optimum growth temperature on artificial media ranges from 6-9°C for psychrophilic *Typhula ishikariensis* var. *canadensis* (Årsvoll and Smith, 1978) to 18-20°C for mesophilic *M. nivale* (Smith, 1987). With the exception of facultative snow molds under natural conditions, most snow mold fungi are dormant at their optimal in vitro temperatures. The inability of the obligate snow molds, *T. incarnata* and *T. ishikariensis* to develop mycelia under their optimal in vitro temperatures has been attributed to microbial antagonism (Matsumoto and Tajimi, 1988). Radial mycelial growth of these fungi at 0°C was about half that at the near-optimal temperature of 10°C (Matsumoto and Tajimi, 1988). However, when cultures were covered with non-sterile soil to introduce natural microbial antagonists, mycelial growth was much greater at 0°C than at 10°C. Thus, these low temperature-tolerant fungi appear to have escaped from the microbial antagonism that occurs at higher temperatures in the absence of snow.

Impaired photosynthesis and snow mold resistance of plants

With increasing duration of snow cover, over wintering plants become less capable of resisting attack by snow molds. Snow-covered host plants therefore represent nutrient-rich resources that are relatively easy to exploit. Nakajima and Abe (1994) correlated reduced etiolated growth of winter wheat seedling with increasing susceptibility to the pink snow mold fungus, *Microdochium nivale*. The amount of etiolated growth represents the amount of carbohydrate reserves available to a plant. Snow mold resistance was reduced by 40% in wheat plants that had exhausted 30-80% of carbohydrate reserves during incubation under snow for 75 days. Thus, accumulation of carbohydrates during cold hardening is critical in terms of snow mold resistance (Mohammad et al., 1997, Yoshida et al., 1998), and the amount of carbohydrates reserved till snowmelt for spring flush is considered to be correlated to varietal differences in winter wheat (Bruehl, 1967, Kiyomoto and Bruehl, 1977).

Recent studies have revealed that plants under snow are not simply dormant but that they have dynamic defense mechanisms against snow molds. The relationship between two types of stresses, i.e., biotic (snow mold infection) and abiotic (freezing damage) stresses, was first elucidated in winter rye. Griffith et al. (1992) reported the production of endogenously produced antifreeze proteins after cold acclimation. These proteins were identical to three classes of pathogenesis-related proteins such as endo-chitinases, endo-β-1,3-glucanases, and thaumatin-like proteins (Hon et al., 1995). They further made comparisons between cold-induced proteins and proteins induced at warm temperatures by infection with *Microdochium nivale* to find that the latter proteins lacked in antifreeze activity (Hiilovaara-Teijo et al., 1999).

Kawakami (2005) subsequently hypothesized as to how snow mold resistance was lowered in winter wheat in terms of fructan metabolism and chitinase activity. Fructans are the main carbon source for survival under persistent snow cover in winter wheat (Yukawa and Watanabe, 1991) and play roles in protection against environmental stresses such as low temperature (Konstantinova et al., 2002). Kawakami (2005) considered the deterioration process of plants under snow as follows; 1) reduction in simple sugars after fructan exhaustion, 2) reduced energy supply, 3) lowered metabolism of plant cells, 4) reduced expression of chitinases and other pathogenesis-related enzymes, and 5) lowered plant resistance to snow molds. In other words, he considered that plant resistance to snow molds was dynamic as was the case with other diseases occurring on growing plants and that carbohydrates were consumed to develop snow mold resistance.

How to exploit limited resources?

Snow cover prevents plant photosynthesis, and consequently, available resources of snow molds, i.e. living plant tissues, do not increase but gradually become depleted during the winter. Snow mold fungi have evolved two different strategies in the utilization of limited resources. Different strains (genotypes) within a species share the resources and can coexist in ascomycetous snow molds such as *Sclerotinia borealis* and *M. nivale*, and possibly *Pythium* spp. *Sclerotinia borealis* spreads extensively by airborne ascospores, and different strains can colonize a single substrate. Their strategy of resource utilization is collectivism.

Basidiomycetous snow molds show strong intraspecies antagonism (Årsvoll and Smith, 1975, Lebeau, 1975, Smith and Årsvoll, 1975, Matsumoto and Tajimi, 1983), and their strategy of resource utilization is individualistic (Rayner et al., 1984). The *Typhula* spp. represent the individualistic snow molds (Matsumoto, 1997c). Each genet (genetic individual) attempts to monopolize the host substrate by excluding others. Matsumoto and Tajimi (1993) identified 31 genets among 72 isolates of *T. ishikariensis* biotype B from 45 mold patches in a 2 × 2 m plot in a turf grass nursery (Matsumoto and Tajimi 1993). Patches in which two or more small plots coalesced to form larger irregular patches, contained two to nine genets. Plants on the junction between constituent small patches often survived infection and formed green lines, apparently zones of inhibition between genets. The three largest genets consisted of 12, 10, and 7 isolates and 19 genets were recovered only once. Low temperature basidiomycetes behave the same way, opposing each other (Lebeau, 1975).

Annual fluctuation

Snow cover is essential for snow mold fungi; however, the duration of persistent snow cover varies year after year, determining their ecological strategy. Some fungi can prevail also in summer, while others exist exclusively under snow. Thus, the mode of adaptation to winter climates varies from species to species.

Facultative snow molds (Table 1) demonstrate another aspect of resource utilization. Some snow mold species can thrive even on growing plants during the normal cropping season in the absence of snow, thus extending their niche. These species tend to have higher optimum growth temperatures. This category is represented by *M. nivale*; isolates obtained in summer from wheat scab caused pink snow mold during the winter, and vice versa (K. Miyajima, personal communication). During the active growing phase of most snow molds, their spatial dispersal is limited by snow cover. This limitation, however, may not affect facultative snow molds which may gain an advantage in both spatial and temporal distribution by their summer activity.

Table 1. Obligate and facultative snow mold fungi*

Species	*References*
Obligate snow mold fungi	
Athelia sp. ('Supponuke' fungus)	Simizu and Miyajima, 1990 Kawakami unpublished**
Coprinus psychromorbidus (low temperature basidiomycete)	Cormack, 1948
Phoma sclerotioides	Sanford, 1933
Pythium iwayamai	Iwayama, 1933
Sclerotinia borealis	Groves and Bowerman, 1955
S. nivalis	Saito, 1997
Typhula incarnata	Tomiyama, 1955
T. ishikariensis	Tomiyama, 1955
T. trifolii	Ylimäki, 1969
Fucaltative snow mold fungi	
Ceratobasidium gramineum	Takamatsu, 1989
Monographella nivalis (*Microdochium nivale*)	Asuyama, 1940
Rhynchosporium secalis	Suzuki and Arai, 1990
S. trifoliorum	Lester and Large, 1958
Pythium graminicola	Hirane, 1960
P. okanoganens	Lipps, 1980
P. paddicum	Hirane, 1960

*Revised from Matsumoto, 1994.
**Inferred from rDNA-ITS sequence.

Other strategies may be found in the species of *Typhula.* In snowy regions, favorable environmental conditions are highly predictable, that is, the beginning and end of persistent snow cover does not vary greatly from year to year (Matsumoto and Tajimi, 1990). Pathogens in such a habitat can wait till plants exhaust their reserves since the fungus can monopolize the resources with few competitors (Matsumoto, 1992, 1994). *Typhula incarnata* is a versatile pathogen that uses its eco-physiological capabilities (Matsumoto and Sato, 1982; Jacobs and Bruehl, 1986) differently in various environments (Matsumoto et al., 1995). In contrast, *T. ishikariensis* has evolved specialized strategies for different environments. The differences in adaptation strategy are reflected in the size variation of sclerotia of the two fungi. The entire range of size variation of sclerotia within the species *T. ishikariensis* occurs within a single strain of *T. incarnata*. Small and large sclerotia produced by a single strain of *T. incarnata* seem to act differently (Matsumoto et al., 1995). Small sclerotia seldom produce basidiospores due to limited energy reserves but develop mycelia on germination, like those of soilborne fungal pathogens (Garrett, 1970). Their inoculum potential is high, but the fungus cannot migrate long distances, only assuring minimum survival. Large sclerotia, on the other hand, produce abundant infectious basidiospores, helping the fungus spread widely but with limited effect where environmental conditions are not favorable because of the limited inoculum potential of basidiospores. As a generalist, *T. incarnata* hedges its bets in its reproductive strategy.

Size variation of sclerotia in *T. ishikariensis* is as great as that in *T. incarnata*, but this is due to isolate variability. Matsumoto and Tajimi (1990) correlated winter climate and sclerotium size in *T. ishikariensis* biotype B. Isolates from snowy regions have large sclerotia, and those from regions with less snow have small sclerotia. Isolates with small sclerotia are more aggressive. In less snowy regions where the beginning and end of persistent snow are difficult to forecast, local populations are very aggressive and attack plants without predisposition under snow (Nakajima and Abe, 1994). They are specialized to be soilborne, attacking roots and lower sheaths in the soil where environmental conditions favor their survival with moisture and constant low temperature (Honkura et al., 1986).

The time of germination is critical to sclerotial snow molds. If sclerotia germinate too early, rather than just before snow cover, the fungus may not survive because infection is unlikely without snow cover. When sclerotia germinate under snow, the fungus may fail to compete with other strains. Long distance dispersal by airborne basidiospores of *T. incarnata* is also hindered by snow. Under cool, moist conditions, sclerotia of isolates from highly predictable habitats where deep snow cover accumulations typically occur, germinate more readily than those from less predictable habitats where light snow accumulations typically occur (Matsumoto and Tajimi, 1990; Matsumoto et al., 1995). Environmental thresholds for sclerotium germination seem to be lower in deep snow regions.

Adaptation to extreme low temperature

The ambient temperature under snow is about 0°C, but on the Canadian prairies (Smith, 1987) and the coastal regions of Norway (Årsvoll, 1973) temperatures fall much lower than 0°C, often causing freezing damage to plants. Hoshino et al. (1997, 1998) studied the adaptation mechanism of the fungus to such extreme low temperatures. In Norway, *T. ishikariensis* group III is considered as an ecotype specialized to low temperatures originating from group I (Matsumoto et al., 1996). In culture, the morphology of these two groups is indistinguishable at 0°C, but at or above 10°C, group III isolates exhibit irregular growth. Two dimensional PAGE reveals that some intracellular proteins of group III are modified at 15°C but not at 4°C (Hoshino et al., 1997), implying that higher temperatures cause aggregation or dissociation of intracellular proteins, leading to the loss of vital properties in group III.

When mycelia of groups I and III were rapidly frozen at –40°C, the regrowth of group I was delayed after transfer to the optimal growth temperature of 10°C, whereas regrowth in group III returned to normal after transfer to its optimal growth temperature of 2°C (Hoshino et al., 1998). Freezing actually enhanced mycelial growth of group III at 10°C and the fungus appeared able to grow in a wider temperature range after freezing stress. Rapid freezing stress destroyed sclerotia of group I whereas group III sclerotia survived the stress. These results indicate that group III strains are more resistant than group I strains to rapid freezing, an adaptation to climatic conditions in the northwest, coastal regions of Norway where cycles of thawing and rapid freezing cause freezing stress to plants (Årsvoll, 1973).

More recent physicochemical analyses provide the clue to adaptation mechanisms of snow molds to low temperatures. Snider et al. (2000) studied the role of ice nucleation and antifreeze activities in pathogenesis and growth of seven snow mold fungi to conclude that ice nucleation activity was unimportant for growth and pathogenesis and that there was no correlation between antifreeze activity and threshold ice nucleation temperatures. According to Terami and Kawakami (2002), growth temperature relations on PDA changed drastically according to water potential of the medium in *Sclerotinia borealis*; its optimum growth temperature ranged between 6-9°C but decreased to 0°C when water potential of PDA (–0.6 to –0.4Mpa) was reduced with sorbitol or KCl (–4.5 to –2.0Mpa), and mycelial growth rate became 1.2-2 times as fast. Investigations on host-parasite interaction especially in freezing environments require further detailed analyses to elucidate how snow mold fungi cause damage to plants at much below 0°C.

Biological control

Winter survival of winter wheat, forage crops, and turf grass depends on snow mold incidence. According to the 2002 statistics of Hokkaido, snow molds occur in approximately 50% of winter wheat fields and damage 9% of them even though fungicides are sprayed in more than 80% of the fields in normal years. When only 48% of the fields received fungicides due to the early start of persistent snow cover in 1997, snow molds damaged 30% of them. On golf course turf grass, chemical control is essential, and creeping bentgrass green, especially, cannot be maintained without chemicals to control snow molds. Thus, there are concerns for environment motivated plant pathologists to develop biocontrol methods (Matsumoto, 1997a).

Biological control of *Typhula* spp. is one of the very few examples of success under field conditions. Most unsuccessful field trials in other diseases may be ascribed to the lack of ecological considerations (e.g. Conway, 1976) prior to field application. Antagonists are, in most cases, difficult to introduce and to establish where pathogens prevail, and antagonist-pathogen interactions are 'thinned' among the complex plant-microbe ecosystem. The habitat of snow molds is practically free from these disadvantages due to its unique habitat under snow cover.

Biological control during the dormant phase in summer

Most obligate snow mold fungi pass the dormant phase in summer in the form of sclerotium. Matsumoto and Tajimi (1985) determined the survival of sclerotia of *T. incarnata* and *T. ishikariensis* biotype A in the field to find that more than 90% of *T. incarnata* sclerotia died due to mycoparasitism and that *T. ishikariensis* biotype A sclerotia mostly survived. Several mycoparasites such as *Coniothyrium minitans, Gliocladium roseum,* and *Trichoderma* spp. were obtained from *T. incarnata* sclerotia, and they were all found parasitic to *T. ishikariensis* biotype A sclerotia as well as those of *T. incarnata* in the laboratory. However, they did not continue subsequent experiments with these mycoparasites due to the following reasons: 1) since basidiospores of *T. incarnata* are effective as inoculum, a few sclerotia survived can cause extensive disease incidence, and 2) reduction in the number of sclerotia in *T. ishikariensis* biotype A sclerotia may directly contribute to the decrease in inoculum potential because the sclerotium is the infection unit in this fungus; however, inoculation with mycoparasites during the dormant phase in summer seemed difficult as was often the case with other sclerotium-forming pathogens (e.g. Trutmann et al., 1982). Biocontrol trials of *Sclerotinia minor* with *Sporidesmium sclerotivorum* may be the only one successful example of biological control in sclerotium-forming pathogens (Adams and Ayers, 1982); the sclerotium serves as an inoculum unit in this fungus, and the antagonist-pathogen interactions are less likely to be affected by other microorganisms because the antagonist *S. sclerotivorum* is highly specialized to parasitize the sclerotium. Such antagonists from *Typhula* sclerotia were not to be found (N. Matsumoto, unpublished). Thus, biological control of snow mold during the summer is difficult, and any successful examples so far are not known.

Biological control during the active phase in winter

Low temperature in the habitat of snow molds limits the number of active micoorganisms under snow, and low temperature-tolerant fungi are easy to establish in the niche blank. On this assumption, Matsumoto and Tajimi (1992) collected fungal isolates just after snowmelt throughout Japan, including *T. incarnata, T. phocorrhiza,* the supponuke fungus, *Trichosporiella* spp., *Trichosporon* sp., and *Acremonium boreale* which was described as a snow mold antagonist (Smith and Davidson, 1979) and evaluated their antagonism against *T. ishikariensis* biotypes A and B using orchardgrass. *T. incarnata* and *T. phocorrhiza* were effective, while others were not. Both *T. incarnata* and *T. phocorrhiza* were antagonistic against biotype B. *T. phocorrhiza* suppressed the disease caused by biotype A, as well. Previous results indicated that *T. incarnata* was not competitive with biotype A on orchardgrass (Matsumoto and Sato, 1983).

T. phacorrhiza also exhibited isolate variability in antagonism against both biotypes, and the variability corresponded to the locality of *T. ishikariensis* biotypes; comparatively less snowy, biotype B area produced biotype B antagonists, and they were invariably not antagonistic against biotype A. Biotype A is exclusively distributed in snowy regions, including Sapporo and the Tenpoku district (Matsumoto et al., 1982), and *T. phocorrhiza* isolates from the biotype A area were antagonistic against biotype A as well as biotype B. Above all, isolates from the Tenpoku district were mostly effective, and their application to perennial ryegrass plots naturally infested with biotype A in Sapporo resulted in yield increase as compared to untreated controls after severe snow mold damage in the third winter. The introduced antagonist was present in the form of sclerotia on plants.

From field trials in the 1960s in Sapporo, Murakami et al. (1965) concluded that ryegrasses including perennial ryegrass were unable to survive many winters and that practical cultivation was possible for 3-4 years at most. However in the Tenpoku district; despite the more prolonged snow cover, perennial ryegrass was found to grow for a longer period than was considered (Tezuka and Komeichi, 1980; Yamagishi, 1988; Ishida et al., 1989). *T. phacorriza* isolates from Sapporo were not as antagonistic against biotype A as those from the Tenpoku district, and biological control by the local population of *T. phacorrhiza* was considered as one of the reasons for successful cultivation of perennial ryegrass (Matsumoto and Tajimi, 1992).

Using *T. phacorrhiza*, Burpee et al. (1987) could suppress gray snow mold of turf grass (creeping bentgrass); 100 and 200 g/m^2 of *T. phacorrhiza* inoculum resulted in 44 and 70% less snow mold, respectively. Their isolate showed residual activity, and its sclerotia were found on the thatch (Lawton and Burpee, 1990). Separate studies revealed that the suppression of snow mold by *T. phacorrhiza* was evident for the first three years but declined by the fourth year after treatment (Hsiang et al., 1999).

Biological control of *Typhula* spp. may be applied to diverse crops with different levels of cultural intensity. Cultivation system is a compromise between inherent productivity of crops, cultural practice, and snow mold damage; snow molds restrict cultivars and species of crops to be planted, and cultural practice affects snow mold incidence. In grasslands cultural practice is the only control measure to enhance the activity of indigenous antagonists and plant resistance. Artificial introduction of antagonists does not pay, and naturally occurring biological control should be exploited in combination with cultural practice. Turf grass in the golf course is contrasting to grasslands, representing the habitat with the most intensive cultural practices. Creeping bentgrass in the golf course green, which is highly susceptible to snow mold, can be free from snow mold if fungicides are used properly, and chemical control pays. Abuse of fungicides, however, may cause pollution and does not gain public acceptance. Biological control on turf grass is promising but still needs improvement for practical use.

Global warming and snow molds

Tomiyama (1955) was the first to describe the relationship between climatic conditions and snow mold incidence. *Sclerotinia borealis* tolerates temperatures of –2 to –3°C at the soil surface under snow, whereas *Typhula* species are more successful at higher temperatures. In Hokkaido, *S. borealis* used to occur in the eastern coastal regions characterized by soil freezing, and *Typhula* spp. in western and central regions with deep snow. However, *S. borealis* has seldom occurred during the last two decades in eastern Hokkaido, and *Typhula* spp., especially, *T. ishikariensis* replaced *S. borealis*.

A similar example may also be found in alfalfa. An extensive survey by Komatsu et al. (1983) in the Tokachi district revealed that yield in mountainous fields with long-lasting snow cover was much lower due to *T. ishikariensis* biotype A than in the plains where biotype B, which was not pathogenic to dicots, prevailed and soil freezing was the principal cause of winter damage. Freezing damage of alfalfa used to be more severe further east: in the Konsen district, alfalfa failed to survive the severe winter due to frost heaving and soil freezing. Agronomists screened 110 cultivars for winter survival to find that cold-tolerant cultivars from Canada often failed to survive (Takeda and Nakashima, 1997a, b). Vigorous autumn growth and resistance to Leptoshaerulina leaf spot caused by *Leptoshaerulina briosiana* were first considered essential in this area: cold tolerance was not critical. The leaf disease reduces carbohydrate level for winter survival. They subsequently selected cultivars for resistance to Leptoshaerulina leaf spot (Takeda and Nakashima, 1997c). During the course of selection, *T. ishikariensis* biotype A badly injured resistant lines in warm winter with deep snow, and resistance to *T. ishikariensis* biotype A had to be included in the breeding scheme. They finally developed the cultivar 'Hisawakaba' with resistance both to *L. briosiana* and *T. ishikariensis* biotype A (Yamaguchi et al., 1995).

A new snow mold was reported from the Abashiri district on winter wheat and referred to as supponuke (Simizu and Miyajima, 1990). The disease was characterized by crown rot, and plant tops were easily pulled. The causal fungus was considered to be *Athelia* sp. on rDNA-ITS region sequence (A. Kawakami, personal communication). According to a four-year survey by Simizu and Miyajima (1992), the disease was restricted to eastern Hokkaido where winter was severe and *S. borealis* was the principal snow mold. The disease became more severe when winter wheat was incubated under snow less than 50 cm thick. They considered that low temperature was essential for the development of supponuke snow mold. Supponuke snow mold disappeared after the leading winter wheat variety, Horoshiri-komugi was replaced with more freeze-tolerant, Chihoku-komugi (Shimizu, 1993). When less freeze-tolerant Hokushin-komugi was subsequently introduced to eastern Hokkaido, supponuke snow mold became prevalent, again (A. Kawakami, personal communication). However, more recently, the disease is difficult to find because mild winter with deep snow cover in this area does not favor supponuke snow mold any more.

FUTURE PERSPECTIVES

Intraspecies variation in *Typhula* spp.

Intraspecies variation is not well-documented in *T. incarnata* or *T. phacorrhiza* but it has been described in *T. ishikariensis*, causing taxonomic controversy. Authors had described *T. ishikariensis* and related taxa without international comparisons of isolates (Imai, 1930; Remsberg, 1940; Ekstrand, 1955) until Bruehl et al. (1975) examined isolates from the U.S.A., Finland and Japan. They regarded *T. idahoensis* Remsberg as a separate species from *T. ishikariensis* based on inter-sterility between the two as well as on morphological and ecological differences. Årsvoll & Smith (1978) reported inter-fertility between isolates from North America and Norway and regarded *T. ishikariensis* as a single species consisting of three varieties: var. *ishikariensis*, var. *idahoensis*, and var. *canadensis*. The varieties *ishikariensis* and *idahoensis* differ in rind cell patterns of sclerotia, and var. *canadensis* is characterized by small sclerotia. Matsumoto et al. (1982) found two inter-sterility groups (biotypes) within Japanese isolates but included them in a single species, *T. ishikariensis*, since these biotypes could be genetically related through an American taxa (Matsumoto et al., 1983). Matsumoto et al. (1996) distinguished three groups in isolates from Norway based on culture morphology and mating patterns between them and the Japanese taxa.

Typhula ishikariensis has not been studied carefully in the former U.S.S.R. despite the fact that this area occupied half the potential habitat of snow mold. Tkachenko et al. (1997) mated isolates from eastern Europe with isolates from Japan, Norway and North America. and found that east European isolates were uniform in culture morphology and in mating patterns, being compatible only with Japanese biotype A and Norwegian group I. The most striking feature of these isolates is that they can cause damage on subterranean parts of hop (Kuznetzova, 1953, Hoshino et al., 2004) and tulip (Procenko, 1967). Their sclerotia produce one, two or, occasionally, more secondary sclerotia, and Tkachenko (1995) regard this habit as an adaptation to a subterranean environment in unpredictable habitat. Japanese biotype A does not attack plant roots (Matsumoto, 1989). In culture, the morphology of isolates from western Siberia resembled that of the eastern European isolates (Matsumoto and Tkachenko, unpublished). However, it is too early to conclude that there is no differentiation into inter-sterility groups in the former U.S.S.R. Tkachenko (unpublished) has recently found that *T. ishikariensis* isolates from western Siberia were exclusively compatible with Norwegian group III isolates.

Matsumoto (1997b) reviewed the literature on mating results and considered that the *T. ishikariensis* complex was composed of two biological species. Biological species I included Japanese biotype A, Norwegian groups I and III, var. *ishikariensis* and var. *idahoensis*. Biological species II consisted of biotype B, group II, and var. *canadensis*. Genetic relationships between the two biological species varied from absolute inter-sterility between biotypes (Matsumoto et al., 1983) to practical inter-fertility between varieties (Årsvoll and Smith, 1978) and between groups (Matsumoto et al.,

1996). Further studies reinforced by molecular comparisons should provide more information on phylogenetic relationships to end the taxonomic confusion in the *T. ishikariensis* complex.

Surveys in polar regions

Surveys in the North and South Poles were initiated by Hoshino et al. (1999) who found patches in moss colonies of *Sanionia uncinata* after snow melt in Svalbard, northern islands of Norway. *Pythium* sp., morphologically similar to *P. ultimum* var. *ultimum*, was isolated from moribund tissues. The fungus showed different growth temperature response to low temperatures as compared to a *P. ultimum* var. *ultimum* isolate from a temperate region: the fungus grew at 0°C. Another *Pythium* isolate from Spitsbergen Island was freezing tolerant (Kida et al., 2005). They (Tojo et al., 2002a, Hakoda et al., 2003) later found at least eight *Pythium* spp. on the moss in the same region and considered that many species of undescribed *Pythium* were involved in the phenomenon (M. Tojo, personal communication). A heterothallic *Pythium* sp. was present on dying moss of *S. uncinata* also in King George Island, Antarctica (Tojo et al., 2002b). Tojo et al. (2002c) found infection of the moss by sterile, non-sclerotial basidiomycetes, as well. The taxonomy and pathogenicity of these fungi are yet to be determined. Surveys in the North and South Poles have just started and should provide us with novel findings and insights on diverse aspects on global warming. Continuous observations are under way to monitor snow mold occurrence on the moss in Spitsbergen Island (Tojo et al., 2006).

CONCLUSIONS

The incidence and severity of snow molds are largely affected by winter climate and, to a lesser extent, by summer and autumn climates. Individual snow mold fungi react differently to environmental conditions, and this governs their distribution and year-to-year occurrence. The recent tendency towards global warming may reduce the impact of snow molds if the present agricultural systems are retained. This may not be the case, however. Warmer environmental conditions may permit the cultivation of more productive but less winter hardy plants and cultivars. The accumulated knowledge of the ecology of snow molds, along with physiological studies of plants, should provide clues for developing counter measures when new agricultural systems are introduced.

Snow molds are cold tolerant and have unique physiological features. Low temperature enzymes and antifreeze proteins should further be studies for practical use. They are also good materials on which to test ecological theories postulated from other organisms and to present new hypotheses on ecology and physiology.

References

Adams, P.B. and Ayers, W.A. 1982. Biological control of Sclerotinia lettuce drop in the field by *Sporidesmium sclerotivorum*. Phytopathology, 72: 485-488.

Anonymous. 1980. Common Names of Economic Plant Diseases in Japan II. Phytopathological Society of Japan, Tokyo (in Japanese).

Årsvoll, K. 1973. Winter damage in Norwegian grasslands, 1968 1971. Meldinger fra Norges landbrukshøgskole, 52(3).

Årsvoll, K. 1975. Fungi Causing Winter Damage on Cultivated Grasses in Norway. Meldinger fra Norges landbrukshøgskole, 54(9).

Årsvoll, K. and Smith, J.D. 1975. Mutual Antagonism Between Isolates of *Typhula ishikariensis* and *Typhula incarnata*. Meldinger fra Norges landbrukshøgskole, 55(19).

Årsvoll, K. and Smith, J.D. 1978. *Typhula ishikariensis* and its varieties, var. *idahoensis* comb. nov. and var. *canadensis* var. nov. Canadian Journal of Botany, 56: 348-364.

Asuyama, H. 1940. Leaf blotch of wheat caused by *Fusarium nivale* (Fr.) Ces. Annals of Phytopathological Society of Japan, 10: 51-54 (in Japanese).

Bruehl, G.W. 1967. Effect of plant size on resistance to snowmold of winter wheat. Plant Disease Reporter, 51: 815-819.

Bruehl, G.W., Sprague, R., Fischer, W.B., Nagamitsu, M., Nelson, W.L. and Vogel, O.A. 1966. Snow Molds of Winter Wheat in Washington. Washington Agricultural Experiment Station Bullutin, 677.

Bruehl, G.W., Machtmes, R. and Kiyomoto, R. 1975. Taxonomic relationships among *Typhula* species as revealed by mating experiments. Phytopathology, 65: 1108-1114.

Burpee, L.L., Kaye, L.M., Goulty, L.G. and Lawton, M.B. 1987. Suppression of gray snow mold on creeping bentgrass by an isolate of *Typhula phacorrhiza*. Plant Disease, 71: 97-100.

Conway, G. 1976. Man versus pests, pp. 257-281. In: The Theoretical Ecology and Applications, R.M. May (ed.), Blackwell, Oxford, UK.

Cormack, M.W. 1948. Winter crown rot or snow mold of alfalfa, clovers, and grasses in Alberta I. Occurrence, parasitism and spread of the pathogen. Canadian Journal of Research, C 26: 71-85.

Ekstrand, H. 1955. Höstädens och vallgräsens övervintring. Statens Växtskyddanst. Meddelande, 67 (in Swedish with English summary).

Garrett, S.D. 1970. Pathogenic Root Infecting Fungi. Cambridge University Press, London, UK.

Griffith, M., Ala, P., Yang, D.S.C., Hon, W.-C. and Moffatt, B.A. 1992. Antifreeze protein produced endogenously in winter rye leaves. Plant Physiology, 100: 593-596.

Groves, J.W. and Bowerman, C.A. 1955. *Sclerotinia borealis* in Canada. Canadian Journal of Botany, 33: 591-594.

Hakoda, A., Tojo, M., Fujii, H., Hoshino, T., Kanda, H. and Ohki, S. 2003. Analysis of amplified ribosomal DNA restriction fragment of *Pythium* spp. isolated from mosses on Spitsbergen Islands. Abstracts XXVI Symposium on Polar Biology, 78.

Hiilovaara-Teijo, M., Hannukkaka, A., Griffith, M., Yu, X.-M. and Pihakaski-Maunsbach, K. 1999. Snow mold-induced appoplastic proteins in winter rye leaves lack antifreeze activity. Plant Physiology, 121: 665-674.

Hirane, S. 1960. Studies on Pythium snow rot on winter cereals with special reference to taxonomy of the pathogens. Transactions of Mycological Society of Japan, 2: 82-87 (in Japanese).

Hon, W.-C., Griffith, M., Mlynarz, A., Kwok, Y.C. and Yang, D.S.C. 1995. Anti-freeze proteins in winter rye are similar to pathogenesis-related proteins. Plant Physiology, 109: 879-889.

Honkura, R., Matsumoto, N. and Inoue, T. 1986. *Typhula ishikariensis* biotype C, a snow mold fungus, can complete its life cycle without snow cover. Transactions of Mycological Society of Japan, 27: 207-210.

Hoshino, T., Tronsmo, A.M., Matsumoto, N., Ohgiya, S. and Ishizaki, K. 1997. Effects of temperature on growth and intracellular proteins of Norwegian *Typhula ishikariensis* isolates. Acta Agriculturae Scandinavica, Section B, Soil and Plant Science, 47: 185-189.

Hoshino, T., Tronsmo, A.M., Matsumoto, N., Araki, T., Georges, F., Goda, T., Ohgiya, S. and Ishizaki, K. 1998. Freezing resistance among isolates of a psychrophilic fungus, *Typhula ishikariensis*, from Norway. Proceeding NIPR Symposium on Polar Biology, 11: 12-118.

Hoshino, T., Tojo, M., Okada, G., Kanda, H., Ohgiya, S. and Ishizaki, K. 1999. A filamentous fungus, *Pythium ultimum* Trow var. *ultimum*, isolated from moribund moss colonies from Svalbard, northern islands of Norway. Polar Bioscience, 12: 68-75.

Hoshino, T., Tkachenko, O.B., Kiriaki, M., Yumoto, I. and Matsumoto, N. 2004. Winter damage caused by *Typhula ishikariensis* biological species I on conifer seedlings and hop roots collected in the Volga-Ural regions of Russia. Canadian Journal of Plant Pathology, 26: 391-396.

Hsiang, T., Wu, C. and Cook, S. 1999. Residual efficacy of *Typhula phacorrhiza* as a biocontrol agent of grey snow mold on creeping bentgrass. Canadian Journal of Plant Pathology, 21: 382-387.

Imai, S. 1930. On the Clavariacea of Japan. II. Transactions of Sapporo Natural History Society, 11: 70-77.

Iwayama, S. 1933. On a new snow rot of winter cereals under snow in Toyama 1. Toyama Agricutural Experiment Station Bulletin, 1-20 (in Japanese).

Jacobs, D.L. and Bruehl, G.W. 1986. Saprophytic ability of *Typhula incarnata, T. idahoensis*, and *T. ishikariensis*. Phytopathology, 76: 695-698.

Ishida, S., Sumiyoshi, M., Nakamura, K., Kawasaki, T. and Ogura, N. 1989. Management of perennial ryegrass pasture mixed with orchard grass and ladino clover in Tenpoku district. Journal of Hokkaido Grassland Science, 23: 22-26 (in Japanese).

Kawakami, A. 2005. Studies on resistance to speckled snow mold in winter wheat. PhD thesis, Hokkaido University, Japan (in Japasese with English summary).

Kida, K., Hoshino, T. and Tojo, M. 2005. Freezing resistance of *Pythium* sp. from Spitsbergen Island. Abstracts XXVIII Symposium on Polar Biology, 58.

Kiyomoto, R.K. and Bruehl, G.W. 1977. Carbohydrate accumulation and depletion by winter cereals differing in resistance to *Typhula idahoensis*. Phytopathology, 67: 206-211.

Komatsu, T., Kubo, M., Tsuchiya, F. and Maruyama, S. 1983. Damage to alfalfa due to *Typhula ishikariensis* and its characteristics from the second survey in the Tokachi district (III). Journal of Hokkaido Grassland Science, 17: 140-144 (in Japanese).

Konstantinova, T., Parvanova, D., Atanassov, A. and Dijilianov, D. 2002. Freezing tolerant tobacco, transformed to accumulate osmoprotectants. Plant Science, 163: 157-164.

Kuznetzova, A. 1953. New species of the fungus *Typhula humulina* A. Kuzn. on subterranean stems of hop. Botanicheskie Materialy Otdela Cporovyh Rastenii 9: 142-145 (in Russian).

Lawton, M.B. and Burpee, L.L. 1990. Effect of rate and frequency of application of *Typhula phacorrhiza* on biological control of Typhula blight of creeping bentgrass. Phytopathology, 80: 70-73.

Lebeau, J.B. 1975. Antagonism between isolates of a snow mold pathogen. Phytopathology, 65: 877-880.

Lester, E. and Large, E.A. 1958. Surveys of clover rot with incidental observations on eel worm in clover: England and Wales 1953-55. Plant Pathology, 7: 115-124.

Lipps, P.E. 1980. A new species of *Pythium* isolated from wheat beneath snow in Washington. Mycologia, 72: 1127-1133.

Matsumoto, N. 1989. Autecology of the pathogenic species of *Typhula*. Research Bulletin of Hokkaido National Agricultural Experiment Stastion. 152: 91-162 (in Japanese with English Summary).

Matsumoto, N. 1992. Evolutionary ecology of the pathogenic species of *Typhula*. Transactions of Mycological Society of Japan, 33: 269-285.

Matsumoto, N. 1994. Ecological adaptations of low temparature plant pathogenic fungi to diverse winter climates. Canadian Journal of Plant Pathology, 16: 237-240.

Matsumoto, N. 1997a. Biological control of snow mold, pp. 343-350. In: Plant Cold Hardiness. P. Li and T. Chen (eds.), Plenum, New York, USA.

Matsumoto, N. 1997b. Evolution and adaptation in snow mold fungi. Soil Micro-organisms 50: 13-19 (in Japanese).

Matsumoto, N. 1997c. Population structure of individualistic snow mold fungi: reflections of ecological strategy and adaptations to environmental disturbance. Proceedings of International Workshop on Plant Microbe Interactions at Low Temperatures Under Snow, Sapporo, 125-131.

Matsumoto, N. and Sato, T. 1982. The competitive saprophytic abilities of *Typhula incarnata* and *T. ishikariensis*. Annals of Phytopathogical Society of Japan, 48: 419-424.

Matsumoto, N. and Sato, T. 1983. Niche separation in the pathogenic species of *Typhula*. Annals of Phytopathogical Society of Japan, 49: 293-298.

Matsumoto, N. and Tajimi, A. 1983. Intra and intertaxon interactions among dikaryons of *Typhula incarnata* and *T. ishikariensis* biotypes A, B, and C. Transactions of Mycological Society of Japan, 24: 250-253.

Matsumoto, N. and Tajimi, A. 1985. Field survival of sclerotia of *Typhula incarnata* and of *T. ishikariensis* biotype A. Canadian Journal of Botany, 63: 1126-1128.

Matsumoto, N. and Tajimi, A. 1988. Life history strategy in *Typhula incarnata* and *T. ishikariensis* biotypes A, B, and C as determined by sclerotium production. Canadian Journal of Botany, 66: 2485-2490.

Matsumoto, N. and Tajimi, A. 1990. Continuous variation within isolates of *Typhula ishikariensis* biotypes B and C associated with habitat differences. Canadian Journal Botany, 68: 1768-1773.

Matsumoto, N. and Tajimi, A. 1992. Biological control of *Typhula ishikariensis* on perennial ryegrass. Annals of Phytopathogical Society of Japan, 58: 741-751.

Matsumoto, N. and Tajimi, A. 1993. Effect of cropping history on the population structure of *Typhula incarnata* and *Typhula ishikariensis*. Canadian Journal of Botany, 71: 1434-1440.

Matsumoto, N., Sato, T. and Araki, T. 1982. Biotype differentiation in the *Typhula ishikariensis* complex and their allopatry in Hokkaido. Ann. Phytopath. Soc. Japan, 48: 275-280.

Matsumoto, N., Sato, T., Araki, T. and Tajimi, A. 1983. Genetic relationships within the *Typhula ishikariensis* complex. Transactions of Mycological Society of Japan, 24: 313-318.

Matsumoto, N., Abe, J. and Shimanuki, T. 1995. Variation within isolates of *Typhula incarnata* from localities differing in winter climate. Mycoscience, 36: 155-158.

Matsumoto, N., Tronsmo, A.M. and Shimanuki, T. 1996. Genetic and biological characteristics of *Typhula ishikariensis* isolates from Norway. European Journal of Plant Pathology, 102: 431-439.

Mohammad, F., Windes J.M. and Souza, E. 1997. Total non-structural carbohydrates in winter wheat populations segregating for snow mold resistance. Crop Science, 37: 108-112.

Murakami, K., Kaneko, K., Kojima, S. and Sekijo, T. 1965. Varietal trials on ryegrass (*Lolium* sp.) Research Bulletin of Hokkaido National Agricultural Experiment Station, 85: 68-76 (in Japanese).

Nakajima, T. and Abe, J. 1994. Development of resistance to *Microdochium nivale* in winter wheat during autumn and decline of the resistance under snow. Canadian Journal of Botany, 72: 1211-1215.

Nissinen, O. 1996. Analyses of Climatic Factors Affecting Snow Mould Injury in First year Timothy (*Phleum pratense* L.) with Special Reference to *Sclerotinia borealis*. Acta Universitatis Ouluensis A 289, 115 pp.

Ozaki, M. 1979. Ecological study of Sclerotinia snow blight disease of orchardgrass. Bull. Hokkaido Prefectural Agricultural Experiment Station, 42: 55-65 (in Japanese with English summary).

Procenko, E.P. 1967. *Typhula borealis* Ekstrand infecting tulips in the U.S.S.R. Mikologiya i fitopatologiya 1: 107-109 (in Russian with English summary).

Rayner, A.D.M., Coates, D., Ainsworth, A.M., Adams, T.J.H., Williams, E.N.D. and Todd, N.K. 1984. The biological consequences of the individualistic mycelium, pp. 509-540. In: The Ecology and Physiology of the Fungal Mycelium. D.H. Jennings and A.D.M. Rayner. (eds.), Cambridge University Press, Cambridge, UK.

Remsberg, R. 1940. Studies in the genus *Typhula*. Mycologia, 45: 75-87.

Saito, I. 1997. *Sclerotinia nivalis*, the pathogen of snow mold of herbaceous dicots in northern Japan. Mycoscience, 38: 227-236.

Sanford, G.B. 1933. A root rot of sweet clover and related crops caused by *Plenodomus meliloti* Dearness & Sanford. Canadian Journal of Research, C8: 337-348.

Schadt, C.W., Martin, A.P., Lipson, D.A. and Schmidt, S.K. 2003. Seasonal dynamics of previously unknown fungal lineages in tundra soils. Science, 301: 1359-1361.

Smith, J.D. 1987. Winter hardiness and overwintering diseases of amenity turfgrasses with special reference to the Canadian Prairies. Agriculture Canada Technical Bulletin, 1987, 12E.

Smith, J.D. and Årsvoll, K. 1975. Competition between basidiomycetes attacking turfgrasses. Journal of Sports Turf Research Institute, 51: 46-51.

Smith, J.D. and Davidson, J.G.N. 1979. *Acremonium boreale* n. sp., a sclerotial, low-temperature-tolerant, snow mold antagonist. Canadian Journal of Botany, 57: 2122-2139.

Simizu, M. 1993. Control of "Supponuke" snow blight on winter wheat. Hokunou, 60: 298-303 (in Japanese).

Simizu, M. and Miyajima, K. 1990. The incidence of supponuke on winter wheat. Annals of Phytopathogical Society of Japan, 56: 141-42 (abstract in Japanese).

Simizu, M. and Miyajima, K. 1992. Distribution of supponuke. Annals of Phytopathogical Society of Japan, 58: 147-8 (abstract in Japanese).

Snider, C.S., Hsiang, T., Zhao, G., Griffith, M. 2000. Role of ice nucleation and antifreeze activities in pathogenesis and growth in snow molds. Phytopathology, 90: 354-361.

Suzuki, H. and Arai, M. 1990. Occurrence and control of barley scald. Shokubutsuboueki, 44: 214-219 (in Japanese).

Takamatsu, S. 1989. A new snow mold of wheat and barley caused by foot rot fungus, *Ceratobasidium graminearum*. Annals of Phytopathogical Society of Japan, 55: 233-237.

Takeda, Y. and Nakashima, K. 1997a. Characteristics of alfalfa (*Medicago sativa* L.) cultivars adapted to Konsen district 1. Difference among cultivars in winter hardiness of the first winter and associated traits. Grassland Science, 43: 144-149 (in Japanese).

Takeda, Y. and Nakashima, K. 1997b. Characteristics of alfalfa (*Medicago sativa* L.) cultivars adapted to Konsen district 2. Difference among cultivars in winter hardiness after the 2nd year and associated traits. Grassland Science, 43: 150-156 (in Japanese).

Takeda, Y. and Nakashima, K. 1997c. Characteristics of alfalfa (*Medicago sativa* L.) cultivars adapted to Konsen district 3. Difference among cultivars in Lepto-leaf spot under the natural infection. Grassland Science, 43: 157-163 (in Japanese).

Terami, F. and Kawakami, A. 2002. Water potential of culture media affect the optimum growing temperature of *Sclerotinia borealis*. Japanese Journal of Phytopathology, 68: 194 (abstract in Japanese).

Tezuka, M. and Komeichi, M. 1980. Varietal differences in winter survival and productivity in late autumn of perennial ryegrass in Tenpoku district. Bulletin of Hokkaido Prefectural Agricultural Experiment Station, 44: 52-61 (in Japanese).

Tkachenko, O.B. 1995. Adaptation of the fungus *Typhula ishikariensis* Imai to the soil inhabitance. Mikologiya i fitopatologiya, 29: 14-19 (in Russian with English summary).

Tkachenko, O.B., Matsumoto, N. and Shimanuki, T. 1997. Mating patterns of east European isolates of *Typhula ishikariensis* with isolates from distant regions. Mikokogiya i fitopatologiya, 31: 68-72.

Tojo, M., Hakoda, A., Fujii, H., Hoshino, T., Kanda, H. and Ohki, S. 2002a. Distribution of a heterothallic *Pythium* sp. on Spitsbergen Islands, Svalbard. Abstracts XXV Symposium on Polar Biology, 64.

Tojo, M., Mühlhauser, H.A., Hoshino, T., Fujii, H., Hakoda, A., Kanda, H. and Ohki, S. 2002b. A preliminary report on a heterothallic *Pythium* sp. from King George Island, Antarctica. Abstracts XXV Symposium on Polar Biology, 65.

Tojo, M., Mühlhauser, H.A., Hoshino, T., Kanda, H. and Ohki, S. 2002c. Occurrence of sterile basidiomycetes causing dead rings in moss mat on King George Island, Antarctica. Abstracts XXV Symposium on Polar Biology, 65.

Tojo, M., Kida, K., Hoshino, T., Toyoda, Y., Tokura, K. and Kanda, H. 2006. Succession of populations of moss pathogenic fungi in a open top chamber (OTC) on moss colonies (*Sanionia uncinata*) at the north side cliff of Japanese Ny-Ålesund observatory, Spitsbergen Island from 2003 to 2006. Abstracts XXIX Symposium on Polar Meteorology and Glaciology Polar Biology, 166.

Tomiyama, K. 1955. Studies on the snow blight disease of winter cereals. Hokkaido National Agricultural Experiment Station Report 47 (in Japanese with English summary).

Trutmann, P., Keane, P.J. and Merriman, P.R. 1982. Biological control of *Sclerotinia sclerotiorum* on aerial parts by the mycoparasite *Coniothyrium minitans*. Transactions of British Mycological Society, 78: 521-529.

Wiese, M.V. 1977. Compendium of Wheat Diseases. American Phytopathological Society, St. Paul, Minnesota, USA.

Yamagishi, Y. 1988. Utilization of perennial ryegrass for grazing in northern Rumoi. Journal of Hokkaido Grassland Science, 22: 225-228 (in Japanese).

Yamaguchi, H., Uchiyama, K., Sawai, A., Gau, M., Ueda, S., Maki, Y., Matsuura, M., Suginobu, K., Sato, R., Takeda, Y., Nakashima, K., Chiba, K., Ochi, H., Sawada, Y. and Tamakake, H. 1995. Breeding of 'Hisawakaba' alfalfa and its characteristics. Research Bulletin of Hokkaido National Agricultural Experiment Station, 161: 17-31.

Ylimäki, A. 1969. Typhula blight of clovers. Annales Agriculturae Fenniae, 8: 30-37.

Yoshida, M., Abe, J., Moriyama, M. and Kuwabara, T. 1998. Carbohydrate levels among winter wheat cultivars varying in freezing tolerance and snow mold resistance during autumn and winter. Physiologia Plantarum, 103: 8-16.

Yukawa, T. and Watanabe, Y. 1991. Studies on fructan accumulation in wheat (Tritichum aestivue L.) I. Relationship between fructan concentration and overwintering ability from aspect on the pedigree. Japanese Journal of Crop Science, 60: 385-391.

8

Fungi from High Nitrogen Environments—Ammonia Fungi: Eco-Physiological Aspects

Akira Suzuki

Department of Biology, Faculty of Education
Chiba University, 1-33 Yayoi-cho, Inage-ku,
Chiba 263-8522, Japan
E-mail: asmush@faculty.chiba-u.jp

Abstract

Ammonia fungi are a chemoecological group of fungi in which reproductive structures appear after the addition of ammonia, urea, or other nitrogenous materials which react as bases, or of alkalis. The appearance of reproductive structures of ammonia fungi generally proceeds in the sequence: saprobic anamorphic fungi, saprobic fungi in Ascomycota, saprobic fungi in Basidiomycota having smaller fruit-bodies, and followed by mostly biotrophic and a few possibly saprobic fungi in Basidiomycota having larger fruit-bodies. The former three groups are described as early phase fungi (EP fungi) and the latter group described as late phase fungi (LP fungi). Spores germination of most ammonia fungi are stimulated by 10-300 mM NH_4-N at pH 7-10. Early-stage EP fungi grow well at pH 7-9 and late-stage EP fungi at pH 6-8, whereas the LP fungi show optimum growth at pH 5-7. Ammonia fungi would be divided into three groups in terms of adaptation to ammonium-ion concentration, namely, high concentration of ammonium-ion adapted species, composed of some EP fungi, wide range concentrations of ammonium-ion adapted species, composed of some EP fungi, and high concentration of ammonium-ion non-adapted species, composed of some EP fungi and LP fungi. Reproductive structure formation of most ammonia fungi is accelerated through the vegetative growth stimulated of ammonium-nitrogen under weak alkaline to neutral conditions, although the fruiting of the late-stage EP fungi *Coprinopsis* spp. is directly induced by the presence of the ammonia or urea. Most early-stage EP fungi intermingle with each other whereas late-stage EP fungi inhibit the growth of early-stage EP fungi or invade into the territories of early-stage EP fungi, irrespective of pH conditions. LP fungus *Hebeloma vinosophyllum* invades into the territories of early-stage EP fungi under acidic condition, but not into those of late-stage EP at any pH conditions. In the five species cultures (early-stage EP fungi *Amblyosporium botrytis* and *Ascobolus denudatus*, late-stage EP fungi *Tephrocybe tesquorum* and *Coprinopsis phlyctidospora*, and LP fungus *He. vinosophyllum*), fruit-body formation of *As. denudatus*, *T. tesquorum* and *C. phlyctidospora*, and *He. vinosophyllum* are reduced, but not *Am. botrytis*. These suggest that successive occurrence of ammonia fungi is caused by the interactions among ammonia fungi as well as by the physiological characteristic of each fungus associates with conditions of its inhabiting soils such as pH and nitrogen concentration. Most ammonia fungi have clear cellulolytic and faint ligninolytic activities. The pH optima for the cellulolytic enzymes of EP fungi are between 6.8 and 9.0, and those for the cellulolytic enzymes of LP fungi are between 5.5 and 6.8. In ecosystems, ammonia fungi would have a role in keeping the carbon cycles instead of pre-inhabiting fungi after the addition of a large amount of nitrogenous materials, and ammonia fungi immobilize nitrogen derived from animal wastes such as urine, feces, and dead bodies. This replacement described may be view as a kind of 'compensation process' in nutrient cycle.

INTRODUCTION

The ecosystem comprises of various biotic communities such as the communities of plants, animals, fungi, and bacteria. Each fungal community is composed of different kinds of ecological groups of fungi that have been categorized based on various ecological standpoints such as their colonizing substrates sometimes including the kinds of decomposing compounds, and their responses to the kinds of disturbance. Litter-decomposing fungi (litter-inhabiting fungi; Frankland et al., 1982), wood-rotting fungi (lignicoulous fungi, lignin-decomposing fungi, ligninolytic fungi, wood-attacking fungi, wood-decaying fungi; Frankland et al., 1982; Lisiewska, 1992; Dix and Webster, 1995; Kirk et al., 2008), fungicolous fungi (Hirsch and Braun, 1992; Kirk et al., 2008), and coprophilous fungi (coprophilic fungi; dung fungi; Dickinson and Pugh, 1974; Hudson, 1980; Frankland et al., 1982; Cooke and Rayner, 1984; Richards, 1987; Lisiewska, 1992; Dix and Webster, 1995; Kirk et al., 2008), chitinolytic fungi (Hudson, 1980), keratinophilic fungi (keratinolytic fungi; keratinophilous fungi,

Hudson, 1980; Richards, 1987; Gams, 1992) are ecological groups of fungi which have been defined based on the former categorization, and the pyrophilous fungi (anthracophilous fungi, carbonicolous fungi, carbophilous fungi, fire place fungi, phoenicoid fungi; Hudson, 1980; Lisiewska, 1992; Dix and Webster, 1995; Kirk et al., 2008), are those of fungi which had been defined based on the latter categorization. 'Ammonia fungi' are one of those ecological groups of fungi that were first recognized by the experimental observation of Sagara and Hamada (1965), Sagara (1973), and Sagara (1975). In other words, 'ammonia fungi' are a chemoecological group of fungi that were recognized from their appearance on the disturbed habitats by high concentration of ammonium-nitrogen (Sagara, 1975). In 1992, Sagara proposed a natural ecological group of fungi, 'postputrefaction fungi', which are recognized from the observation of their occurrences from animal waste decomposing sites such as urine, feces, and dead bodies (Sagara, 1975, 1976a, 1976b, 1976c, 1977, 1981, 1984, 1989a, 1992, 1995a; Sagara et al., 1985, 2000, Miller and Hilton, 1986; Wang and Sagara, 1997; Fukiharu et al., 2000a, 2000b; Harmaja, 2002; Kasuya, 2002; Tibbett and Carter, 2003) or deserted midden (latrines) of mammalian animals (Sagara 1976a, 1976c, 1978, 1989a, 1989b, 1992, 1995a, 1995b, 1998, 1999; Sagara et al., 1988, 1989, 1993a, 1993b, 2006, 2008; Sagara and Abe, 1993; Tibbett and Carter, 2003). Until now, more than 60 fungus species have been recorded as postputrefaction fungi by the application of urea and/or other nitrogenous material(s) both in the field and in the laboratory, and the observation at deserted middens of mammalians such as moles (cf., Table 1; Laiho, 1970; Sagara, 1975, 1981, 1989a, 1992, 1995a; Lehmann and Hudson, 1977; Miller and Hilton, 1986; Clémançon and Hongo, 1994; Fukiharu and Hongo, 1995; Sagara et al., 2000, 2008; Imamura, 2001; Harmaja, 2002; Kasuya, 2002; Nagao et al., 2003; Suzuki et al., 2002a; Tibbett and Carter, 2003). In order to elucidate the physiological background of colonization, establishment, successive appearance of reproductive structures (= occurrence), and ecological role of postputrefaction fungi, the physiological characteristics of postputrefaction fungi have been examined except for the fungus species such as *Hebeloma radicosum*, of which colonization was only observed from the deserted midden (Tibbett and Carter, 2003). In the following discussion, the physiology of 'ammonia fungi' is focused, since no physiological research has been done on the fungus species belonging to 'postputrefaction fungi' such as *Hebeloma radicosum* which has not been recorded as 'ammonia fungi.'

Table 1. Succession and nutritional mode of ammonia fungi

Fungus species	*Nutritional mode*	*Succession*	*Reference**
Anamorphic fungi			
Amblyosporium botrytis	Saprotrophy	early stage EP	1, 3, 5-9, 11, 12, 14, 16
Cladorrhinum foecundissimum	Saprotrophy	early stage EP	4, 12
Ascomycota			
Ascobolus denudatus	Saprotrophy	early stage EP	1, 3, 6-8, 10-12, 14, 15
Ascobolus hansenii	Saprotrophy	early stage EP	3, 4, 15
Pseudombrophila petrakii	Saprotrophy	early stage EP	1, 2, 5-8, 11, 12, 15, 16
Peziza moravecii	Saprotrophy	early stage EP	1-3, 5, 8, 10, 12, 14, 15
Peziza urinophila	Saprotrophy	early stage EP	15
Thecoteus urinamans	Saprotrophy	early stage EP	13
Humaria velonovskyi	Saprotrophy	late stage EP	3, 12
Basidiomycota			
Coprinopsis cinerea	Saprotrophy	late stage EP	4, 9
Coprinopsis echinospora	Saprotrophy	late stage EP	1-3, 5-7, 16
Coprinopsis neolagopus	Saprotrophy	late stage EP	1, 11, 15
Coprinopsis phlyctidospora	Saprotrophy	late stage EP	4, 8-12, 16
Coprinopsis stercorea	Saprotrophy	late stage EP	3
Crucispora rhombisperma	Saprotrophy	late stage EP	3, 4, 9
Tephrocybe tesquorum	Saprotrophy	late stage EP	1-3, 5-7, 10-12, 15, 16

Contd.

Table 1 continued

Fungus species	*Nutritional mode*	*Succession*	*Reference**
Collybia cookei	Nectrotrophy (fungicolous)	LP	3, 7
Lepista sordida	?Saprotrophy	LP	3
Tephrocybe ambusta	?Saprotrophy	LP	1-3, 16
Alnicola lactariolens	Biotrophy (EM)	LP	4, 15
Calocybe leucocephala	?Saprotrophy	LP	3
Hebeloma aminophilum	?Biotrophy (?EM)	LP	10
Hebeloma luchuense	?Biotrophy (?EM)	LP	4
Hebeloma radicosoides	Biotrophy (EM)	LP	3, 4, 15, 16
Hebeloma spoliatum	Biotrophy (EM)	LP	3, 9, 11, 12, 15
Hebeloma vinosophyllum	Biotrophy (EM)	LP	2, 3, 8, 11, 15
Laccaria amethystina	Biotrophy (EM)	LP	3, 15
Laccaria bicolor	Biotrophy (EM)	LP	1-3, 5, 7, 15, 16
Rhizopogon sucossus	?Biotrophy (?EM)	LP	1-3
Suillus bovinus	Biotrophy (EM)	LP	1-3

1) Sagara, 1975; 2) Sagara, 1992; 3) Sagara, 1995a; 4) Fukiharu and Hongo, 1995; 5) Yamanaka, 1995a; 6) Yamanaka, 1995b; 7) Yamanaka, 1995c; 8) Sato and Suzuki, 1997; 9) Fukiharu et al., 1997; 10) Suzuki et al., 1998; 11) Suzuki, 2000; 12) Suzuki et al., 2002b; 13) Nagao et al., 2003; 14) He and Suzuki, 2004; 15) Imamura and Yumoto, 2004; 16) Sagara et al., 2008.

Ecological background of ammonia fungi

1. Definition of ammonia fungi

In 1975, 'ammonia fungi' were defined by Sagara as a new category of an ecological group of fungi. Namely, 'ammonia fungi' are a chemo-ecological group of fungi, which sequentially develop reproductive structures exclusively or relatively luxuriantly on the soil after the sudden addition of ammonia, or of some other nitrogenous material that reacts as a base or of alkalis (cf., Fig. 1; Sagara, 1975). Many ammonia fungus species have been recognized as the member of other ecological groups of fungi such as fungicolous fungi, coprophilous fungi, and pyrophilous fungi, since each fungus species would be categorized based on various ecological standpoints (characteristics), (cf., Fig. 2; Sagara, 1973, 1975, 1992; Suzuki, 2004).

2. Sequential occurrence of ammonia fungi in the field

The sequential appearance of reproductive structures (= succession) of ammonia fungi generally proceeds as follows: saprobic anamorphic fungi → saprobic cup fungi in Ascomycota → saprobic agaric fungi in Basidiomycota having smaller fruit-bodies (the fungi belonging to these three phases are described as early phase fungi: EP fungi) → mostly biotrophic and a few possibly saprobic fungi in Basidiomycota having larger fruit-bodies (the fungi belonging to this phase is described as late phase fungi: LP fungi (cf., Fig. 1; Table 1; Sagara, 1975, 1992, 1995a; Imamura and Yumoto, 2004; Tibbett and Carter, 2003).

Biogeographical distribution of ammonia fungi

Ammonia fungi have been recorded in Eastern Asia, Southern Asia, Oceania (Australia, New Zealand, Hawaii), Europe, North Africa, East Africa, North America, and northern South America (cf., Suzuki et al., 2003). Most collection records of ammonia fungi in Japan and New Zealand have been obtained from the surveys for the urea-treated plots placed in different geographical areas, whereas those of ammonia fungi in other regions have been obtained from the very restricted geographical areas with or without urea-treatment (Fukiharu and Horigome, 1996; Suzuki et al., 2003). Fungal community of ammonia fungi in each habitat would comprise fungus species having different biogeographical distributing areas. Namely, each fungal community of ammonia fungi would consist of both saprobic and ectomycorrhizal species having different biogeographical distribution (cf., Fukiharu and Horigome, 1996; Suzuki et al., 2003). To elucidate the commu-

Fig. 1. Occurrence of LP fungus (Ectomycorrhizal ammonia fungus) *Hebeloma spoliatum* on the urea-treated plot in Kiyosumi, Japan. Urea (800 g/m^2) was applied on the forest floor encircled by strings. The size of sub-plot encircled by a string is 1 m × 1 m. The fruit-bodies appeared about 6 months after the urea application .

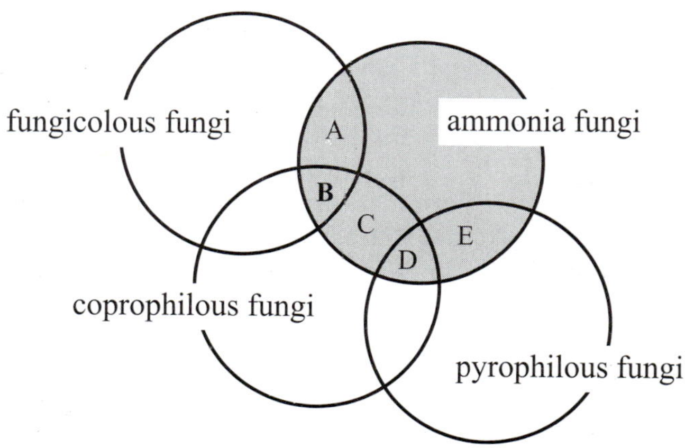

A: *Collybia cookei;* B: *Amblyosporium botrytis;* C: *Cladorrhinum foecundissimum, Ascobolus denudatus, Coprinopsis cinerea, Coprinopsis stercorea;* D: Not recorded; E: *Coprinopsis phlyctidospora, Laccaria bicolor*

Fig. 2. Ammonia fungus species belonging to more than two ecological categories

nity structure of ammonia fungi in each habitat, further surveys for the mycobiota of ammonia fungi at a worldwide scale are required, since surveys for the ammonia fungi by the urea application have been done in pinpoint places even in the surveyed areas of Europe, Australia, and North America.

PHYSIOLOGY OF AMMONIA FUNGI

Spore germination

Spores germination of most saprobic and ectomycorrhizal ammonia fungi was stimulated by 10-300 mM NH_4-N at pH 7-10 (Table 2; Suzuki, 1978, 1989, 1992, 2004, 2006; Suzuki et al., 1982; Deng and Suzuki, 2008). Conidium germination of early-stage EP fungus (saprobic ammonia fungus) *Amblyosporium botrytis*, which appears first with other anamorphic ammonia fungi in succession, is markedly stimulated by 300 mM NH_4-N at pH 8. Ascospore germination of early-stage EP fungus (saprobic ammonia fungus) *Ascobolus denudatus*, which appears second in succession, is remarkably stimulated with 10-30 mM NH_4-N at pH 8-10. Ascospore germination of another early-stage EP fungus (saprobic ammonia fungus) *Peziza moavecii* is remarkably stimulated by 10 mM NH_4-N at pH 8-10. Basidiospore germination of late-stage EP fungi (saprobic ammonia fungi) *Coprinopsis* spp. and that of LP fungi (ectomycorrhizal ammonia fungi) *Hebeloma* spp. are drastically stimulated by about 100 mM NH_4-N at about pH 8 (Table 2). The stimulation of spore germination was also observed by sterilized water extracts of soils collected at different times after urea application (Fig. 3). This means that the spores of ammonia fungi, both saprobic and ectomycorrhizal species, germinate at a high percentage in the short days after an application of a large amount of nitrogenous material(s) in the field. In other words, it is expected that no strong inhibitory factors for spore germination are present in the forest soils, and that pH and ammonium-nitrogen are principal factors for the stimulation for the spore germination of ammonia fungi in the field. This also suggests that most ammonia fungi germinate at the same time in the field after a sudden disturbance by a large amount of nitrogenous material(s), irrespective of the sequence of the succession.

Temperature optima for spore germination of early-stage EP fungus *Amblyoporium botrytis*, and LP fungi *Hebeloma radicosoides* and *Hebeloma spoliatum* are somewhat lower than those for spore germination of early-stage EP fungus *Ascobolus denudatus* and late-stage EP fungi *Coprinospsis neolagopus* and *Coprinopsis phlyctidospora*, and LP fungus *Hebeloma vinosophyllum*. Among them, temperature optimum for spore germination of *C. neolagopus* is definitely higher than that for spore germination of other

Table 2. Optima of NH_4-N, pH, and temperature for spore germination of ammonia fungi

Fungus species	*Spore form*	*NH_4-N (mM)*	*Tested reagent*	*pH*	*Temperature (°C)*	*Source*
Amblyosporium botrytis	conidia	300	NH_4Cl	8	10-15	Licyayo et al. (unpublished data)
Ascobolus denudatus	ascospore	10-30	NH_4Cl	8-10	15-30	Suzuki et al. (unpublished data)
Peziza moravecii	ascospore	10	NH_4Cl	8-10	(15)	Suzuki et al. (unpublished data)
Coprinopsis cinerea	basidiospore	200	$(NH_4)_2HPO_4$	7- 9	(30)	Suzuki et al. (unpublished data)
Coprinopsis echinospora	basidiospore	200	$(NH_4)_2HPO_4$	8	(25)	Suzuki et al. (unpublished data)
Coprinopsis neolagopus	basidiospore	30-150	NH_4Cl	8-10	30-35	Suzuki et al. (unpublished data)
Coprinopsis phlyctidospora	basidiospore	200	$(NH_4)_2HPO_4$	8	15-25	Suzuki et al. (unpublished data)
Coprinopsis stercorea	basidiospore	200	$(NH_4)_2HPO_4$	8	(25)	Suzuki et al. (unpublished data)
Hebeloma radicosoides	basidiospore	100-300	NH_4Cl	8.0-8.5	5-20	Suzuki et al. (unpublished data)
Hebeloma spoliatum	basidiospore	60-200	$(NH_4)_2HPO_4$	8	10-20	Suzuki et al. (unpublished data)
Hebeloma vinosophyllum	basidiospore	100	NH_4Cl	8.0-8.5	25	Deng and Suzuki, 2008

Numerical values in parenthesis express the tested condition and not examined the optimum condition.

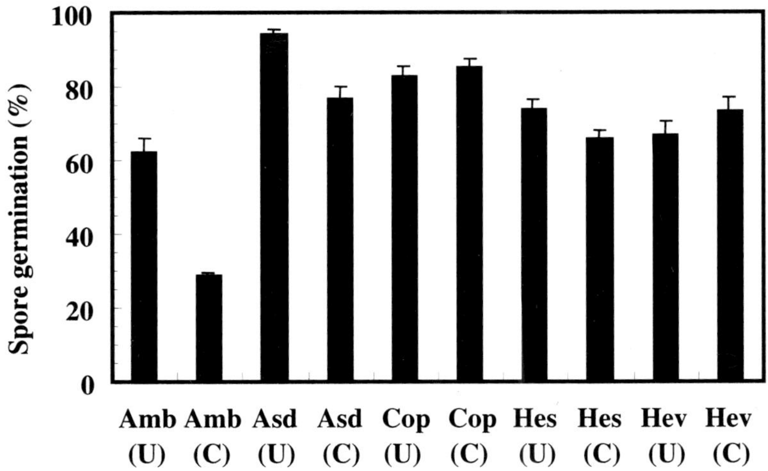

Fig. 3. Spore germination of ammonia fungi stimulated by water extracts of urea-treated soils. Bar: Standard error, (U): Water extract of urea-treated soil, (C): Water solution of following chemicals (10 mM NH_4Cl adjusted at pH 9.0 for *Ascobolus denudatus* and 100 mM $(NH_4)_2HPO_4$ for other 4 fungus species). Spore germinations of the five fungus species are not stimulated by water extract of non-urea-treated soil or by pure water. Detailed information about the experiment is shown in Suzuki (2006; Table 2). Amb: Conidia of *Amblyosporium botrytis*; Asd: Ascospores of *Ascobolus denudatus*; Cop: Basidiospores of *Coprinopsis phlyctidospora*; Hes: Basidiospores of *Hebeloma spoliatum*; Hev: Bsidiospores of *Hebeloma vinosophyllum*. (Based on data from Suzuki, 2006)

tested fungi (Table 2). In the field, *C. neolagopus* occurs when the urea treatment is conducted in summer, but not in winter. In laboratory experiments, *C. neolagopus* also occurs when urea application is performed above 20°C (Sagara, 1975, 1976a). These phenomena give us the idea that the influence of temperature during urea application described above would be partially explained by the difference in the optimum temperature for the spore germination of each ammonia fungus.

Spore longevity

Spore longevity is one of the important factors to speculate the colonization strategy of fungi in the field. Spore longevity of ammonia fungi has not been examined except for LP fungus (ectomycorrhizal ammonia fungus) *Hebeloma spoliatum*. Germination percentage of basidispore of *He. spoliatum* gradually decreased with the increment of storage period. The tendency of this decline is more prominent at 25°C than 5°C. The longevity of

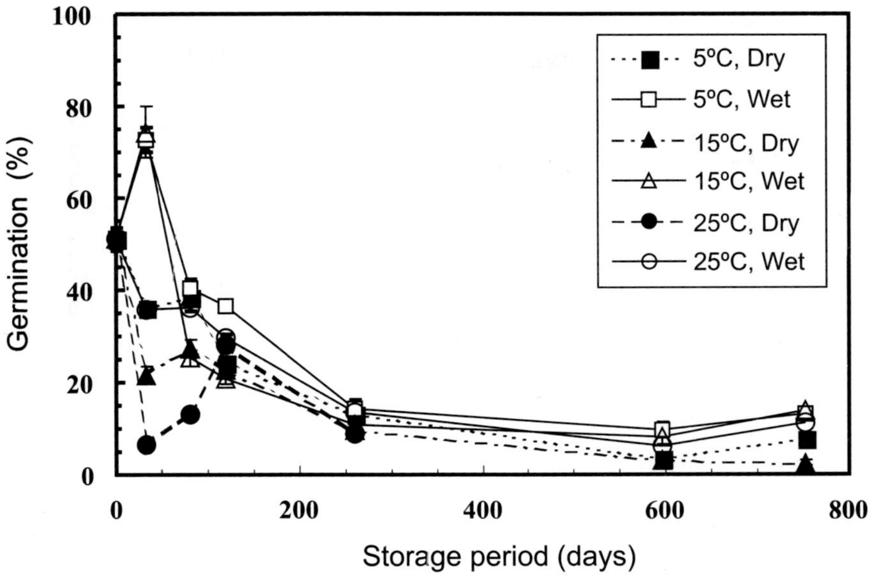

Fig. 4. Longevity of basidiospores of *Hebeloma spoliatum*. Values in the graph are shown as average with standard error. (Modified from Suzuki, 2009, Mycoscience 50: 44, fig. 1 with permission from Springer Ltd. and the Mycological Society of Japan). Basidiomata of *He. spoliatum* was collected from a urea-treated plot in the field. Basidiospores were collected from spore prints obtained separately at 5°C, 15°C and 25°C from the pilei of the harvested fruit-bodies. Basidiospores collected at different temperatures were mixed with roughly equal volumes. The mixed basidiospores were stored at 5°C under dry condition, 5°C under wet condition, 15°C under dry condition, 15°C under wet condition, 25°C under dry condition, and 25°C under wet condition, respectively. Dry condition was kept by the storage with silica gel and wet condition was kept by water suspension of the mixed basidiospores and both sample spores were stored in darkness. At designated time, a small amount of the stored spores were suspended in 100 mM NH_4Cl adjusted at pH 8.0 with 1 M KOH. The spore density was adjusted at the optimum conditions for the germination. The spore suspensions were incubated at 15°C in the dark and collected at 6, 15 and 30 days after the incubation. The percentage germination of the spores was expressed as (Number of germinated spores/500 spores) × 100. The germination percentage in the graph is shown by the maximum germination percentage obtained among 6, 15 and 30 days' incubations

basidiospores of *He. spoliatum* becomes shorter when they were exposed to dry conditions. It is expected that basidiospores of *He. spoliatum* in the field would maintain their germination ability for more than three years, when they are not exposed to extremely dry conditions (Fig. 4). This supports the assumption that the invasion and colonization of ammonia fungi would initiate from spores as well as hyphae (cf., Sagara, 1976b; Suzuki, 2006; Sagara et al., 2008).

Vegetative growth

Early-stage EP fungi (saprobic ammonia fungi) grow well at pH 7-8 and late-stage EP fungi (saprobic ammonia fungi) at pH 6-7, whereas the LP fungi (ectomycorrhizal ammonia fungi) show optimum growth at pH 5-6 (Yamanaka, 2003). pH of most Japanese forest soils without urea application aligns at 3.5-6.5 (Sagara, 1975, 1992; Yamanaka, 1995a, 1995b, 1995c; Fukiharu et al., 1997; Sato and Suzuki, 1997; Suzuki, 2000; Suzuki et al., 2002b; He and Suzuki, 2004). These suggest that both EP and LP fungi (saprobic and ectomycrrhizal ammonia fungi) adapt to the weak alkaline and neutral conditions at the vegetative growth stage.

EP fungi grow well in both inorganic nitrogen such as NH_4-N and NO_3-N and organic nitrogen such as l-asparagine and urea. Early-stage EP fungus *Amblyosporium botrytis*, and late-stage EP fungi *Tephrocybe tesquorum* and *Coprinopsis echinospora* grow vigorously in ammonium salts, l-asparagine, and urea, but not in nitrate. In bovine serum albumin, *T. tesquorum* and *C. echinospora* also grow well but *Am. botrytis* is not able to grow (Yamanaka, 1999). In contrast, early-stage EP fungi *Peziza urinophila* and *Pseudombrophila petrakii* (= *Pseudombrophila deerata*) do not grow in any of the above nitrogen sources (Yamanaka 1999). LP fungus *Hebeloma vinosophyllum* grows well in nitrogen sources such as NH_4Cl, KNO_3, KNO_2, and urea. pH optima for vegetative growth of *He. vinosophyllum* using each of these four nitrogen sources are 7-8, 5-6, 8, and 4-6, respectively. *He. vinosophylum* also grows well in l-asparagine and bovine serum albumin. Another LP fungus *Laccaria bicolor* also grows well in ammonium salts, nitrate, and urea, but not in l-asparagine and bovine serum (Yamanaka 1999). Early-stage EP fungi *Ascobolus denudatus* and *Ps. petrakii,* late-stage EP fungi *Coprinopsis phlyctidospora* and *Coprinopsis* sp. (allied species of *C. phlyctidospora* in Oceania (Suzuki et al., 2002a)), and LP fungi *He. vinosophyllum* and *He. aminophilum* have vegetative growth maxima at 0.3-0.1 M NH_4Cl and 0.01-0.1 M NH_4Cl, respectively. The upper limit concentration of NH_4Cl for their vegetative growth is 0.6 M. Late-stage EP fungus *Humaria velenovskyi* has growth optima in a wider range of NH_4Cl concentrations, whereas early-stage EP fungi *Am. botrytis* and *Peziza moravecii* have growth optima at 0.03-0.3 M NH_4Cl. The upper limit concentrations for the vegetative growth of these three ammonia fungi and early-stage EP fungus *Peziza moravecii* are between 1.1-1.6 M NH_4Cl. *Pe. moravecii* grows vigorously even at 0.0001 M NH_4Cl. These indicate that ammonia fungi would be divided into three groups in terms of adaptation to ammonium-

High ammonium-nitrogen concentration adapted species

Amblyosporium botrytis

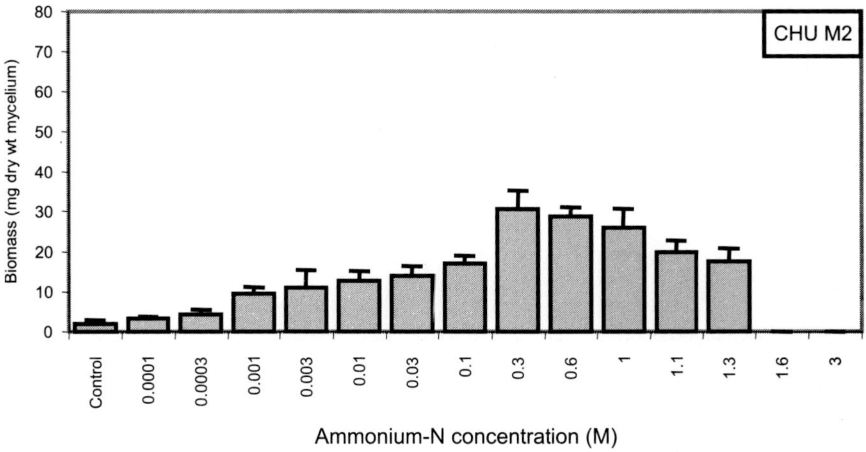

Peziza moravecii

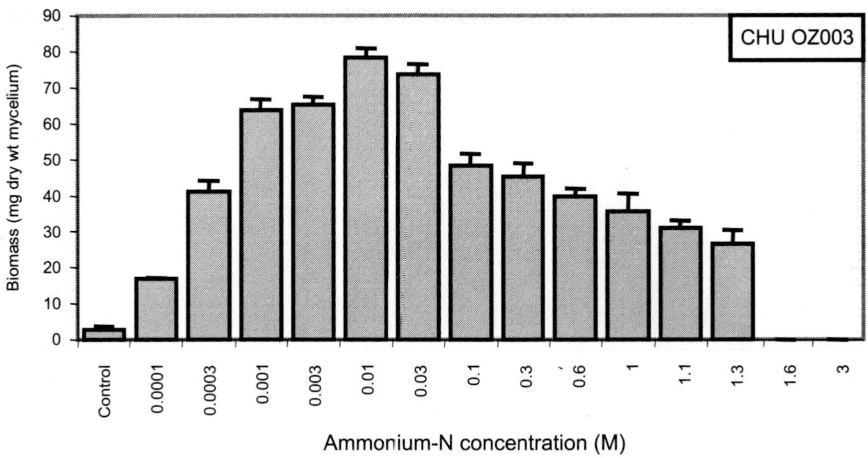

Fig. 5. The effect of ammonium-nitrogen concentration on the vegetative growth of ammonia fungi. Bar: standard error. *Peziza moravecii* is categorized as high ammonium-nitrogen concentration adapted species, although this fungus was once categorized as wide range ammonium-nitrogen concentration adapted species. (After Licyayo and Suzuki, 2006, with permission)

High ammonium-nitrogen concentration non-adapted species

Ascobolus denudatus

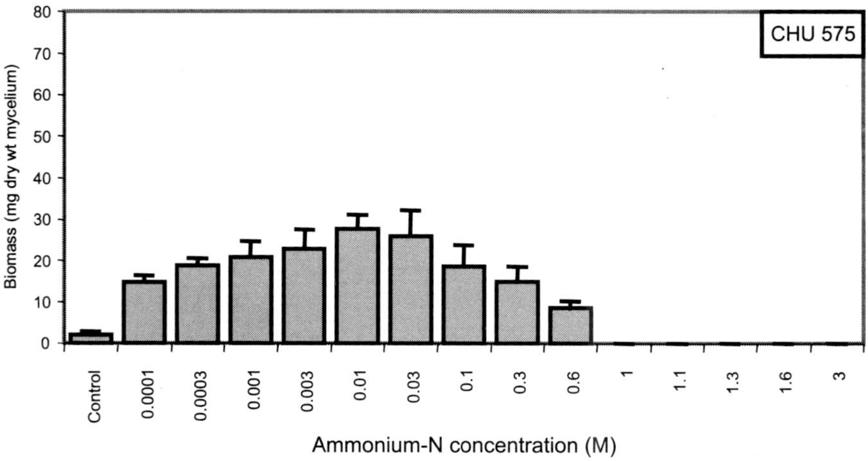

Coprinopsis phlyctidospora

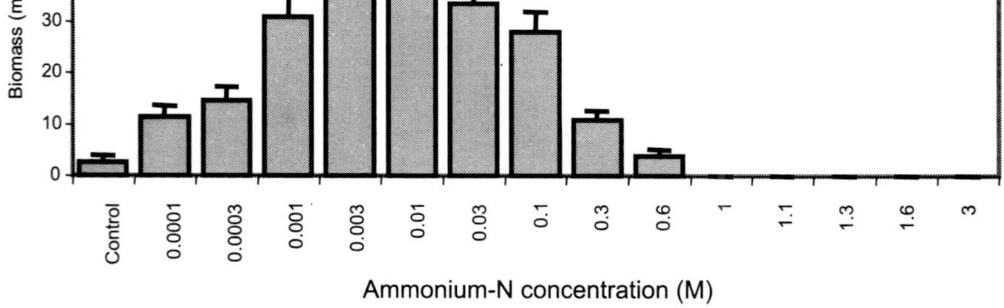

Fig. 5 continued. The effect of ammonium-nitrogen concentration on the vegetative growth of ammonia fungi. Bar: standard error. *Peziza moravecii* is categorized as high ammonium-nitrogen concentration adapted species, although this fungus was once categorized as wide range ammonium-nitrogen concentration adapted species. (After Licyayo and Suzuki, 2006, with permission)

High ammonium-nitrogen concentration non-adapted species

Hebeloma vinosophyllum

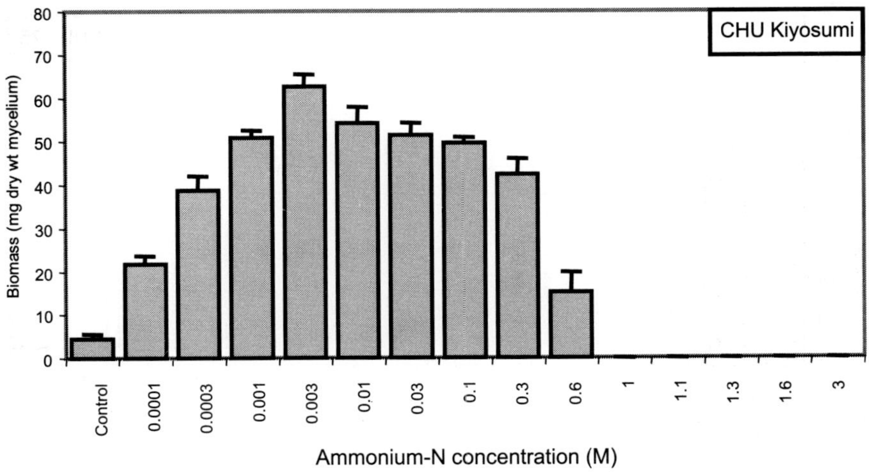

Wide range ammonium-nitrogen concentration adapted species

Tephrocybe tesquorum

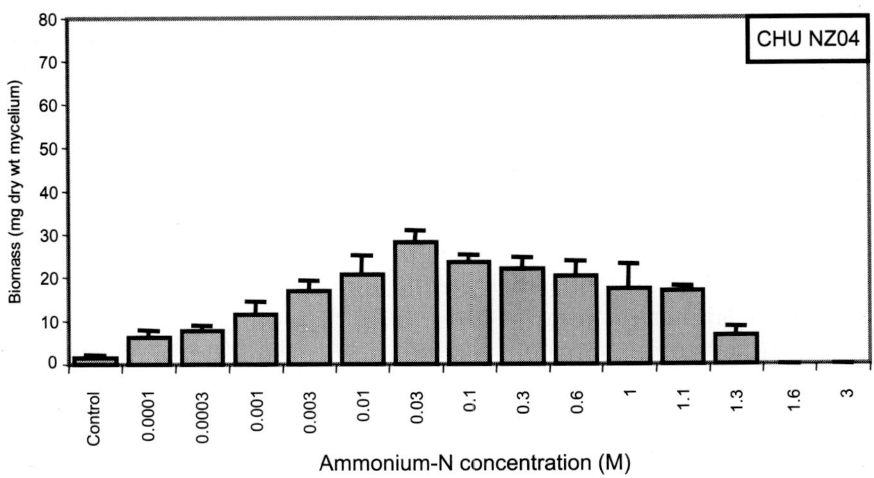

Fig. 5 continued. The effect of ammonium-nitrogen concentration on the vegetative growth of ammonia fungi. Bar: standard error. *Peziza moravecii* is categorized as high ammonium-nitrogen concentration adapted species, although this fungus was once categorized as wide range ammonium-nitrogen concentration adapted species. (After Licyayo and Suzuki, 2006, with permission)

Wide range ammonium-nitrogen concentration adapted species

Humaria velenovskyi

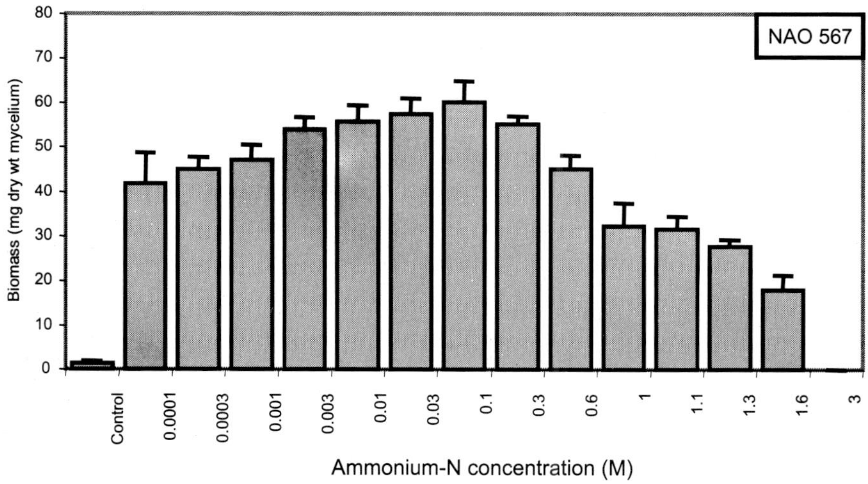

Fig. 5 continued. The effect of ammonium-nitrogen concentration on the vegetative growth of ammonia fungi. Bar: standard error. *Peziza moravecii* is categorized as high ammonium-nitrogen concentration adapted species, although this fungus was once categorized as wide range ammonium-nitrogen concentration adapted species. (After Licyayo and Suzuki, 2006, with permission)

ion concentration, namely, 1) high concentration of ammonium-ion non-adapted species, composed of some EP fungi and LP fungi, i.e., some saprobic ammonia fungi and ectomycorrhizal ammonia fungi, 2) wide range concentrations of ammonium-ion adapted species, composed of some EP fungi (saprobic ammonia fungi), and 3) high concentration of ammonium-ion adapted species, composed of some EP fungi (saprobic ammonia fungi), (Fig. 5; Licyayo and Suzuki, 2006).

The major nitrogen form just after the urea-treated soils is ammonium-nitrogen and urea gradually changes into nitrate-nitrogen and reaches maximum concentration at around the period of the first flush of LP fungi (cf., Yamanaka, 1995a, 1995b; Suzuki, 2000; Suzuki et al., 2002b). The changes in pH, concentrations of ammonium-nitrogen, and that of nitrate-nitrogen are as follows: pH and ammonium-nitrogen concentration rapidly increased just after urea application and then gradually decrease during the occurrence periods of both EP and LP fungi. These results suggest that the major nitrogen source for vegetative growth of ammonia fungi would be ammonium-nitrogen, irrespective of their nutritional modes under neutral to weakly alkaline conditions, and nitrogen source utilized by the ectomycorrhizal ammonia fungi would gradually be replaced from ammonium-nitrogen to nitrate-nitrogen associated with the decline of pH in the urea-treated soil.

Reproductive structure formation

Late-stage EP fungus (saprobic ammonia fungus) *Coprinopsis cinerea* forms fruit body primordium even in darkness when urea or ammonium-nitrogen is applied as the sole nitrogen source to the potato-dextrose agar medium. While, the fungus does not fruit in water solutions of non-basic ammonium salts. The presence of the ammonium-nitrogen (0.05-0.4 g NH_3/liter) at least for 24 hours is enough to induce its primordium formation. (Morimoto et al., 1981). Late-stage EP fungus (saprobic ammonia fungus) *Coprinopsis stercorea* fruits in light, when urea (0.25-1.0 g urea/liter) is applied to a potato-dextrose-agar medium (Morimoto et al., 1982). Another late-stage EP fungus (saprobic ammonia fungus) *Coprinopsis phlyctidospora* fruits both in light and darkness when urea (0.2-1.7 g urea/liter) is applied to a liquid media (He and Suzuki, 2003). These results suggest that urea would be an effective stimulant for fruiting of saprobic ammonia fungi *in vitro*. However, urea would not be one of the major determinants for the occurrence of ammonia fungi in the field, since the concentration of urea in the soil following the urea application is probably lower than that of urea which is effective for primordium formation of the *Coprinopsis* spp. *in vitro*.

Vegetative growth and fruit-body formation of a late-stage EP fungus (saprobic ammonia fungus) *Tephrocybe tesquorum* are stimulated by gamma-rays sterilized forest soil having been collected from Ao horizon and incubated for five days after application of urea, namely when ammonium-nitrogen concentration in the soil increased (Yamanaka, 2001).

In monoculture, both early-stage EP fungi *Amblyosporium botrytis* and *Ascobolus denudatus*, and late-stage EP fungi *Tephrocybe tesquorum* and *Coprinopsis phlyctidospora*, and LP fungus *Hebeloma vinosophyllum* grow well and form reproductive structures on the urea-treated forest soils sterilized by gamma-rays, when they are cultured on the soils collected at around the occurrence period of each species in the field (Fig. 6; Suzuki, 2006). The vegetative growth rate of mycelium of each ammonia fungus increases when it is cultured in the soils collected at around the occurrence period of each species in the field.

In the five species culture, fruit-body formation of early-stage EP fungus *Ascobolus denudatus*, late-stage EP fungi *Tephrocybe tesquorum* and *Coprinopsis phlyctidospora*, and LP fungus *Hebeloma vinosophyllum* are reduced, but not that of early-stage EP fungus *Amblyosporium botrytis*. This tendency is more drastic in *T. tesquorum*, namely its fruiting is absolutely

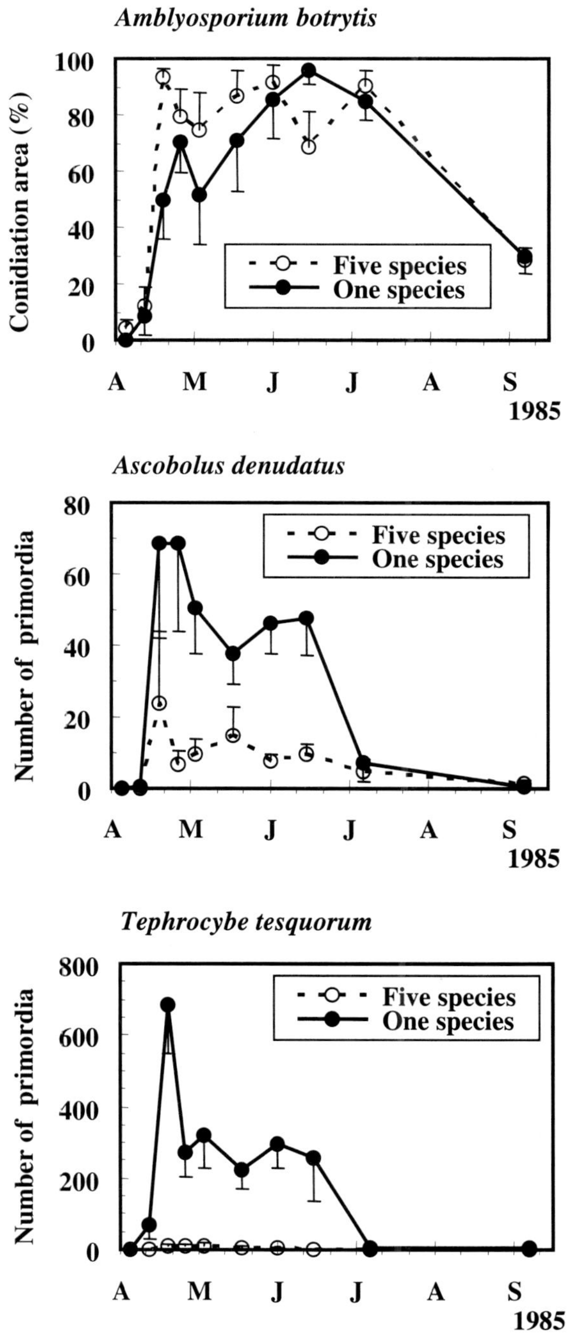
Amblyosporium botrytis
Conidiation area (%)
100
80
60
40
20
0
Five species
One species
A
M
J
J
A
S
1985
Ascobolus denudatus
Number of primordia
80
60
40
20
0
Five species
One species
A
M
J
J
A
S
1985
Tephrocybe tesquorum
Number of primordia
800
600
400
200
0
Five species
One species
A
M
J
J
A
S
1985

Fig. 6

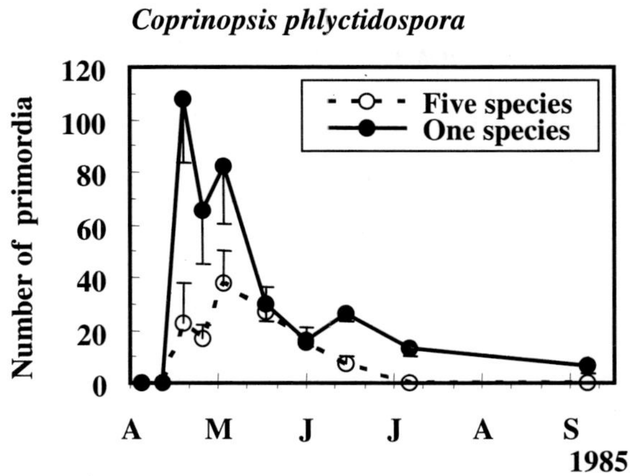

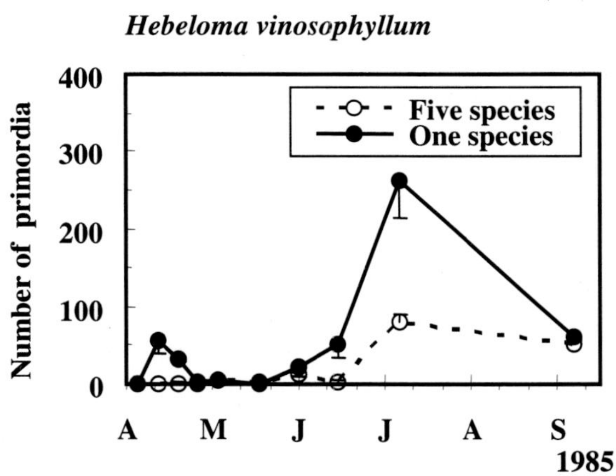

Fig. 6 continued. Structure formation of ammonia fungi on sterilized soils. Urea (800 g/m^2) was applied on April 6, 1985. Isolate(s) inoculated on the sterilized soils collected at different days after urea application was (were) incubated at 20°C. Occurrence period of each ammonia fungus was as follows. *Amblyosporium botrytis* (from April 17, 1985 to June 12, 1985), *Ascobolus denudatus* (from May 1, 1985 to May 15, 1985), *Tephrocybe tesquorum* (from May 8, 1985 to June 26, 1985), *Coprinopsis phlyctidospora* (from May 29, 1985 to September 18, 1985), *Hebeloma spoliatum* (July 17, 1985 to December 22, 1985 and in the autumn of 1986). Detailed information about the experiment is shown in Suzuki, 2006. (From Suzuki, 2006, Mycoscience **47**: 8, Fig. 6 with permission from Springer Ltd. and the Mycological Society of Japan)

inhibited in the five species cultures except for the cultivation of the five fungus species in the urea-treated soils collected at their fruiting periods in the field (Fig 6; Suzuki, 2006).

Co-cultures on malt extract-yeast extract agar media with pH 5.5, 7.0, 8.0 and 9.0 among the early-stage EP fungi *Amblyosporium botrytis*, *Ascobolus denudatus*, *Pseudombrophila petrakii*, and *Peziza moravecii* generally do not inhibit the reproductive structure formation of the opposed EP ammonia fungi. Among the early-stage EP fungi, *Am. botrytis*, *As. denudatus*, and *Pe. moravecii* intermingle with each other, irrespective of pH conditions. The late-stage EP fungus *Tephrocybe tesquorum* inhibits the growth of most of EP fungi under any pH conditions. Another late-stage EP fungus *Coprinopsis phlyctidospora* shows invasion ability to other EP fungi, irrespective of pH conditions, but does not deeply invade into the territories of early-stage EP fungi. LP fungus *Hebeloma vinosophyllum* invades into the territories of early-stage EP fungi at pH 5.5 but is inhibited to grow by late-stage EP fungi, at any pH conditions. *He. vinosophyllum* tends to accelerate fruit-body formation of *C. phlyctidospora* at pH 5.5 and 9.0. *He. vinosophyllum* forms the highest numbers of fruit-bodies at pH 5.5 (Licyayo et al., 2007). These results show that successive occurrence of ammonia fungi is caused by the interspecific interactions among ammonia fungi as well as by the physiological characteristic of each fungus associates with conditions of its inhabiting soils such as pH and nitrogen concentration.

The results obtained from the five species cultures and the co-cultures suggest that the occurrence period of each ammonia fungus in the field partly derived from the interactions with other ammonia fungi. In addition, the establishment of symbiosis with *Hebeloma vinosophyllum* and plant roots (i.e., the formation of ectomycorrhizae) is a propagation strategy of *He. vinosophyllum* to overcome potentially competitive interactions with soil microbes, especially with saprobic ammonia fungi (Fig. 7).

Enzyme activities

Most early-stage EP fungi (saprobic ammonia fungi) have faint ligninolytic activities as well as clear cellulolytic, proteolytic, and lipolytic activities. Most late-stage EP fungi (saprobic ammonia fungi) have clear cellulolytic, ligninolytic, chitinolytic, proteolytic, and lipolytic activities. LP fungi (ectomycorrhizal ammonia fungi) *Hebeloma* spp. have faint cellulolytic, ligninolytic, chitinolytic, proteolytic, and lipolytic activities, but another LP fungus *Laccaria bicolor* has no chitinolytic activities (cf., Enokibara et al., 1993; Yamanaka, 1995b; Soponsathatien, 1998a, 1998b; Ikehata et al., 2004). The pH optima for the cellulolytic enzymes of EP fungi are between 6.8 and 9.0 and those for the cellulolytic enzymes of LP fungi are between 5.5 and 6.8. (Enokibara et al., 1993). Generally speaking, cellulolytic and ligninolytic activities of late-stage EP fungi are stronger than those of LP fungi (Table 3). These results suggest that, in the urea-treated sites, cellulose would be mainly decomposed by the activities of some saprobic ammonia fungi, and lignin would be faintly decomposed by the activities of some saprobic ammonia fungi. These also suggest that the remarkable decomposition of lignin would happen only after the colonization of pre-inhabiting fungi which gradually displace again to ammonia fungi in the site with progress of time after the disturbance by ammonium-nitrogen. Early-stage EP fungi such as *Amblyosporium botryt*is and *Ascobolus denudatus* would probably utilize sugars without decomposition of plant materials treated with a large amount of nitrogenous material(s), namely they may grow as sugar fungi. On the other hand, late-stage EP fungi such as *Coprinopsis echinospora* and *Tephrocybe tesquorum* would be principal decomposers of the litter treated with a large amount of nitrogenous material(s). LP fungi *Hebeloma* spp. probably utilize plant materials by their faint decomposing activities as well as sugars produced by decomposing activities of other microbes, and slowly colonize into the territories of early-stage EP fungi. The *Hebeloma* spp., however, soon establish mycorrhizal symbiosis to evade the competition with other microbes including late-stage EP fungi (saprobic ammonia fungi).

EPILOGUE

The ammonia fungi can happen to colonize into litter just after a disturbance by a high concentration of ammonium-nitrogen. The ammonia fungi probably initiate to colonize from spores and/or small mycelia having existed prior to the disturbance by ammonium-nitrogen (Sagara, 1976b; Suzuki, 2006; Sagara et al., 2008). Each ammonia fungus establishes its territory during high concentration of ammonium-nitrogen amended habitats accompanying weak alkaline to neutral conditions as a consequence of the interactions between other ammonia fungi, since competition with other microbes would be suppressed by the high ammonium-nitrogen concentration accompanying weak alkaline to neutral conditions. The vigorous growth of ammonia fungi causes the declining of pH and ammonium-nitrogen concentration in their inhabiting substratum as a result of the utilization of ammonium-nitrogen. In other words, a large amount of nitrogen is immobilized into the mycelia of ammonia fungi (Sagara, 1992, 1995a). This may accelerate the successive invasion and colonization of ammonia fungi. Sequential occurrence of ammonia fungi would be explained mainly by the time required for reproductive structures of each ammonia fungus and interaction among ammonia fungi (cf., Suzuki, 1989, 2006; Suzuki et al., 2002b; Licyayo et al., 2007). EP fungi, namely saprobic ammonia fungi, would be principal litter and wood decomposers

Table 3. Cellulolytic and ligninolytic enzyme activities of ammonia fungi

	Cellulolytic enzyme			*Ligninolytic enzymes*			*References*
Fungus species	*Avicelase*	*CMase*	*ß-glucosidase*	*Laccase*	*Tyrosinase*	*Peroxidase*	
Anamorphic fungi							
Amblyosporium botrytis			+++	(+)	–	(+)	2, 4
Doratomyces puteredinis				–	–	(+)	2
Ascomycota							
Ascobolus denudatus			++	–	–	(+)	2, 4
Pseudombrophila petrakii			–	–	–	++	2
Peziza moravecii			++	(+)	–	+	2, 4
Basidiomycota							
Coprinopsis cinerea	++(AL)	+++(AL)	++(AL)	(+)	–	(+)	1, 2
Coprinopsis echinospora	+(AL)	+(AL)	+(AL)	+	–	++	1, 2
Coprinopsis neolagopus				(+)	–	(+)	2
Coprinopsis phlyctidospora	++(AL)	+(AL)	+(AL)	++	(+)		1, 2
Crucispora rhombisperma			+	+	–		2, 3
Tephrocybe tesquorum	++(AL)	+(AL)	++(AL)	(+)	–	(+)	1, 2
Collybia cookei				–	–	+	2
Lepista sordida			+	++	–		2, 3

Contd.

Table 3 continued

Fungus species	*Cellulolytic enzyme*			*Ligninolytic enzymes*			*References*
	Avicelase	*CMase*	*ß-glucosidase*	*Laccase*	*Tyrosinase*	*Peroxidase*	
Alnicola lactariolens				+	–		2, 3
Hebeloma radicosoides				(+)	(+)	+	2
Hebeloma spoliatum	–	(+)(WA)	–	+	(+)	++	1, 2
Hebeloma vinosophyllum	+(WA)	+(WA)	–	+	(+)	(+)	1, 2
Laccaria bicolor	+(WA)	–	–	–	–	+	1, 2

+: slightly; '++: weak, +++: rather strong; –: no activity; (+): A few strains of the same species have slight activity; Blank: not examined. AL: Active at weak alkaline range (pH 6.8-9.0); WA: Active at weak acidic and neutral range (pH 5.5-6.8). 1) Enokibara et al. 1993; 2) Soponsathien, 1998a; 3) Soponsathien, 1998b; 4) Yamanaka, 1995b.

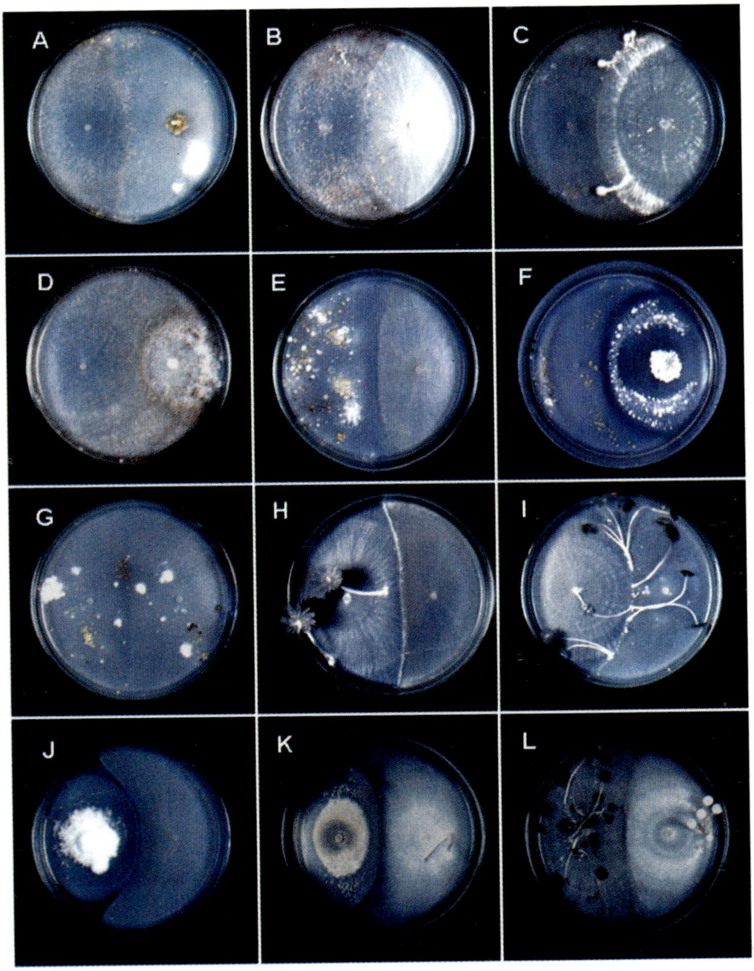

Fig. 7. Interactions between ammonia fungi. Two discs from different fungal species grown on MY agar media were inoculated 55 mm apart on a MY agar plate adjusted at different pH 8.0. Slow-growing species were inoculated earlier than the fast-growing species (Detailed cultivation schedule are shown in Licyayo et al., 2007). **A:** *Amblyosporium botrytis* (left) and *Ascobolus denudatus* (right); **B:** *Am. botrytis* (left) and *Pseudombrophila petrakii* (right); **C:** *Am. botrytis* (left) and *Coprinopsis phlyctidospora* (right); **D:** *Am. botrytis* (left) and *Peziza moravecii* (right); **E:** *As. denudatus* (left) and *Ps. petrakii* (right), **F:** *As. denudatus* (left) and; *Tephrocybe tesquorum* (right) **G:** *As. denudatus* (left) and *As. denudatus* (right); **H:** *C. phlyctidospora* (left) and *Ps. petrakii* (right); **I:** *C. phlyctidospora* (left) and *As. denudatus* (right); **J:** *T. tesquorum* (left) and *Ps. petrakii* (right); **K:** *T. tesquorum* (left) and *Hebeloma vinosophyllum* (right); **L:** *C. phlyctidospora* (left) and *He. vinosophyllum* (right)

under high concentration of ammonium-nitrogen with neutral to weak alkaline conditions that were caused by a sudden application of nitrogenous material(s) (Suzuki, 2006). On the other hand, ectomycorrhizal ammonia fungi are pioneer fungi of secondary fungus succession (Suzuki, 2002). Namely, ectomycorrhizal ammonia fungi first establish mycorrhizal symbiosis in the habitat where the mycorrhizal symbiosis had been once destroyed by the high concentration of ammonium-nitrogen and, instead of the pre-inhabiting mycorrhizal fungi, and they would sustain the nutrients cycle during the habitat having a high concentration of ammonium-nitrogen with neutral to weak alkaline conditions. In 1995, Sagara (Sagara, 1995a) proposed cleaning symbiosis for the ecological role of ectomycorrhizal postputrefaction fungi which colonize in mammalians deserted middens and this concept comprises of a nutrient cycle in animal waste decomposing sites. In other words, in an ecosystem, ammonia fungi would have a role in keeping the carbon cycles instead of pre-inhabiting fungi after the addition of a large amount of nitrogenous material(s) (Suzuki, 2006). These indicate that the nutrient cycle in an ecosystem is not interrupted even after a sudden application of a large amount of nitrogenous material(s) in the field, since the nutrient cycle conducting by litter-decomposing fungi and wood-rotting would be replaced by postputrefaction fungi comprising ammonia fungi. The replacement described may be viewed as a kind of 'compensation process' in a nutrient cycle (Suzuki, 2002, 2006).

References

Clémançon, H. and Hongo, T. 1994. Notes on three Japanese Agaricales, Mycoscience, 36: 53-58.

Cooke, R.C. and Rayner, A.D.M. 1984. Ecology of Saprotrophic Fungi, Longman, London, UK.

Deng, Z. and Suzuki, A. 2008. Effects of pH, NH_4-N concentration, temperature, and storage period on basidiospore germination in an ectomycorrhizal ammonia fungus *Hebeloma vinosophyllum*. Mycoscience, 49: 178-184.

Dickinson, C.H. and Pugh, G.J.F. (eds.) 1974. Biology of Plant Litter Decomposition, Vol. I, II, Academic Press, New York, USA.

Dix, N.J. and Webster, J. 1995. Fungal Ecology, Chapman & Hall, London, UK.

Enokibara, S., Suzuki, A., Fujita, C., Kashiwagi, M., Mori, N. and Kitamoto, Y. 1993. Diversity of pH spectra of cellulolytic enzymes in Basidiomycetes. Transactions of the Mycological Society of Japan, 34: 221-228 (In Japanese with English summary).

Frankland, J.C., Hedger, J.N. and Swift, M.J. (eds.) 1982. Decomposer Basidiomycetes: Their Biology and Ecology. Cambridge University Press, Cambridge, UK.

Fukiharu, T. and Hongo, T. 1995. Ammonia fungi of Iriomote Island in the southern Ryukyus, Japan and a new ammonia fungus, *Hebeloma luchuense*. Mycoscience, 36: 425-430.

Fukiharu, T. and Horigome, R. 1996. Ammonia fungi in the Abukuma Mountains and its biogeographical distribution around Japan, Memoirs of the National Museum, Tokyo, No. 29: 105-112 (In Japanese with English summary).

Fukiharu, T., Sato, Y. and Suzuki, A. 1997. The occurrence of ammonia fungi, and changes in soil conditions and decay rate of bamboo in response to application of a large amount of urea in a bamboo grove in Chiba Prefecture, central Japan, The Bulletin of the Faculty of Education. Chiba University 45 (III: Natural Sciences): 61-67.

Fukiharu, T., Osaku, K., Iguchi, K. and Asada, M. 2000a. Occurrence of ammonia fungi on the forest ground after decomposition of a dog carcass. Natural History Research 6 (1): 9-14.

Fukiharu, T., Yokoyama, G. and Oba, T. 2000b. Occurrence of *Hebeloma vinosophyllum* on the forest ground after decomposition of crow carcass. Mycoscience, 41: 401-402.

Gams, W. 1992. The analysis of communities of saprophytic microfungi with special reference to soil fungi, pp. 183-223. In: Fungi in Vegetation Science, Handbook of Vegetation Science Vol. 19/1, W. Winterhoff (ed.), Kluswer Academic Publishers, Dordrecht, The Netherlands.

Harmaja, H. 2002. *Alciphila vulgaris*, a new genus and species of Deutromycetes, Karstenia, 42: 33-38.

He, X.M. and Suzuki, A. 2003. Effect of nitrogen resources and pH on growth and fruit body formation of *Coprinopsis phlyctidospora*. Fungal Diversity No. 12: 35-44.

He, X.M. and Suzuki, A. 2004. Effects of urea treatment on litter decomposition in *Pasania edulis* forest soil. Journal of Wood Science, 50: 266-270.

Hirsch, G. and Braun, U. 1992. Communities of parasitic microfungi pp. 225-250. In: Fungi in Vegetation Science. W. Winterhoff (ed.), Kluswer Academic Publishers, Dordrecht, The Netherlands.

Hudson, H.J. 1980. Fungal Saprophytism. 2nd edition, Edward Arnold, London, UK.

Ikehata, K., Buchanan, I.D. and Smith, D.W. 2004. Extracellular peroxidase production by *Coprinus* species from urea-treated soil. Canadian Journal of Microbiology, 50: 57-60.

Imamura, A. 2001. Report on *Laccaria amethystina*, newly confirmed as an ammonia fungus. Mycoscience, 42: 623-625.

Imamura, A. and Yumoto, T. 2004. The time of urea treatment and its effects on the succession of ammonia fungi in two warm temperate forests in Japan. Mycoscience, 45: 123-130.

Kasuya, T. 2002. Reports on fruiting of ammonia fungi on the ground after decomposition of animal wastes. Nippon Kingakukai Kaiho, 43: 99-104 (In Japanese with English summary).

Kirk, P.M., Cannon, P.F., David, J.C. and Stalpers, J.A. 2008. Ainsworth & Bisby's Dictionary of the Fungi, 10th edition. CAB International, Oxon, UK.

Laiho, O. 1970. *Paxillus involutus* as a mycorrhizal symbiont of forest trees. Acta Forestalia Fennica, 106: 1-72.

Lehmann, P.F. and Hudson, H.J. 1977. The fungal succession on normal and urea-treated pine needles. Transactions of the British Mycological Society, 68: 221-228.

Licyayo, D.C. and Suzuki, A. 2006. Growth responses of ammonia fungi to different concentration of ammonium-nitrogen. Mushroom Science and Biotechnology, 14: 145-156.

Licyayo, D.C.M., Suzuki, A. and Matsumoto, M. 2007. Interactions among ammonia fungi on MY agar medium with varying pH. Mycoscience, 48: 20-28.

Lisiewska, M. 1992. Macrofungi on special substrates, pp. 151-182. In: Fungi in Vegetation Science, Handbook of Vegetation Science Vol. 19/1. W. Winterhoff (ed.), Kluswer Academic Publishers, Dordrecht, The Netherlands.

Miller, O.K. Jr, and Hilton, R.N. 1986. New and interesting agarics from Western Australia. Sydowia, 39: 126-137.

Morimoto, N., Suda, S. and Sagara, N. 1981. Effect of ammonia on fruit-body induction of *Coprinus cinereus* in darkness. Plant and Cell Physiology, 22: 247-254.

Morimoto, N. Suda, S. and Sagara, N. 1982. The effects of urea on the vegetative and reproductive growth of *Coprinus stercorarius* in pure culture. Transactions of the Mycological Society of Japan, 23: 79-83.

Nagao, H., Udagawa, S., Bougher, N.L., Suzuki A. and Tommerup I.C. 2003. The genus *Thecotheus* (Pezizales) in Australia: *T. urinamans* sp. nov. from urea-treated jarrah (*Eucalyptus marginata*) forest. Mycologia, 95: 688-693.

Richards, B.N. 1987. The Microbiology of Terrestrial Ecosystems. Longman Scientific & Technical, Essex, UK.

Sagara, N. 1973. Proteophilous fungi and fire place fungi (A preliminary report) Transactions of the Mycological Society of Japan, 14: 41-46.

Sagara, N. 1975. Ammonia fungi—a chemoecological grouping of terrestrial fungi. Contributions from the Biological Laboratory, Kyoto University, 24: 205-276, 7 plates.

Sagara, N. 1976a. Growth and reproduction of the ammonia fungi, pp. 153-178. In: Ecology of Microorganisms (3). K. Biseibutsuseitai-Kenkyukai (ed.). University of Tokyo Press, Tokyo, Japan (In Japanese).

Sagara, N. 1976b. Supplement to the studies of ammonia fungi (I). Transactions of the Mycological Society of Japan 17: 418-428 (In Japanese with English summary).

Sagara, N. 1976c. Presence of a buried mammalian carcass indicated by fungal fruiting bodies. Nature, 262: 816.

Sagara, N. 1977. Detecting dead bodies in 1718 and 1976. Nature, 269: 284.

Sagara, N. 1978. The occurrence of fungi in association with wood mouse nests, Transactions of the Mycological Society of Japan, 19: 201-214.

Sagara, N. 1981. Occurrence of *Laccaria proxima* in the grave site of a cat. Transactions of the Mycological Society of Japan, 22: 271-275.

Sagara, N. 1984. On "Corpse finder." McIlvainea, 6(2): 7-9.

Sagara, N. 1989a. Mushrooms and Animals, Tsukiji-shokan, Tokyo, Japan (In Japanese).

Sagara, N. 1989b. European record of the presence of a mole's nest indicated by a particular fungus. Mammalia, 53: 301-305.

Sagara, N. 1992. Experimental disturbances and epigeous fungi, pp. 427-454. In: The Fungal Community—Its Organization and Role in the Ecosystem. 2nd edition. G.C. Carroll and D.T. Wicklow (eds.), Marcel Dekker. New York, USA.

Sagara, N. 1995a. Association of ectomycorrhizal fungi with decomposed animal wastes in forest habitats: a cleaning symbiosis?, Canadian Journal of Botany 73 (Supplement 1): S1423-S1433.

Sagara, N. 1995b. A request—pinpoint marking of *Hebeloma radicosum* fruiting sites. Mycologist, 9: 128.

Sagara, N. 1998. Methods for studying the nesting ecology of moles through observation of mushroom fruiting. Honyurui Kagaku (Mammalian Science) 38: 271-292 (In Japanese with English summary).

Sagara, N. 1999. Mycological approach to the natural history of talpid moles: a review with new data and proposed of "habitat-cleaning symbiosis," pp. 33-55. In: Recent Advances in the Biology of Japanese Insectivore. Y. Yokohata and S. Nakamura (eds.), Hiba Society of Natural History, Shoubara, Japan.

Sagara, N. and Hamada, M. 1965. Responses of higher fungi to some chemical treatments of forest ground. Transactions of the Mycological Society of Japan, 6: 72-74.

Sagara, N. and Abe, H. 1993. A case of late breeding in the mole *Mogera kobeae* and its nest. Journal of the Mammalogical Society of Japan, 18: 53-59.

Sagara, N., Abe, H. and Okabe, H. 1993a. The persistence of moles in nesting at the same site as indicated by mushroom fruiting and nest reconstruction, Canadian Journal of Zoology, 71: 1690-1693.

Sagara, N., Honda, S., Kuroyanagi, E. and Takayama, S. 1981. The occurrence of *Hebeloma spoliatum* and *Hebeloma radicosum* on the dung-deposited burrows of *Utrotrichus talpodes* (shrew mole). Transactions of the Mycological Society of Japan, 22: 441-455.

Sagara, N., Kitamoto, Y., Nishio, R. and Yoshimi, S. 1985. Asssociation of two *Hebeloma* species with decomposed nests of vespine wasps. Transactions of the British Mycological Society, 84: 349-352.

Sagara, N., Murakami, Y. and Clémançon, H. 1988. Association of *Hebeloma radicosum* with a nest of the wood mouse *Apodemus*. Mycologia Helvetica 3 (1): 27-36.

Sagara, N., Kobayashi, S., Ota, H., Itsubo, T. and Okabe, H. 1989. Finding *Euroscaptor mizura* (Mammalia: Insectivora) and its nest from under *Hebeloma radicosum* (Fungi: Agaricales) in Ashiu, Kyoto, with data of possible contiguous occurrence of three talpine species in this region. Contributions from the Biological Laboratory, Kyoto University, 27: 261-272.

Sagara, N., Okabe, H. and Kikuchi, I. 1993b. Occurrence of an agaric fungus *Hebeloma* on the underground nest of wood mouse. Transactions of the Mycological Society of Japan, 34: 315-322.

Sagara, N., Hongo, T., Murakami, Y., Hashimoto, T., Nagamatsu, H., Fukiharu, T. and Asakawa, Y. 2000. *Hebeloma radicosoides* sp. nov., an agaric belonging to the chemoecological group ammonia fungi. Mycological Research, 104: 1017-1024.

Sagara, N., Senn-Irlet, B. and Marstad, P. 2006. Establishment of the case of *Hebeloma radicosum* growth on the latrine of the wood mouse. Mycoscience, 47: 263-268.

Sagara, N., Yamanaka, T. and Tibbett, M. 2008. Soil fungi associated with graves and latrines: Toward a forensic mycology, pp. 67-107. In: Soil Analysis in Forensic Taphonomy-Chemical and Biological Effects of Buried Human Remains. M. Tibbett and D.O. Carter (eds.), CRC Press, New York, USA.

Sato, Y. and Suzuki, A. 1997. The occurrence of ammonia fungi, and changes in soil conditions and wood decay rate in response to application of a large amount of urea in a *Quercus serrata* dominated mixed forest in Meguro. Tokyo, The Bulletin of Faculty of Education, Chiba University 45 (III: Natural Sciences): 53-59.

Soponsathien, S. 1998a. Some characteristics of ammonia fungi 1. In relation to their ligninolytic enzyme activities. Journal of General and Applied Microbiology, 44: 337-345.

Soponsathien, S. 1998b. Study on the production of acetyl esterase and side-group cleaving glycosidases of ammonia fungi. Journal of General and Applied Microbiology, 44: 389-397.

Suzuki, A. 1978. Basidiospore germination by aqua ammonia in *Hebeloma vinosophyllum*, Transactions of the Mycological Society of Japan, 19: 362.

Suzuki, A. 1989. Analyses of factors affecting the occurrence and succession of the ammonia fungi, pp. 275-279. In: Recent Advances in Microbial Ecology (ISME 5). T. Hattori, Y. Ishida, Y. Maruyama, R.Y. Morita, and A. Uchida (eds.), Japan Scientific Societies Press, Tokyo, Japan.

Suzuki, A. 1992. Physiological characteristics of the ammonia fungi and their role in bioscience. Proceedings of the Asian Mycological Symposium, Seoul, ROK. pp. 238-247.

Suzuki, A. 2000. A survey of species assemblage of ammonia fungi. Journal of Plant History—Planta. No. 68: 27-35 (In Japanese).

Suzuki, A. 2002. Fungal succession at different scale. Fungal Diversity No. 10: 11-20.

Suzuki, A. 2004. Physiology and ecology of ammonia fungi and their biogeographical distribution, pp. 5-12. In: Natural History of Fungi. T. Kobayashi, and H. Takahashi (eds.), The Museum of Hokkaido University, Asahitosho Kanko Center, Sapporo, Japan (In Japanese).

Suzuki, A. 2006. Experimental and physiological ecology of ammonia fungi: studies using natural substances and artificial media. Mycoscience, 47: 3-17.

Suzuki, A. 2009. Propagation strategy of ammonia fungi. Mycoscience, 50: 39-51.

Suzuki, A., Motoyoshi, N. and Sagara, N. 1982. Effects of ammonia, ammonium salts, urea, and potassium salts on basidiospore germination in *Coprinus cinereus* and *Coprinus phlyctidosporus*. Transactions of the Mycological Society of Japan, 23: 217-224.

Suzuki, A., Tommerup, I.C., and Bougher, N.L. 1998. Ammonia fungi in the jarrah forest of western Australia and parallelism with other geographical regions of the world, Proceedings of the 2nd International Conference on Mycorrhiza. Upsala, Sweden.

Suzuki, A., Tanaka, C., Bougher, N., L., Tommerup, I.C., Buchanan, P.K., Fukiharu, T., Tsuchida, S., Tsuda, M., Oda, T., Fukada, J. and Sagara, N. 2002a. ITS rDNA variation of the *Coprinopsis phlyctidospora* (Syn.: *Coprinus phlyctidosporus*) complex in the Northern and Southern Hemispheres, Mycoscience, 43: 229-238.

Suzuki, A., Uchida, M. and Kita, M. 2002b. Experimental analyses of successive occurrence of ammonia fungi in the field. Fungal Diversity No. 10: 141-165.

Suzuki, A., Fukiharu, T., Tanaka, C., Ohono, T. and Buchanan, P.K. 2003. Saprobic and ectomycorrhizal ammonia fungi in the Southern Hemisphere, New Zealand. Journal of Botany, 41: 391-406.

Tibbett, M. and Carter, D.O. 2003. Mushrooms and taphonomy: the fungi that mark woodland graves. Mycologist News 17 (Part 1): 20-24.

Wang Y.-Z., and Sagara, N. 1997 *Peziza urinophila*, a new ammonophilic discomycete. Mycotaxon, 65: 447-452.

Yamanaka, T. 1995a. Nitrification in a Japanese pine forest soil treated with a large amount of urea. Journal of Japanese Forest Society, 77: 232-238.

Yamanaka, T. 1995b. Changes in organic matter composition of forest soil treated with a large amount of urea to promote ammonia fungi and the abilities of these fungi to decompose organic matter. Mycoscience, 36: 17-23.

Yamanaka, T. 1995c. Changes in soil conditions following treatment with a large amount of urea to enhance fungal fruiting-body production in a Japanese red pine forest. Bulletin of Japanese Society of Microbial Ecology, 10: 67-72.

Yamanaka, T. 1999. Utilization of inorganic and organic nitrogen in pure cultures by saprotrophic and ectomycorrhizal fungi producing sporophores on urea-treated forest floor. Mycological Research, 103: 811-816.

Yamanaka, T. 2001. Fruit-body production and mycelial growth of *Tephrocybe tesquorum* in urea-treated forest soil. Mycoscience, 42: 333-338.

Yamanaka, T. 2003. The effect of pH on the growth of saprotrophic and ectomycorrhizal ammonia fungi *in vitro*. Mycologia, 95: 584-589.

9

Prospecting for Novel Enzyme Activities and Their Genes in Filamentous Fungi from Extreme Environments

Helena Nevalainen[1, 2], **Junior Te'o**[1, 2] **and Ron Bradner**[1, 3]

[1]Department of Chemistry and Biomolecular Sciences
Macquarie University, Sydney, NSW 2109, Australia
E-mail: hnevalai@els.mq.edu.au

[2]Applimex Systems Pty Ltd.
Level 2, 4 Park Drive, Macquarie University,
Sydney, NSW 2109 Australia

[3]Grain Foods CRC Ltd.
Riverside Corporate Park, 1 Rivett Road,
North Ryde, NSW 2113 Australia

Abstract

Some 4000 enzymes are known today and of these about 200 are in commercial use with an estimated value of US$2.3 billion in 2007. The majority of today's industrial enzymes are microbial in origin and about 40% of these are produced from fungal hosts. Prospecting for filamentous fungi exhibiting novel bioactivities or genes with economic potential has not been as vigorous as with bacteria, for example, but is expanding rapidly. Current investigations reveal that the more unusual and extreme regions of the planet play host to a previously unrecognized wealth of microfungal material, in particular those with the potential to yield cold-active and thermophilic enzymes. Recent oil crises and increasing prices at the local pump stations have promoted global research into alternative sources for fuel. In particular, this has invigorated the search for fungi capable of producing a set of enzymes that can effectively convert plant biomass to fermentable sugars. A complement to classical prospecting by cultivation is the molecular approach of screening for enzyme-encoding DNA sequences. This PCR based strategy combined with chromosomal walking PCR has been successfully used for the isolation and heterologous expression of a number of enzyme-encoding genes. This latter approach can be further supported with the expanding volume of global sequence data becoming available from the various fungal genome sequencing projects, and from recent developments in proteomics and mass spectrometry for protein identification.

INTRODUCTION

There are nearly 4000 enzymes known today and of these, about 200 are in commercial use (Sharma et al., 2001). The hydrolases, which feature carbohydrases (amylases, cellulases and hemicellulases), esterases (lipases) and proteases represent at least 75% of all industrial enzymes and are used for the degradation of various natural substances (Godfrey and West, 1996). Subsequently, the commercial value of industrial enzymes applied worldwide was estimated at US$2.3 billion by the end of 2007 (http://bccresearcg.com/report/BIO030E.html). The majority of today's industrial enzymes are microbial in origin and it has been estimated that some 40% of the production is derived from fungal sources.

Prospecting for novel genes and bioactivities has been extensively carried out with bacteria across varying environments including different extreme sites such as Antarctic soils and lakes, hot pools, caves, hypersaline locations and deep sea vents (Bergquist et al., 1987; Cavicchioli et al., 2002; Farrell et al., 2003; Gilbert et al., 2004; Gomes and Steiner, 2004; Guezennec, 2002; Park et al., 2004) but to a notably lesser amount with filamentous fungi. Reports on the environmental prospecting for genes and gene products in microorganisms describe studies targeting cellulase enzyme producers in soil (reviewed in Ohmiya et al., 2003), finding novel bioactivities in aqueous environments (Liu et al., 2002; Wagner-Döbler et al., 2002) and screening Antarctic fungal isolates for various cold-active enzymes of industrial interest (Bradner et al., 1999a). The prospecting activities reflect the broader range of habitats that bacteria are capable of living in, for example, in terms of temperature. Where there are quite a few bacteria and archaea that can grow in temperatures in excess of 70°C, one would be hard pressed to find fungi that can tolerate temperatures above 55°C (Maheshwari et al., 2000).

The main strategies applied in gene prospecting have also changed considerably over the last 10 years. The traditional approach to prospecting with microorganisms is based on culturing the organisms, either isolated from a particular location of interest or lifted from a culture collection and

Table 1. Milestones in the development of *Trichoderma*

Year	*Development*	*Reference*
1940-41	Discovery of *Trichoderma*	http://www.thebioenergysite.com/news/617/the-discovery-of-trichoderma-reesei
1950	The "C_1-C_X" concept for cellulose hydrolysis	Reese et al. (1950)
1957	Production of cellulases in liquid culture	Mandels and Reese (1957)
1971-81	Isolation of hypercellulotyic mutant strains	Reviewed in Nevalainen et al. (1994)
1983	Cloning of the *cbh1* gene encoding the major cellulase	Shoemaker et al. (1983); Teeri et al. (1983)
1986	Two domain sructure of cellobiohydrolase I	Van Tilbeurgh et al. (1986)
1987	Genetic transformation of *Trichoderma*	Penttilä et al. (1987)
1990	3D structure of the CBHI core	Bergfors et al. (1990)
1993	Targeted integration in transformation	Karhunen et al. (1993); Suominen et al. (1993)
1993	*Trichoderma* heterologous host	Nyyssönen et al. (1993)
2001	First published study in *Trichoderma* proteomics	Lim et al. (2001)
2003	*Trichoderma* genome	http://genome.jgi-psf.org/Trire2/Trire2.home.html

proceeding from there. The modern way engages direct PCR amplification of particular target genes which may, for example, originate from genomic biomass in the environment. Further studies and potential industrial application of the PCR-amplified DNA would require cloning the DNA into suitable vectors and expressing the desired gene(s) in a surrogate host(s). This latter approach has circumvented the requirement to first culture a microorganism or to construct and test an entire gene library for the identification and further exploitation of the gene(s) of interest. A new source for gene prospecting is the genome, several of which have emerged from the global genome sequencing programs, now also featuring several filamentous fungi (see Table 1). Therefore, a considerable amount of gene prospecting is currently being carried out *in silico* in association with various genome databases intent on finding suitable targets and to aid in primer design.

Environmental prospecting of fungi: A historical view

Fungi are abundant in soil where they degrade recalcitrant natural polymers such as lignocelluloses and parasitize other organisms such as insect larvae, nematodes and other microorganisms (e.g. Hay et al., 2002; Rabinovich et al., 2004; Scholte et al., 2004; Martinez et al., 2005). This rather aggressive lifestyle is made possible by their excellent ability to produce and secrete various hydrolytic enzymes, for example cellulases, xylanases, ligninases, proteases, chitinases and lipases, all of which are also of considerable industrial interest. A number of soil fungi produce volatile compounds; 2-methyl-1-propanol, 3-methyl-1-butanol and the malodorous 2-methyl-isoborneol and geosmin are some common volatiles from fungi such as *Aspergillus* sp., *Fusarium* sp. and *Penicillium* sp. (Schnurer et al. 1999). Antibiotics are also produced, which contribute to overall competitiveness of fungi in the soil environment. A classical example of an antibiotic-producing fungus is *Penicillium chrysogenum*, originally discovered as a contaminant on a plate of *Staphylococcus* culture (reviewed in Demain and Elander, 1999). Fungi are also avid spore-forming organisms which helps them survive in less favorable environments and over temporary adverse conditions or seasonal variations. Recently, a phylogenetic study into microbial communities in tundra soil revealed high biodiversity and three novel fungal clades that constitute new groups of fungi across the major fungal groups (Schadt et al., 2003). A novel species of a filamentous fungus, *Embellisia* sp. has been isolated from a bryophyte collected in the Antarctica (Bradner et al., 2000), showing that there is plenty of room for discovery concerning fungi and their activities in extreme environments.

On the lookout for cellulose-degrading fungi

Prospecting for cellulose degrading enzymes and genes encoding them has, until recently, been based on the premise that the candidate producers are cultivable. The big boost into finding efficient cellulases came from the oil crisis in the early 1970s. This emergency initiated global research into renewable sources for fuel, especially plant biomass, and organisms capable of degrading the cellulosic biomass to glucose (reviewed in Enari and Markkanen, 1977; Virendra et al., 1981). This line of prospecting has since paved the way for the design of some of the highest enzyme-secreting strains and most powerful cell factories currently available (reviewed in Mäntylä et al., 1998). The discovery of the cellulolytic fungus *Trichoderma reesei* during World War II led first to a better understanding of the mechanism for cellulose degradation followed by production of improved cellulase secreting strains, isolation of the cellulase-encoding genes and their regulatory regions, fungal transformation and to solving the three dimensional structures of the main cellulases (Table 1).

To date, there are over a hundred cellulase-encoding genes isolated from fungi and related information (e.g. at the protein level) on a number of the corresponding gene products which can be found at http://srs.ebi.ac.uk, with key words such as 'endoglucanase and fungi' or 'cellulase and fungi' used as input in the search window. Also present are many more sequences which have been tentatively labelled as cellulase-encoding, but only by sequence similarity comparison.

A good amount of current knowledge on the molecular biology and mechanisms of cellulose degradation has originated from the studies of the ascomycetous soft rot fungus *Trichoderma reesei* (*Hypocrea jecorina*) (reviewed in Teeri et al., 1992; Koivula et al., 1998; Nevalainen and Penttilä, 2004), *Penicillium pinophilum* (e.g. Wood et al., 1989), the thermophilic fungi *Humicola grisea* (Schülein et al., 1993; Takashima et al., 1996; Takashima et al., 1999) and *H. insolens* (Varrot et al., 1999; Davies et al., 2000; Boisset et al., 2001) and the white rot fungus *Phanerochaete chrysosporium* (reviewed in Kirk and Cullen, 1998; Munoz et al., 2001). Cellulase-encoding genes have also been isolated from edible basidiomycetes such as the common mushroom *Agaricus bisporus* (Yagüe et al., 1997 and references therein) and anaerobic rumen fungi, for example *Neocallimastix patriciarum* (Reymond et al., 1991; Denman et al., 1996; see references in Garcia-Vallvé et al., 2000).

The new wave of prospecting for genes from cultivable cellulolytic fungi has targeted particular characteristics such as alkaline cellulases from *Humicola insolens* (Schülein et al., 1993), *Myceliophora thermophila* (von der Osten and Schülein, 1999) and *Chrysosporium lucknowense* (van Zeijl et al., 2001) or neutral cellulases from *Melanocarpus albomyces* (Miettinen-Oinonen et al., 2004) which have found applications in e.g. enzymatic denim washing, biostoning (Miettinen-Oinonen, 2004). In many of these cases, the studies have progressed to gene isolation and improved expression of the gene product of interest either in the original host or a heterologous expression system.

In addition to cellulolytic activities, cultivable filamentous fungi are currently being searched for other enzyme activities and genes involved in lignocellulose degradation, for example, extracellular lignin peroxidases (reviewed in Eriksson, 1997) and genes encoding enzymes that degrade hemicellulose for applications in the pulp and paper industry (Suurnäkki et al., 1997). Molecular approaches are discussed further below.

Fungi in cold environments, cold active enzymes

The microbiota present in the more frigid regions of the planet, in particular the Arctic and Antarctic have attracted some level of attention in the past, however a detailed examination of the filamentous fungi present in these environments has been largely overlooked. The modest research carried out so far indicates that fungi represent a large and so far untapped bioresource of gene products with the potential to produce a broad spectrum of novel cold-adapted enzymes. The means by which these fungi tolerate and survive in the extreme environmental conditions which, in addition to the intense cold, can also include low water availability, high salinity, broad ranges in pH and the potential damaging effects of ultra violet radiation, are highly complex. The response to the stresses imposed on the fungi may lead to changes to the intracellular trehalose and polyol concentrations and membrane lipids or to the synthesis of new anti-freeze and other proteins and cold active enzymes (reviewed in Robinson, 2001).

The distribution of fungi in Antarctica is most often associated with strong biotic influences such as birds (penguin rookeries) and vegetation (bryophyte and lichen communities). Within the Antarctic environment, moss is one of the microhabitats richest in fungi (Table 2; Tosi et al., 2002). About 230 different species of fungi have been recorded in maritime and continental Antarctica excluding sub-Antarctic (Onofri, 1998). From these, some 50 odd appear to be new Antarctic species, the remainder are cosmopolitan (Onofri et al., 1999).

Antarctic microfungi have generally been found to be psychrotrophic (mesophilic psychrotolerant) rather than psychrophilic (Azmi and Seppelt, 1997; Kerry, 1990; Zucconi et al., 1996), which, in view of the apparent cosmopolitan origins of many of the species isolated, is suggestive of adaptations to their cold environment.

There is a growing interest in cold adapted enzymes due to their potential in industrial applications. Enzymes produced by cold-adapted microorganisms display a high catalytic efficiency that offers considerable potential to the biotechnology industry, making it possible to reduce the amount of enzyme needed in catalytic processes (Demirjian et al., 2001). The low temperature optimum and temperature stability makes them preferable for use in reactions performed at low temperatures that require an easy inactivation (Yamamura et al., 1999).

From the biotechnological point of view, cold-adapted enzymes, which

Table 2. Fungal genera reported to be associated with Antarctic bryophytes

Acremonium[2, 3]	*Alternaria*[1]	*Arthrobotrys*[3, 4]
Arthroderma[2]	*Aspergillus*[1]	*Aureobasidium*[1, 4]
Cadophora[4]	*Chaetomium*[4]	*Chalara*[2]
Chaunopycnis[2]	*Chrysosporium*[1, 3]	*Cladosporium*[1, 4]
Conidiobolus[4]	*Cylindrocarpon*[2]	*Epicoccum*[1]
Geomyces[2, 3, 4]	*Monascella*[2]	*Monocillium*[2]
Mortierella[1, 2, 3, 4]	*Mucor*[1, 3]	*Mycelia sterilia*[2, 3]
Myrioconium[2]	*Nectria*[3]	*Paecilomyces*[2, 4]
Penicillium[1, 2, 3, 4]	*Phaeosphaeria*[2]	*Phaeoseptoria*[2]
Phialophora[1, 2, 3]	*Phoma*[2, 3, 4]	*Phomopsis*[2]
Pycnostysanus[2]	*Rhodotorula*[4]	*Scolecobasidium*[4]
Thelebolus[2, 3, 4]	*Tolypocladium*[2]	*Trichoderma*[1]
Verticillium[4]		

[1]Greenfield and Wilson, 1981; [2]Möller and Dreyfuss, 1996; [3]Azmi and Seppelt, 1998; [4]Tosi et al., 2002; [5]Bradner et al., 2000; [6]Bradner, 2003.

Antarctic fungi can potentially supply, could in the future replace their mesophilic counterparts and help to establish new improved bioprocesses under low temperature conditions. These enzymes can help to save energy, to save labile or volatile compounds, to prevent the growth of mesophilic contaminants at low temperature, and could be easily inactivated by moderate temperatures.

Enzyme activities from Antarctic fungal isolates

As already indicated above, the absence of any inventory of Antarctic filamentous fungi has resulted in a concomitant paucity of information on the extent or potential of their enzymes. Fenice and his colleagues (1997) investigated the extracellular enzymatic production of some 33 Antarctic fungal strains isolated from sites in the general region of the Italian Antarctic base at Terra Nova (Victoria Land). We have also looked at several enzyme activities produced and secreted by a number of Antarctic fungal isolates recovered from ornithogenic soils (Windmill Islands, Eastern Antarctica), bryophytes (Dry Valleys/Ross Island and McMurdo Sound, Southern Antarctica Victoria Land) and soils contaminated with petroleum (Ross Island, Marble Pt and McMurdo Dry Valleys) (Bradner et al., 1999a; Bradner et al., 1999b; Bradner, 2003). Based on the analyses, isolates from a penguin colony would provide a good source for example of cold-active chitinases and mannanases (Fig. 1). It also appears that fungi isolated from the petroleum-contaminated soil produce considerable amounts of cold-

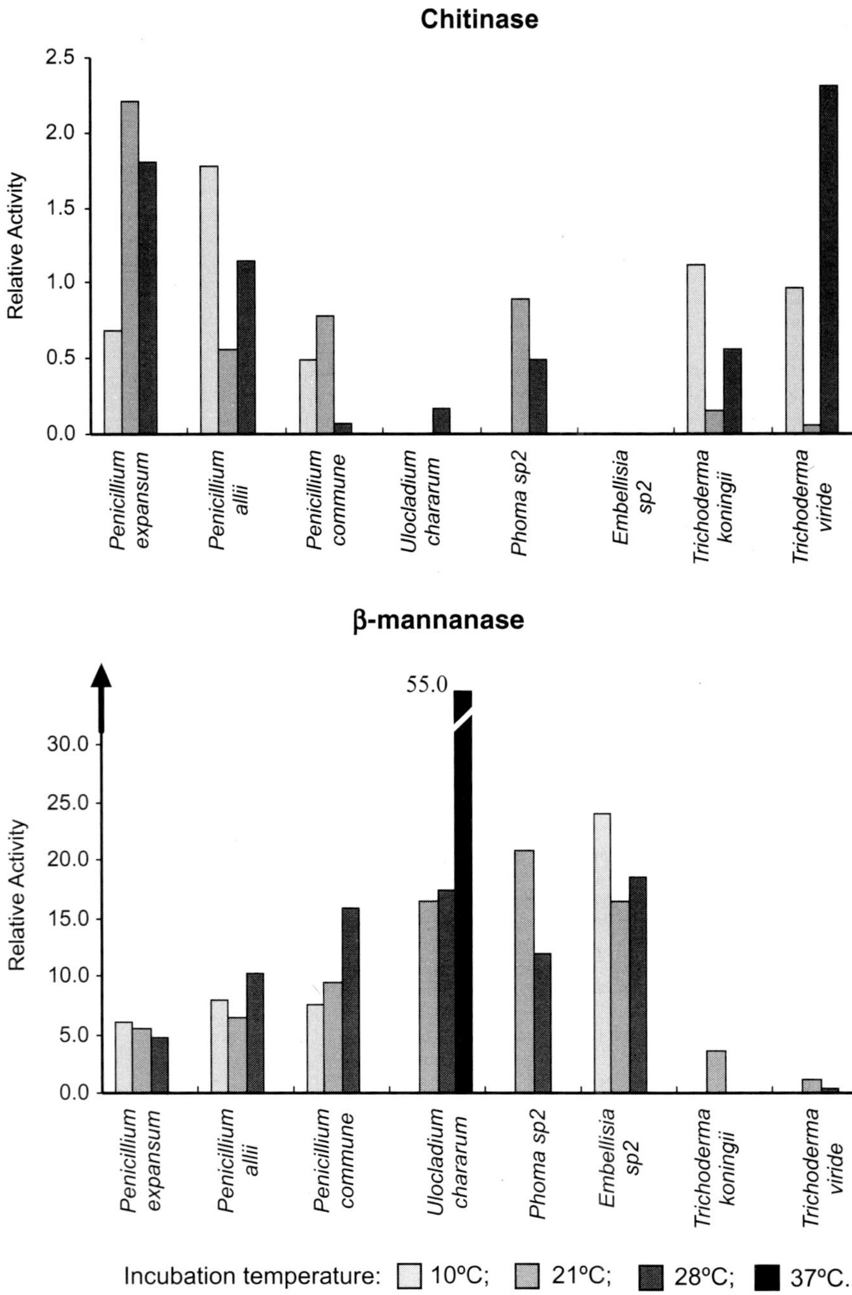

Fig. 1. Chitinase and mannanse activities from selected Antarctic fungi isolated from ornithogenic soils. Adapted from Bradner et al. (1999a)

active chitinases and some are capable of utilizing 6% crude oil as a carbon source at low temperatures in the range of 4-15°C (unpublished).

Further studies into the pH and temperature behavior of the fungi from ornithogenic soils clearly demonstrated that the temperature optimum for e.g. of hemicellulases had shifted by about 10-40°C towards a colder direction. For example, maximum β-mannanase activity was detected at 40-60°C compared to 70°C for the reference mesophile strain (*Trichoderma reesei*) also *Penicillium expansum* and *Phoma* sp. retained more than 40% β-mannanase activity at 0°C (Bradner et al., 2000). The cold-adapted hemicellulases also exhibited activity over 50% broader temperature range than the reference strain.

A similar investigation of an Antarctic strain of *Verticillium* (Fenice et al., 1998) indicated chitinolytic activity at 5°C that was four times higher than that obtained from two *Trichoderma harzianum* reference strains. Additionally, the chitinolytic enzyme was found to be active over a broad range of temperatures (5-60°C) and at 5°C its relative activity was still 50% of that recorded at the optimal temperature of 40°C. Therefore, it seems clear that novel cold-active hydrolases and their genes are available from Antarctic fungal isolates.

Thermophilic fungi

The highest temperatures at which some fungi have been found to grow optimally are between 45 and 50°C with some species capable of surviving into the high 50s (Maheshwari et al., 2000). These temperatures are much higher than can be tolerated by most other eukaryotic organisms. Despite this feature and the fact that thermophilic fungi are a rich source for enzymes that degrade plant biomass, they have not provoked a great amount of research attention. There are, however, indications that interest in these fungi is increasing as evidenced by recent increases in the amount of literature and the range of species being investigated (Li et al., 2005; Ko et al., 2005; Badhan et al., 2004; Li et al., 2003). The thermophilic fungus *Thermomyces lanuginosus* currently enjoys widespread application as a potential source of stable, high temperature (50-80°C) cellulase-free β-xylanases with activity over a broad pH range (3-12) (reviewed in Singh et al., 2003). Although the mesophilic *Trichoderma reesei* has recently been utilized as an expression host for a thermostable β-glucosidase gene isolated from the thermophilic fungus *Talaromyces emersonii* (Murray et al., 2004), thermophilic fungi themselves would be ideally suited to act as production hosts for thermophilic and thermolabile proteins. So far, no high-yielding expression systems developed for thermophilic fungi have been reported.

Marine fungi

The seas and oceans of the world cover in excess of 70% of the surface of the planet and contain within them a vast range of habitats and a diversity of life forms. Over the last few decades there has been an increasing recognition of the wealth and complexity of the microbiota present in the seas and the crucial role they play in ocean ecosystems. As a consequence, there has been a steady increase in interest in marine microorganisms and an intensity in the level of investigation into their genetic diversity and their potential to provide a wealth of new and novel commercial products (reviewed in DeLong and Karl, 2005). Although not known for their abundance, fungi nevertheless do inhabit the aqueous environment and some notable discoveries of therapeutic compounds have been identified as originating from them. By way of example, fungi isolated from marine invertebrates have become a dominant source of novel metabolites some with potent antibacterial and anticancer activities (D'Souza et al., 2005; Bernan et al., 2004; McDonald et al., 1999). Elsewhere, an antiparasitic agent has been isolated from a strain of *Aspergillus carneus* isolated from an estuarine sediment (Capon et al., 2003) and antifungal antibiotics have been produced by a marine *Phoma* sp. (Nagai et al., 2002).

Molecular approaches into gene prospecting

Molecular screening of enzyme-encoding sequences for the desired type of gene is largely based on gene alignments for the purpose of identifying specific areas of conserved DNA sequence that can be used for the design of PCR primers. The PCR based strategy combined with chromosomal walking PCR (Morris et al., 1998) has been successfully used for the isolation and cloning of a number of enzyme-encoding genes from thermophilic microorganisms, mainly bacteria and archaea (Peek et al., 1992), and also from unculturable microorganisms in a mixed environmental sample (Bell et al., 2002). The PCR based approach described by Morris et al. (1998) has scope for the isolation of variants of the same type of genes from other microorganisms, and even higher organisms.

An example combining Antarctic fungal isolates as a source of novel genes and the chromosomal walking PCR approach featured isolation of a lipase-encoding gene from *Penicillium allii* (Bradner et al., 2003). Lipases are serine hydrolases with an active site formed by a catalytic triad consisting of serine, a carboxylic acid and a histidine residue (Brzozowski et al., 2000; Jaeger et al., 1994). A region in the vicinity of the active-site serine and a second region corresponding to the oxyanion hole (Herggård et al., 2000; Jaeger et al., 1994) has been found to be relatively well conserved in several investigations into the amino acid sequences of bacterial and fungal lipases. These two sites, which in filamentous fungi appear to be consistently separated by around 72 aa (216 bp), were targeted as the most promising for the design of PCR primers (Table 3). Due to the significantly reduced level of similarity that existed at the nucleotide level, a set of degenerate PCR primers was designed using the CODEHOP design principle (Rose et al., 1998).

Table 3. Degenerate oligonucleotide primers designed to identify fungal lipases. Adapted from Bradner et al. (2003)

Function	*Region targeted*	*Primer name*	*Sequence (5′ - 3′)*
Forward primers	Oxyanion hole	FoxF1	atc gtt ctg gYN KtN MgN gg
		FoxF2	att gtc ctt KcN KtN MgN gg
		FoxF3	att tac att KYN ttN MgN gg
		FoxF4	atc ggc atc RSN ttN MgN gg
Reverse primers	Active site	FacR1	tgc ccc tcc NaK Nga Rtg NSc
		FacR2	tgc gcc Ncc NaK Rct Rtg NSc

Base abbreviations: R = a/g; Y = c/t; M = a/c; K = g/t; S = g/c; N = a/c/g/t.

Isolation of lipase genes provides an example of 'difficult-to-clone genes' because of the notably varying nucleotide sequences between different lipases. In contrast to this, an example of a relatively straightforward approach into fungal gene isolation is provided by xylanase-encoding genes. Xylanases across species belong to two main families, F or G (Beguin, 1990), which correspond to families 10 and 11 of the glycosyl hydrolases (Henrissat, 1991), respectively. A high degree of sequence conservation can be found at the amino acid level (Henrissat, 1991) and consensus primers have been designed and used successfully, for example, in the isolation of a xylanase gene fragment from the thermophilic bacterium *Dictyoglomus thermophilum* (Morris et al., 1998).

Other techniques into molecular prospecting for genes related to a known function include using heterologous genes (homologues) as hybridization probes and functional expression of cDNA libraries in a suitable host such as the bakers yeast (Saloheimo et al., 1997; Dalboge, 1997).

Genomics approaches

To date, over 25 fungal genomes (e.g. http://www.fgsc.net/outlink.html) have been or are in the process of being sequenced. Examples feature human and plant pathogens as well as fungi that serve as basic models for molecular and cellular biology. However, as yet there is only a small number of completed and annotated genomes. Examples of the completed sequencing projects include the genomes of *Neurospora crassa,* (Galagan et al., 2003; http://mips.gsf.de/projects/fungi/neurospora.html), *Aspergillus nidulans* (Roe et al., 1999; http://www.broad.mit.edu/annotation/fungi/aspergillus/), the plant pathogens *Magnaporthe grisea* (Dean et al., 2005), *Fusarium graminearum* (http://mips.gsf.de/projects/fungi/Fgraminearum.html) and *Stagonospora nodorum* (http://ohioline.osu.edu/ac-fact/0002.html), the opportunistic human pathogen *Aspergillus fumigatus* (Brookman and Denning, 2000), the ligninolytic *Phanerochaete chrysosporium* (Martinez et al., 2004; http://genome.jgi-psf.org/whiterot1/whiterot1.home.html) and the industrially relevant fungi *Aspergillus niger* (http://www.dsm.com/en_US/html/dfs/genomics_aniger.htm and *Trichoderma reesei* (http://genome.jgi-psf.org/Trire2/Trire2.home.html).

With genome sequence in hand, scanning for particular genes of interest is relatively straightforward, provided that there is enough sequence similarity between the query sequences and the genes in the organism of interest. Consequently, only the relevant single gene or a group of genes can be lifted directly from the genome of the target organism. An example of a comparative genomics approach concerning sequences that encode proteins involved in N-linked glycosylation of proteins is given in Table 4.

In addition to full genomic data, several EST (Expressed Sequence Tags) are available for various fungi for further exploration. Mapping gene expression using transcriptomics, e.g. fungal microarrays, broadens the scope into all genes that may be functional under particular conditions of interest, making possible the isolation of several genes contributing to the targeted function such as biomass hydrolysis. For example, analysis of EST libraries made from the cellulolytic *Trichoderma reesei* grown on various cultivation media spiked with simple sugars or on complex carbon sources exposed a number of genes encoding precursors for biomass-degrading enzymes and proteins involved in cellular targeting and vesicle formation. Moreover, homologs for genes involved in protein transport and posttranslational modification of proteins were identified by BLAST searches against databases across different organisms (Diener et al., 2004a). In another genome-wide search into *T. reesei*, 34 different probes derived from genes related to biomass degradation were hybridized to a BAC library providing coverage of 28 genome equivalents resulted in 475 positive BAC recombinants (Diener et al., 2004b). In the sequencing of over 5100 random *T. reesei* cDNA clones, 12 sequences were identified which encode previously unknown enzymes with a likely role in biomass degradation (Foreman et al., 2003).

In summary, the genomic approaches have high discovery value and

Table 4. Enzymes involved in N-linked protein glycosylation in the yeast *Saccharomyces cerevisiae* and their homologues in the filamentous fungus *Aspergillus nidulans*. #http://www.broad.mit.edu/annotation/fungi/aspergillus/.

Gene Locus/Enzyme in S. cerevisiae	*Accession No./ Reference*	*Definition/Function*	*Homolog in A. nidulans*#
CWH41 or GLS1, (α-1,2-glucosidase I)	EMBL: Z36098	ER glucosidase I, removes the terminal α-1,2 linked glucose	AN6606.2
ROT2 or GLS2, (α-1,3-glucosidase II)	EMBL: Z36098	ER Glucosidase II, removes the two α-1,3 linked glucose residues after the glucosidase I enzyme activity	AN8217.2
MNS1, (α-1,2-mannosidase)	EMBL: Z49631	ER α-1,2-mannosidase, specific for removal of one mannose from $Man_9GlcNAc_2$ to $Man_8GlcNAc_2$	AN0551.2
OCH1p	EMBL: Z72560	α-1,6 mannosyltransferase	AN4716.2
MNN9p	EMBL: U44030	has α-1,6 mannosyltransferase activity	AN4716.2
Van1p	EMBL: Z49210	involved in regulation of the phosphorylation of a number of proteins, some of which appear to be important in cell growth control	AN2159.2
Mnn9p	EMBL: U44030	α-1,6 mannosyltransferase	AN4716.2
Anp1p	EMBL: U18779	required to maintain a functional Golgi apparatus and for glycosylation in the Golgi, role in retention of glycosyltransferases in the Golgi	AN2159.2
Mnn10p	EMBL: Z49701	subunit of mannosyltransferase complex	AN7562.2
Mnn11p	EMBL: Z49458	related to Mnn10p	AN1969.2
Hoc1p	EMBL: U62942	forms a complex with Anp1p and Mnn9p in the Golgi membrane	AN4716.2
Mnn2p	EMBL: Z35884	has α-1,2 mannosyltransferase activity	AN6571.2
Mnn5p	EMBL: Z49461	has α-1,2 mannosyltransferase activity	AN6571.2
Mnn4p	EMBL: Z28201	regulates the mannophosphorylation	AN1268.2
Mnn6p	EMBL: U39205	mannosylphosphotransferase	AN2752.2
Mnn1p	EMBL: U18778	α-1,3 mannosyltransferase, required for complex glycosylation of both N- and O-linked oligosaccharides	No significant match

can be further applied for purposes such as strain improvement, engineering the secretion pathway, posttranslational modification of gene products and metabolic engineering. Areas such as fungal pathogenicity and research into pathogen-host interactions are also sourcing targets from the available genome information.

Proteomics

Recent developments in proteomics and mass spectrometry for protein identification have made the approach a powerful tool for gene product discovery. Filamentous fungi have gained relatively little attention so far but the situation is changing rapidly with the increasing availability of genome sequence data. For example, proteomics has been used in search for proteins involved in fungal biological control (Grinyer et al., 2005), mapping out fungal secretomes (Vanden Wymelenberg et al., 2005) and studies of the molecular basis of secretion stress (reviewed in Nevalainen et al., 2005). Identification of proteins separated, for example, by 2-dimensional gel electrophoresis provides basis for peptide sequencing which in turn is the basis for the design of oligonucleotides for isolation of the corresponding gene by PCR and chromosomal walking.

Proteomics may also help in finding novel, strong condition-specific gene promoters for gene expression. For example, in the studies into the proteins associated with the fungal cell envelope, Lim et al. (2001) noticed that over 50% of the protein extracted from fungal hyphae grown on glucose-containing medium represented HEX1, the major protein in a fungal Woronin body. The *hex1* promoter has since been characterized and tested for protein expression (Curach et al., 2004).

Proteomic analyses complement the genome and transcriptome-based approaches in gene prospecting providing insight into the actual gene product, for example its posttranslational modification by proteolysis, glycosylation or inclusion of charged molecules such as phosphate or sulfate (Harrison et al., 1998). It is also worth noticing that not all mRNAs are translated into proteins and that complexity of the material to be explored increases exponentially when proceeding from DNA to proteins. For example, a genome of 22,000 genes may well produce a proteome of 50,000 proteins or more as a result of various post-translational modifications to proteins such as proteolytic processing and glycosylation.

CONCLUSION AND FUTURE ASPECTS

It has been shown that the sources of filamentous fungi, their enzymes and genes are extensive and remain to a large degree untapped. The great potential of the more extreme regions of our planet—the Arctic/Antarctic, the oceans and thermal environments, to suggest just a few, invite further systematic investigation.

The molecular approaches into gene and gene product prospecting in filamentous fungi, including genome mining and proteomics, are clearly taking over the more traditional techniques based on culturing of the microorganisms as a starting point. Microbial culture collections as a source for prospecting have been sidelined by exploration of gene and genome data banks. The application of comparative genomics, where possible, opens up whole new ways for gene prospecting across organism boundaries. New fungal genes are practically being revealed on a daily basis. Undoubtedly, the vast amount of data streaming in, are creating pressure on the handling, storage and correct interpretation of information. New themes emerging from the genomics programs include understanding the evolutionary relationships between fungi (Dean et al., 2005) and systems biology aiming at understanding how organisms work. Considering the estimates that there are still hundreds of thousands if not millions of novel fungi to be discovered, research into fungal metagenomics is likely to attract plenty of attention in the future.

References

Azmi, O.R. and Seppelt, R.D. 1997. Fungi of the Windmill Islands, continental Antarctica. Effect of temperature, pH and culture media on the growth of selected microfungi. Polar Biology, 18: 128-134.

Azmi, O.R. and Seppelt, R.D. 1998. The broad-scale distribution of microfungi in the Windmill Islands region, continental Antarctica. Polar Biology, 19: 92-100.

Badhan, A.K., Chadha, B.S., Sonia, K.G., Saini, H.S. and Bhat, M.K. 2004. Functionally diverse multiple xylanases of thermophilic fungus Myceliophthora sp. IMI 387099. Enzyme and Microbial Technology, 35(5): 460-466.

Beguin, P. 1990. Molecular biology of cellulose degradation. Annual Review of Microbiology, 44: 219-248.

Bell, P.J.L., Sunna, A., Gibbs, M.D., Curach, N., Nevalainen, H. and Bergquist, P.L. 2002. Prospecting for novel lipase genes using PCR. Microbiology, 148: 2283-2291.

Bergfors, T., Rouvinen, J., Lehtovaara, P., Caldentey, X., Tomme, P., Claeyssens, M., Pettersson, G., Teeri, T., Knowles, J. and Jones, T.A. 1990. Three-dimensional structure of cellobiohydrolase II from *Trichoderma reesei*. Science, 249: 380-386.

Bergquist, P.L., Love, D.R., Croft, J.E., Streiff, M.B., Daniel, R.M., and Morgan, W.H. 1987. Genetics and potential biotechnological applications of thermophilic and extremely thermophilic micro-organisms. Biotechnology and Genetic Engineering Reviews, 5: 199-244.

Bernan, V.S., Greenstein, M. and Carter, G.T. 2004. Mining marine microorganisms as a source of new antimicrobials and antifungals. Current Medicinal Chemistry: Anti-Infective Agents, 3(3): 181-195.

Boisset, C., Pétrequin, C., Chanzy, H., Henrissat, B. and Schülein, M. 2001. Optimized mixtures of recombinant *Humicola insolens* cellulases for the biodegradation of crystalline cellulose. Biotechnology and Bioengineering, 72: 339-345.

Bradner, J.R. 2003. Antarctic microfungi as a potential bioresource. PhD Thesis. Biotechnology Research Institute, Macquarie University, Sydney, N.S.W., Australia.

Bradner, J.R., Gillings, M. and Nevalainen, K.M.H. 1999a. Qualitative assessment of hydrolytic activities in antarctic fungi at different temperatures on solid media. World Journal of Microbiology and Biotechnology, 15: 143-145.

Bradner, J.R., Sidhu, R.K., Gillings, G. and Nevalainen, K.M.H. 1999b. Hemicellulase activity of antarctic microfungi. Journal of Applied Microbiology, 87: 366-370.

Bradner, J.R., Sidhu, R.K., Yee, B., Skotnicki, M.L., Selkirk, P.M. and Nevalainen, K.M.H. 2000. A new microfungal isolate, *Embellisia* sp. associated with the Antarctic moss *Bryum argenteum*. Polar Biology, 23: 730-732.

Bradner, J.R., Bell, P.J.L., Te'o, V.S.J. and Nevalainen, K.M.H. 2003. The application of PCR for the isolation of a lipase gene from the genomic DNA of an Antarctic microfungus. Current Genetics, 44: 224-230.

Brookman, J.L. and Denning, D.W. 2000. Molecular genetics of *Aspergillus fumigatus*. Current Opinion in Biotechnology, 3: 468-474.

Brzozowski, A.M., Savage, H., Verma, C.S., Turkenburg, J.P., Lawson, D.M., Svendsen, A. and Patkar, S. 2000. Structural origins of the interfacial activation in *Thermomyces (Humicola) lanuginosa* lipase. Biochemistry, 39: 15071-15082.

Capon, R.J., Skene, C., Stewart, M., Ford, J., O'Hair, R.A.J., Williams, L., Lacey, E., Gill, J.H., Heiland, K. and Friedel, T. 2003. Aspergillicins A-E: Five novel depsipeptides from the marine-derived fungus *Aspergillus carneus*. Organic and Biomolecular Chemistry, 1(11): 1856-1862.

Cavicchioli, R., Siddiqui, K.S., Andrews, D. and Sowers, K.R. 2002. Low-temperature extremophiles and their applications. Current Opinion in Biotechnology, 13: 253–261.

Curach, N.C., Te'o, V.S.J., Gibbs, M.D., Bergquist, P.L. and Nevalainen, K.M.H. 2004. Isolation, characterization and expression of the hex1 gene from *Trichoderma reesei*. Gene, 331: 133-140.

D'Souza, D.T., Tiwari, R., Sah, A.K. and Raghukumar, C. 2005. Enhanced production of laccase by a marine fungus during treatment of colored effluents and synthetic dyes. Enzyme and Microbial Technology, 38: 504-511.

Dalboge, H. 1997. Expression cloning of fungal enzyme genes; a novel approach for efficient isolation of enzyme genes of industrial relevance. FEMS Microbiology Reviews, 1: 29-42.

Davies, G.J., Brzozowski, A.M., Dauter, M., Varrot, A. and Schülein, M. 2000. Structure and function of *Humicola insolens* family 6 cellulases: structure of the endoglucanase, Cel6B, at 1.6Å resolution. Biochemical Jour, 15: 201-207.

Dean, R.A., Talbot, N.J., Ebbole, D.J., Farman, M.L., Mitchell, T.K., Orbach, M.J., Thon, M., Kulkarni, R., Xu, J-R., Pan, H. Read, N.D., Lee, Y-H., Carbone, I., Brown, D., Oh, Y.Y., Donofrio, N., Jeong, J.S., Soanes, D.M., Djonovic, S., Kolomiets, E., Rehmeyer, C., Li, W., Harding, M., Kim, S., Lebrun, M-H., Bohnert, H., Coughlan, S.B.J., Calvo, S., Ma, L-J., Nicol, R., Purcell, S., Nusbaum, C., Galagan, J.E. and Birren, B.W. 2005. The genome sequence of the rice blast fungus *Magnaporthe grisea*. Nature, 434 (7036): 980-986.

DeLong, E.F. and Karl, D.M. 2005. Genomic perspectives in microbial oceanography. Nature, 437: 336-342.

Demain, A.L. and Elander, R.P. 1999. The beta-lactam antibiotics: Past, present, and future. Antonie van Leeuwenhoek, 75: 5-19.

Demirjian, D.C., Morís-Varas, F. and Cassidy, C.S. 2001. Enzymes from extremophiles. Current Opinion in Chemical Biology, 5: 44-151.

Denman, S., Xue, G.P. and Patel, B. 1996 Characterization of a *Neocallimastix patriciarum* cellulase cDNA (*celA*) homologous to *Trichoderma reesei* cellobiohydrolase II. Applied and Environmental Microbiology, 62: 1889-1996.

Diener, S.E., Dunn-Coleman, N., Foreman, P., Houfek, T.D., Teunissen, P.J.M., van Solingen, P., Dankmeyer, L., Mitchell, T.K., Ward, M. and Dean, R.A. 2004a. Characterization of the protein processing and secretion pathways in a comprehensive set of expressed sequence tags from *Trichoderma reesei*. FEMS Microbiology Letters, 230(2): 275-282.

Diener, S.E., Chellappan, M.K., Mitchell, T.K., Dunn-Coleman, N., Ward, M. and Dean, R.A. 2004b. Insight into *Trichoderma reesei*'s genome content, organization and evolution revealed through BAC library characterization. Fungal Genetics and Biology, 41(12): 1077-1087.

Enari, T.M. and Markkanen, P. 1977. Production of cellulolytic enzymes by fungi, pp. 1-24. In: Advances in Biochemical Engineering 5. T.K. Ghose, A. Fiechter, and N. Blakebrough (eds.). Springer-Verlag, Berlin, Gemany.

Eriksson, K-E. 1997. Biotechnology in the pulp and paper industry: An overview. ACS Symposium Series, 687: 2-14.

Farrell, R.L., Rhodes, P.L and Aislabie, J. 2003. Toluene-degrading Antarctic *Pseudomonas* strains from fuel-contaminated soil. Biochemical and Biophysical Research Communications, 312: 235-240.

Fenice, M., Selbmann, L., Zucconi, L. and Onofri, S. 1997. Production of extracellular enzymes by Antarctic fungal strains. Polar Biology, 17: 275-280.

Fenice, M., Selbmann, L., Di Giambattista, R. and Federici, F. 1998. Chitinolytic activity at low temperature of an Antarctic strain (A3) of *Verticillium lecanii*. Research in Microbiology, 149: 289-300.

Foreman, P.K., Brown, D., Dankmeyer, L., Dean, R., Diener, S., Dunn-Coleman, N.S., Goedegebuur, F., Houfek, T.D., England, G.J., Kelley, A.S., Meerman, H.J., Mitchell, T., Mitchinson, C., Olivares, H.A., Teunissen, P.J., Yao, J. and Ward, M. 2003. Transcriptional regulation of biomass-degrading enzymes in the filamentous fungus *Trichoderma reesei*. Journal of Biological Chemistry, 278: 31988-97.

Galagan, J.E., Calvo, S.E., Borkovich, K.A., Selker, E.U., Read, N.D., Jaffe, D., FitzHugh, W., Ma, L.J. Smirnov, S., Purcell, S., Rehman, B., Elkins, T., Engels, R., Wang, S., Nielsen, C.B., Butler, J., Endrizzi, M., Qui, D., Ianakiev, P., Bell-

Pedersen, D., Nelson, M.A., Werner-Washburne, M., Selitrennikoff, C.P., Kinsey, J.A., Braun, E.L., Zelter, A., Schulte, U., Kothe, G.O., Jedd, G., Mewes, W., Staben, C., Marcotte, E., Greenberg, D., Roy, A., Foley, K., Naylor, J., Stange-Thomann, N., Barrett, R., Gnerre, S., Kamal, M., Kamvysselis, M., Mauceli, E., Bielke, C., Rudd, S., Frishman, D., Krystofova, S., Rasmussen, C., Metzenberg, R.L., Perkins, D.D., Kroken, S., Cogoni, C., Macino, G., Catcheside, D., Li, W., Pratt, R.J., Osmani, S.A., DeSouza, C.P., Glass, L., Orbach, M.J., Berglund, J.A., Voelker, R., Yarden, O., Plamann, M., Seiler, S., Dunlap, J., Radford, A., Aramayo, R., Natvig, D.O., Alex, L.A., Mannhaupt, G., Ebbole, D.J., Freitag, M., Paulsen, I., Sachs, M.S., Lander, E.S., Nusbaum, C. and Birren, B. 2003. The genome sequence of the filamentous fungus *Neurospora crassa*. Nature, 422(6934): 859-868.

Garcia-Vallvé, S., Romeu, A. and Palau, J. 2000. Horizontal gene transfer of glycosyl hydrolases of the rumen fungi. Molecular Biology and Evolution, 17:352-361.

Gilbert, J.A., Hill, P.J., Dodd, C.E.R. and Laybourn-Parry, J. 2004. Demonstration of antifreeze protein activity in Antarctic lake bacteria. Microbiology, 50(1): 171-180.

Godfrey, T. and West, S. 1996. Industrial Enzymology. Macmillan Press, London, UK.

Gomes, J. and Steiner, W. 2004. The biocatalytic potential of extremophiles and extremozymes. Food Technology and Biotechnology, 42(4): 223-235.

Greenfield, L. and Wilson, G. 1981. University of Canterbury, Antarctic Expedition No. 19, Report University of Canterbury, Christchurch, New Zealand.

Grinyer, J., Hunt, S., McKay, M., Herbert, B.R. and Nevalainen, H. 2005. Proteomic response of the biological control fungus *Trichoderma atroviride* to growth on the cell walls of Rhizoctonia solani. Current Genetics, 47(6): 381-388.

Guezennec, J. 2002. Deep-sea hydrothermal vents: A new source of innovative bacterial exopolysaccharides of biotechnological interest. Journal of Industrial Microbiology and Biotechnology, 29(4): 204-208.

Harrison, M.J., Nouwens, A.S., Jardine, D.R., Zachara, N.E., Gooley, A.A., Nevalainen, H. and Packer, N.H. 1998. Glycosylation of cellobiohydrolase I from *Trichoderma reesei*. European Journal of Biochemistry, 256: 119-127.

Hay, F.S., Niezen, J.H. Leathwick, D. and Skipp, R.A. 2002. Nematophagous fungi in pasture: Colonisation of sheep faeces and their potential for control of free-living stages of gastro-intestinal nematode parasites of sheep. Australian Journal of Experimental Agriculture, 42(1):7-13.

Henrissat, B. 1991. A classification of glycosyl hydrolases based on amino acid sequence similarities. Biochemical Journal, 280: 309-316.

Herggård, S., Gibas, C.J. and Subramaniam, S. 2000. Role of an electrostatic network of residues in the enzymatic action of the *Rhizomucor miehei* lipase family. Biochemistry, 39: 2921-2930.

Jaeger, K-E., Ransac, S., Dijkstra, B.W., Colson, C., van Heuvel, M. and Misset, O. 1994. Bacterial lipases. FEMS Microbiology Reviews, 15: 29-63.

Karhunen, T., Mäntylä, A., Nevalainen, H. and Suominen, P. 1993. High frequency one-step gene replacement in *Trichoderma reesei*. I. Endoglucanase I overproduction. Molecular and General Genetics, 241: 515-522.

Kerry, E. 1990. Effects of temperature on growth rates of fungi from subantarctic Macquarie Island and Casey, Antarctica. Polar Biology, 10: 423-430.

Kirk, T.K. and Cullen, D. 1998. Enzymology and molecular genetics of wood degradation by white-rot fungi, pp. 273-308. In: Environmentally Friendly Technologies for the Pulp and Paper Industry. R.A. Young, and M. Achtar (eds.), John Wiley and Sons, New York, USA.

Ko, H-S., Yokoyama, Y., Ohno, N., Okadome, M., Amachi, S., Shinoyama, H. and Fujii, T. 2005. Purification and characterization of intracellular and extracellular, thermostable and alkali-tolerant alcohol oxidases produced by a thermophilic fungus, *Thermoascus aurantiacus* NBRC 31693. Journal of Bioscience and Bioengineering, 99(4):348-353.

Koivula, A., Linder, M and Teeri, T.T. 1998. Structure-function relationships in *Trichoderma* cellulolytic enzymes, pp. 3-23. In: *Trichoderma* & *Gliocladium* Vol. 2, Enzymes, biological control and commercial applications. G.R. Harman and C.P. Kubicek (eds.), Taylor and Francis, London, UK.

Li, D-C., Lu, M., Li, Y-L. and Lu, J. 2003. Purification and characterization of an endocellulase from the thermophilic fungus *Chaetomium thermophilum* CT2. Enzyme and Microbial Technology, 33(7): 932-937.

Li, L., Tian, H, Cheng, Y., Jiang, Z. and Yang, S. 2005. Purification and characterization of a thermostable cellulase-free xylanase from the newly isolated *Paecilomyces themophila*. Enzyme and Microbial Technology Sep 8 [ePub].

Lim, D., Hains, P., Walsh, B., Bergquist, P. and Nevalainen, H. 2001. Proteins associated with the cell envelope of *Trichoderma reesei*: A proteomic approach. Proteomics, 1: 899-910.

Liu, Q., Hu, J., Xue, D., Ma, C. and Wang, S. 2002. Bio-active substances derived from marine microorganisms. Chinese Journal of Applied Ecology, 13(7): 901-905.

Maheshwari, R., Bharadwaj, G. and Bhat, M.K. 2000. Thermophilic fungi: their physiology and enzymes. Microbiology and Molecular Biology Reviews, 64(3): 461-488.

Mandels, M. and Reese, E.T. 1957. Induction of cellulase in *Trichoderma viride* as influenced by carbon sources and metals. Journal of Bacteriology, 73: 269-278.

Mäntylä, A., Paloheimo, M. and Suominen, P. 1998. Industrial mutants and recombinant strains of *Trichoderma reesei*, pp. 291-309. In: *Trichoderma* and *Gliocladium* Vol. 2, Enzymes, biological control and commercial applications. G.R. Harman, and C.P. Kubicek, (eds.), Taylor and Francis, London, UK.

Martínez, A.T., Speranza, M., Ruiz-Dueñas, F.J., Ferreira, P., Camarero, S., Guillén, F., Martínez, M.J., Gutiérrez, A. and Del Río, J.C. 2005. Biodegradation of lignocellulosics: Microbial, chemical, and enzymatic aspects of the fungal attack of lignin. International Microbiology, 8(3):195-204.

Martinez, D., Larrondo, L.F., Putnam, N., Gelpke, M.D., Huang, K., Chapman, J., Helfenbein, K.G., Ramaiya, P., Detter, J.C., Larimer, F., Coutinho, P.M., Henrissat, B., Berka, R., Cullen, D., and Rokhsar, D. 2004. Genome sequence of the lignocellulose degrading fungus *Phanerochaete chrysosporium* strain RP78. Nature Biotechnology, (6): 679-680.

McDonald, L.A., Abbanat, D.R., Barbieri, L.R., Bernan, V.S., Discafani, C.M.,

Greenstein, M., Janota, K., Korshalla, J.D., Lassota, P., Tischle, M. and Carter, G.T. 1999. Spiroxins, DNA cleaving antitumor antibiotics from a marine-derived fungus. Tetrahedron Letters, 40(13): 2489-2492.

Miettinen-Oinonen, A. 2004. *Trichoderma reesei* strains for industrial production of cellulases, VTT Publications 550, Finland.

Miettinen-Oinonen, A., Londesborough, J., Joutsjoki, V., Lantto, R. and Vehmaanperä, J. 2004. Three cellulases from *Melanocarpus albomyces* with applications in the textile industry. Enzyme and Microbial Technology, 34: 332–341.

Möller, C. and Dreyfuss, M.M. 1996. Microfungi from Antarctic lichens, mosses and vascular plants. Mycologia, 88: 922-933.

Morris, D.D., Gibbs, M.D., Chin, C.W., Koh, M.H., Wong, K.K., Allison, R.W., Nelson, P.J. and Bergquist, P.L. 1998. Cloning of the *xynB* gene from *Dictyoglomus thermophilum* Rt46B.1 and action of the gene product on kraft pulp. Applied and Environmental Microbiology, 64: 1759-1765.

Munoz, I.G., Ubhayasekera, W., Henriksson, H., Szabo, I., Pettersson, G., Johansson, G., Mowbray, S.L. and Ståhlberg, J. 2001. Family 7 cellobiohydrolases from *Phanoerocahete chrysosporium*: crystal structure of the catalytic module of Cel7D (CBH58) at 1.3 Å resolution. Journal of Molecular Biology, 314(5): 1097-1111.

Murray, P., Aro, N., Collins, C., Grassick, A., Penttilä, M., Saloheimo, M. and Tuohy, P. 2004. Expression in *Trichoderma reesei* and characterisation of a thermostable family 3 β-glucosidase from the moderately thermophilic fungus *Talaromyces emersonii*. Protein Expression and Purification, 38(2): 248-257.

Nagai, K., Kamigiri, K., Matsumoto, H., Kawano, Y., Yamaoka, M., Shimoi, H., Watanabe, M. and Suzuki, K. 2002. YM-202204, a new antifungal antibiotic produced by marine fungus *Phoma* sp. Journal of Antibiotics, 55(12): 1036-1041.

Nevalainen, H. and Penttilä, M. 2004. Molecular biology of cellulolytic fungi, pp. 369-390. In: The Mycota, II, Genetics and Biotechnology, 2nd edition. U. Kück (ed), Springer-Verlag, Berlin-Heidelberg, Germany.

Nevalainen, H., Suominen, P, and Taimisto, K. 1994. On the safety of *Trichoderma reesei*. Journal of Biotechnology, 37: 193-200.

Nevalainen, K.M.H., Te'o, V.S.J. and Bergquist, P.L. 2005. Heterologous protein expression in filamentous fungi. Trends in Biotechnology, 23(9): 468-474.

Nyyssönen, E., Penttilä, M., Harkki, A., Saloheimo, A., Knowles, J.K.C. and Keränen, S. 1993. Efficient production of antibody fragments by the filamentous fungus *Trichoderma reesei*. Bio/Technology, 11: 591-595.

Ohmiya, K., Sakka, K., Kimura, T. and Morimoto, K. 2003. Application of microbial genes to recalcitrant biomass utilization and environmental conservation. Journal of Bioscience and Bioengineering, 95(6): 549-561.

Onofri, S. 1998. Ecology of Antarctic fungi. In: 6th International Mycological Congress (IMC6). Jerusalem, Israel, 23-28 August.

Onofri, S., Pagano, S., Zucconi, L., Cicalini, A.R., Selbmann, L. and Fenice, M. 1999. Microfungi in Antarctic extreme environments. Extracellular enzyme activities. In: IBC's World Congress on Enzyme Technologies. San Francisco, CA, USA, 10-12 March.

Park, S-J., Joo, W-A., Choi, J., Lee, S-H. and Kim, C-W. 2004. Identification and characterization of inosine monophosphate dehydrogenase from *Halobacterium salinarum*. Proteomics, 4(11): 3632-3641.

Peek, K., Ruttersmith, L.D., Daniel, R.M., Morgan, H.W. and Bergquist, P.L. 1992. Thermophilic enzymes as industrial catalysts. International Journal of Biotechnology, 9: 466-470.

Penttilä, M.E., Nevalainen, H., Salminen, E., Rättö, M. and Knowles, J. 1987. Versatile transformation system for the cellulolytic filamentous fungus *Trichoderma reesei*. Gene, 61: 155-164.

Rabinovich, M.L., Bolobova, A.V. and Vasil'chenko, L.G. 2004. Decomposition of natural aromatic structures and xenobiotics by fungi. Prikladnaia biokhimiia i mikrobiologiia, 40(1): 5-23.

Reese, E.T., Siu, R.G.H. and Levinson, H.S. 1950. The biological degradation of soluble cellulose derivatives and its relationship to the mechanism of cellulose hydrolysis. Journal of Bacteriology, 59: 485-497.

Reymond, P., Durand, H., Herbaud, M. and Fevre, M. 1991. Molecular cloning of genes from the rumen anaerobic fungus *Neocallimastix frontalis*: expression during hydrolase induction. FEMS Microbiology Letters, 77: 107-112.

Robinson, C.H. 2001. Cold adaptation in Arctic and Antarctic fungi. New Phytologist, 151: 341-353.

Roe, B.A., Kupfer, D., Zhu, H., Gray, J., Clifton, S., Prade, R., Loros, J. and Dunlap, J. 1999. *Aspergillus nidulans* sequencing project. http://www.genome.ou.edu/fungal.html.

Rose, T.M., Schultz, E.R., Henikoff, J.G., Pietrokovski, S., McCallum, C.M. and Henikoff, S. 1998. Consensus-degenerate hybrid oligonucleotide primers for amplification of distantly related sequences. Biotechnology and Applied Biochemistry, 26: 1628-1635.

Saloheimo, M., Nakari-Setälä, T., Tenkanen, M. and Penttilä, M. 1997. cDNA cloning of a *Trichoderma reesei* cellulase and demonstration of endoglucanase activity by expression in yeast. European Journal of Biochemistry, 429: 584-591.

Schadt, C.W., Martin, A.P., Lipson, D.A. and Schmidt, S.K. 2003. Seasonal dynamics of previously unknown fungal lineages in tundra soils. Science, 301: 1359-1361.

Schnurer, J., Olsson, J. and Borjesson, T. 1999. Fungal volatiles as inodicators of food and feeds spoilage. Fungal Genetics and Biology, 27: 209-217

Scholte, E-J., Knols, B.G.J., Samson, R.A. and Takken, W. 2004. Entomopathogenic fungi for mosquito control: A review. Journal of Insect Science, 4(19): 1-24.

Schülein, M., Tikhomirov, D.E. and Schou, C. 1993. *Humicola insolens* alkaline cellulases. In: Proceedings of the 2nd Tricel Meeting, Majvik, Finland T. Reinikainen and P. Suominen (eds.), Foundation for Biotechnical and Industrial Fermentation Research, 8: 109-116.

Sharma, R., Chisti, Y. and Banerjee, U.C. 2001. Production, purification, characterization, and applications of lipases. Biotechnology Advances, 19: 627-662.

Shoemaker, S., Schweickart, V., Ladner, M., Gelfand, D., Kwok, S., Myambo, K. and Innis, M. 1983. Molecular cloning of the exo-cellobiohydrolase I derived from *Trichoderma reesei* strain L27. Bio/Technology, 1: 691-696.

Singh, S., Madlala, A.M. and Prior, B.A. 2003. *Thermomyces lanuginosus*: properties of strains and their hemicellulases. FEMS Microbiology Reviews, 27(1): 3-16.

Suominen, P., Mäntylä, A., Karhunen, T. and Nevalainen, H. 1993b. High frequency one-step gene replacement in *Trichoderma reesei*. II. Effects of deletion of individual cellulase genes. Molecular and General Genetics, 241: 522-530.

Suurnäkki, A., Tenkanen, M., Buchert, J. and Viikari, L. 1997. Hemicellulases in the bleaching of chemical pulps. Advances in Biochemical Engineering and Biotechnology, 57:261-87.

Takashima, S., Nakamura, A., Hidaka, M., Masaki, H. and Uozumi, T. 1996. Cloning, sequencing and expression of the cellulase genes of *Humicola grisea* var. *thermoidea*. Journal of Biotechnology, 50: 137-147.

Takashima, S., Iikura, H., Nakamura, A., Hidaka, M., Masaki, H. and Uozumi, T. 1999. Comparison of gene structures and enzymatic properties between two endoglucanases from *Humicola grisea*. Journal of Biotechnology, 67: 85-97.

Teeri, T.T., Salovuori, I. and Knowles, J. 1983. The molecular cloning of the major cellulose gene from *Trichoderma reesei*. Bio/Technology, 1: 696-699.

Teeri, T.T., Penttilä, M., Keränen, S., Nevalainen, H. and Knowles, J.K.C. 1992. Structure, function, and genetics of cellulases, pp. 417-445. In: Biotechnology of Filamentous Fungi. D.B. Finkelstein, and C. Ball (eds.), Butterworth-Heinemann, Boston, USA.

Tosi, S., Casado, B., Gerdol, R. and Caretta, G. 2002. Fungi isolated from Antarctic mosses. Polar Biology, 25: 262-268.

Vanden Wymelenberg, A., Sabat, G., Martinez, D., Rajangam, A.S., Teeri, T.T., Gaskell, J., Kersten, P.J. and Cullen, D. 2005. The *Phanerochaete chrysosporium* secretome: Database predictions and initial mass spectrometry peptide identifications in cellulose-grown medium. Journal of Biotechnology, 118: 17-34.

Van Tilbeurgh, H., Toome, P. Claeyssens, M., Bhikhabhai, R. and Pettersson, G. 1986. Limited proteolysis of the cellobiohydrolase I from *Trichoderma reesei*. FEBS Letters, 204: 223-227.

van Zeijl, C., Punt, P., Emalfarb, M., Burlinghame, R., Sinitsyn, A., Parriche, M., Bousson, J.C. and van den Hondel, C.A.M.J.J. 2001. *Chrysosporium lucknowense*, a new fungal host for protein production. Fungal Genetics Newsletter 48, Suppl. p. 89.

Varrot, A., Halstrup, S., Schülein, M. and Davies, G.J. 1999. Crystal structure of the catalytic core domain of the family 6 cellobiohydrolase II, Cel 6A, from *Humicola insolens*, at 1.92Å resolution. Biochemical Journal, 15: 297-304.

Virendra, S., Bisaria, V.S. and Ghose, T.K. 1981. Biodegradation of cellulosic materials: substrates, microorganisms, enzymes and products. Enzyme and Microbial Technology, 3: 90-104.

von der Osten, C. and Schülein, M. 1999. Alkaline cellulases. US Patent 5,912,157.

Wagner-Döbler, I., Beil, W., Lang, S., Meiners, M. and Laatsch, H. 2002. Integrated approach to explore the potential of marine microorganisms for the production of bioactive metabolites. Advances in Biochemical Engineering/Biotechnology, 74: 207-238.

Wood, T.M., McCrae, S.I. and Bhat, M. 1989. The mechanism of fungal cellulase action. Biochemical Journal, 260: 37-43.

Yagüe, E., Mehak-Zunic, M., Morgan, L., Wood, D.A. and Thurston, C.F. 1997. Expression of CEL2 and CEL4, two proteins from *Agaricus bisporus* with similarity to fungal cellobiohydrolase I and β-mannanse, respectively, is regulated by the carbon source. Microbiology, 143: 239-244.

Yamamura, A., Sakaguchi, T., Murakami, Y., Yokoyama, K. and Tamiya, E. 1999. Purification and characterization of cold-active L-glutamate dehydrogenase of NAD(P) and oxygen. Journal of Biochemistry, 125: 760-769.

Zucconi, L., Pagano, S., Fenice, M., Selbmann, L., Tosi, S. and Onofri, S. 1996. Growth temperature preferences of fungal strains from Victoria Land, Antarctica. Polar Biology, 16: 53-61.

10

The Cuckoo Fungus 'Termite ball' Mimicking Termite Eggs: A Novel Insect-fungal Association

Kenji Matsuura* and Toshihisa Yashiro
Laboratory of Insect Ecology,
Graduate School of Environmental Science,
Okayama University, Okayama 700-8530, Japan
*E-mail: kenjijpn@cc.okayama-u.ac.jp

Abstract

Unlike the over 200 years long history of research on the symbiosis between fungus-growing termites and *Termitomyces* fungi (Wilson, 1971), the research history of the termite egg mimicking fungi is only 10 years. To date, termite balls have been found in five *Reticulitermes* species. Climatic factors play much more important roles in restricting the distribution of termite ball fungi than do geographic distances and phylogeny of host termites. Phylogenetic analysis illustrated that termite ball fungi isolated from different termite species were all very similar, with no significant molecular differences among host species or geographic locations. The evidences to date indicate long-distance gene flow and horizontal transmission of the termite ball fungi. This seems impossible to explain without spore formation, although inducing spore formation in laboratory conditions has so far not succeeded. It is most likely that this fungus has a free-living sexual stage independent of termites. Evidences for the strict morphological and chemical egg-mimicry by the termite ball fungi were provided. The net outcome of the interaction seems most likely to be negative for termites. In contrast, the termite ball fungus profits from this interaction by receiving competitor-free habitat in termite nests and protection from termites. Thus it can be concluded that termite balls parasitically mimic termite eggs to be harbored in termite nests.

INTRODUCTION

The nesting and feeding ecology of termites exposes colony members to a great variety of microbes including bacteria, fungi, protozoa, viruses, spirochetes and nematodes (Rosengaus and Traniello, 2001). Thus, it can be said that termites live in the world of microorganisms. Therefore, one of the most important selection pressures on termites is how they cope with various microorganisms, resulting in the evolution of behavioral and physiological adaptations (Matsuura, 2003). The interactions between termites and microorganisms vary from total dependence of the termite on the microorganism for food at one end of the spectrum to total dependence of the microorganism on the termite at the other end, with many degrees of mutualism in between. The best-known termite-fungal association is the mutualistic symbiosis between fungus-growing termites and *Termitomyces* fungi (Wood and Thomas, 1989; Aanen et al., 2002; Aanen and Boomsma, 2005). This chapter introduces a novel termite-fungus interaction, that is, the termite-egg mimicry by a sclerotium-forming fungus, whereby the fungus gains competitor-free habitat in termite nests.

Mimicry has evolved in a wide range of organisms and encompasses diverse tactics for defence (e.g., Lindstrom et al., 2004; Ito et al., 2004), foraging (e.g., Allan et al., 2002), pollination (e.g., Singer et al., 2004) and social parasitism (e.g., Kistner, 2000). Mimicry has long fascinated evolutionary biologists, because it provides a model system for the study of the 'evolutionary arms race' between hosts and parasites (Wickler, 1968; Malcolm, 1990). Egg protection is an essential behavior in social animals, and egg mimicry is a well-known tactic in avian brood parasitism (Rothstein, 1990; Servedio and Lande, 2003). The phenomenon of termites harboring brown fungal balls alongside their eggs was found in 1997 and reported for the first time in 2000 (Matsuura et al., 2000). When termite workers see the eggs laid by queens, they bring the eggs together and heap them in order to care for them. In nests of *Reticulitermes* termites, workers make several piles of eggs in this way (Fig. 1A). Brown balls (named 'termite balls') that are similar to eggs in their size and smooth texture are frequently found within these egg piles (Fig. 1A). DNA sequence analysis of the internal transcribed spacers (ITS) and the 5.8S ribosomal RNA gene of the nuclear ribosomal repeat unit identified these balls as sclerotia of a corticioid fungus (Basidiomycotina: Aphyllophorales), an undescribed species of the genus *Fibularhizoctonia*, which is phylogenetically closest to decay fungi, *Athelia* spp. (Matsuura et al., 2000). Based on the phylogeny of fungi (James et al., 2006), the origin of the termite ball fungus is completely independent of the symbiotic *Termitomyces* fungi. The sclerotia are tough balls of densely packed filaments (Fig. 1B) that germinate into fungal colonies under favorable conditions (Fig. 1C).

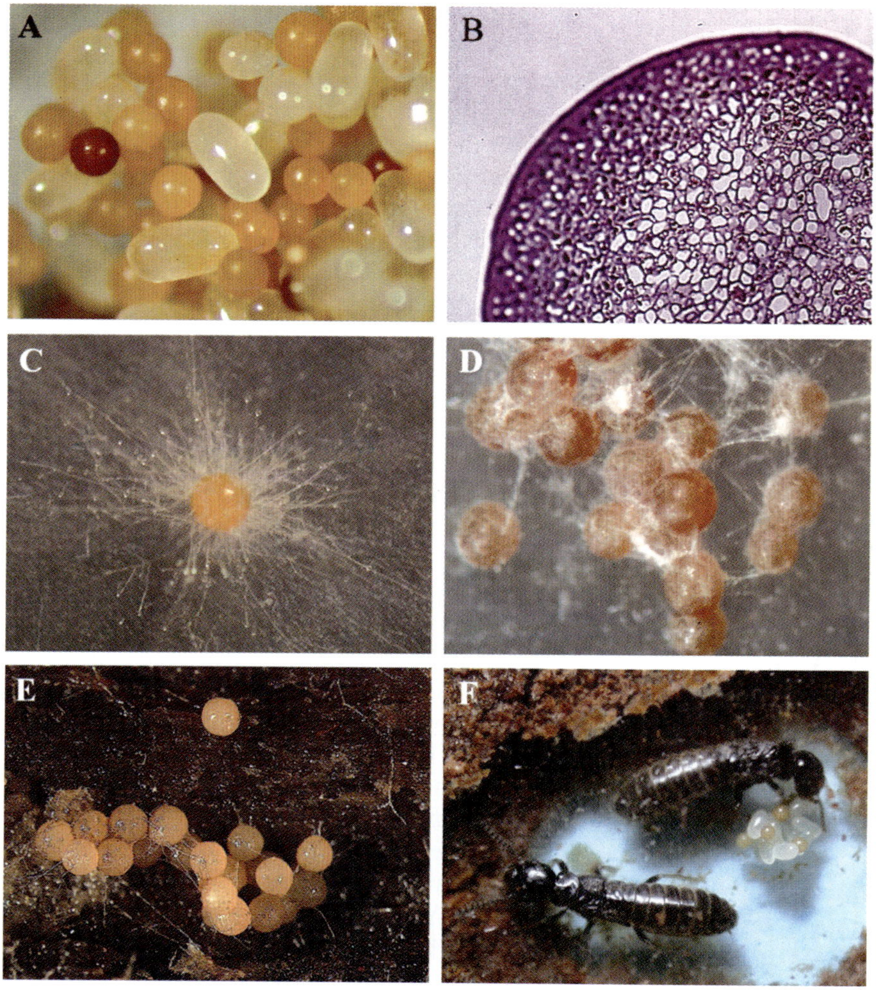

Fig. 1. The termite balls. (A) A pile of eggs. Termite eggs are transparent and oval, whereas termite balls are brown and spherical. (B) A cross section of a termite ball. (C) Germination of a termite ball on a PDA plate. (D) The fungus forming termite balls on a PDA plate. (E) The fungus forming termite balls in a termite nest. (F) A founding pair of termites tending the termite balls given experimentally under 24h light condition. (Photos by K. Matsuura and T. Yashiro)

Distribution and phylogenetic analysis of termite balls

To identify factors influencing the distribution of termite balls among their host termite species, wide-range sampling were conducted in Japan, Taiwan and the United States. To date, termite balls have been found in five *Reticulitermes* species (*R. speratus*, *R. kanmonensis*, *R. amamianus* in Japan, *R. flavipes* and *R. virginicus* in the United States) (Matsuura, 2005; Yashiro and Matsuura, 2007). On the other hand, no termite ball has been found in the nest of *R. okinawanus* in Okinawa Island, Japan or *R. hesperus* in California, USA (Matsuura, 2005). Therefore, the distribution of termite-ball fungi can not be explained simply by geographic barriers.

Analyses of termite ball distribution among nine *Reticulitermes* species (Fig. 2) suggested that climate is the most important factor restricting the distribution of the termite ball fungi (Yashiro and Matsuura, in press). It is notable that termite balls have been found only in *Reticulitermes* species distributed in temperate zones (Fig. 2). In Amami-Oshima Island, *R. amamianus* colonies had termite balls only in temperate highland regions, but no termite balls were found in its subtropical lowland region. Kitade and Hayashi (2002) suggested that *R. kanmonensis* was introduced artificially from southern China to the Kanmon area in Japan a few hundred years ago, where an international port, Shimonoseki, is located. Phylogenetic analysis showed that *R. kanmonensis* in the Kanmon area did not diverge from *R. flaviceps* in Taiwan (Yashiro and Matsuura, 2007) (Fig. 3). Termite balls were frequently found in the egg piles of *R. kanmonensis* as well as in the sympatric *R. speratus* in the Kanmon area, whereas no termite balls were found in the egg piles of *R. flaviceps* in Taiwan (Fig. 2). This suggests that the history of the interaction between *R. kanmonensis* and the termite ball fungus is likely to be short: a few hundred years at the longest. Matsuura (2006) showed that even *R. okinawanus*, which has no natural association with termiteball fungus in Okinawa, tended termite balls along with its eggs when they were experimentally provided with termite balls. Therefore, the difference of termite species, at least in the genus *Reticulitermes*, cannot explain the distribution of termite balls. These data strongly suggest that climatic factors play much more important roles in restricting the distribution of termite ball fungi than do geographic distances and phylogeny of host termites.

Tree flora also affects the distribution of the termite ball fungi in conjunction with climate. The percentage of colonies possessing termite balls differed significantly between the tree species that the termites harvested. All of the *R. speratus* colonies collected from rotten pine woods (*Pinus densiflora*) had termite balls in their egg piles, although only 40% of the colonies nesting in rotten Japanese cedar (*Cryptomeria japonica*) had termite balls (Matsuura et al., 2000). The colonies nesting in cedar wood

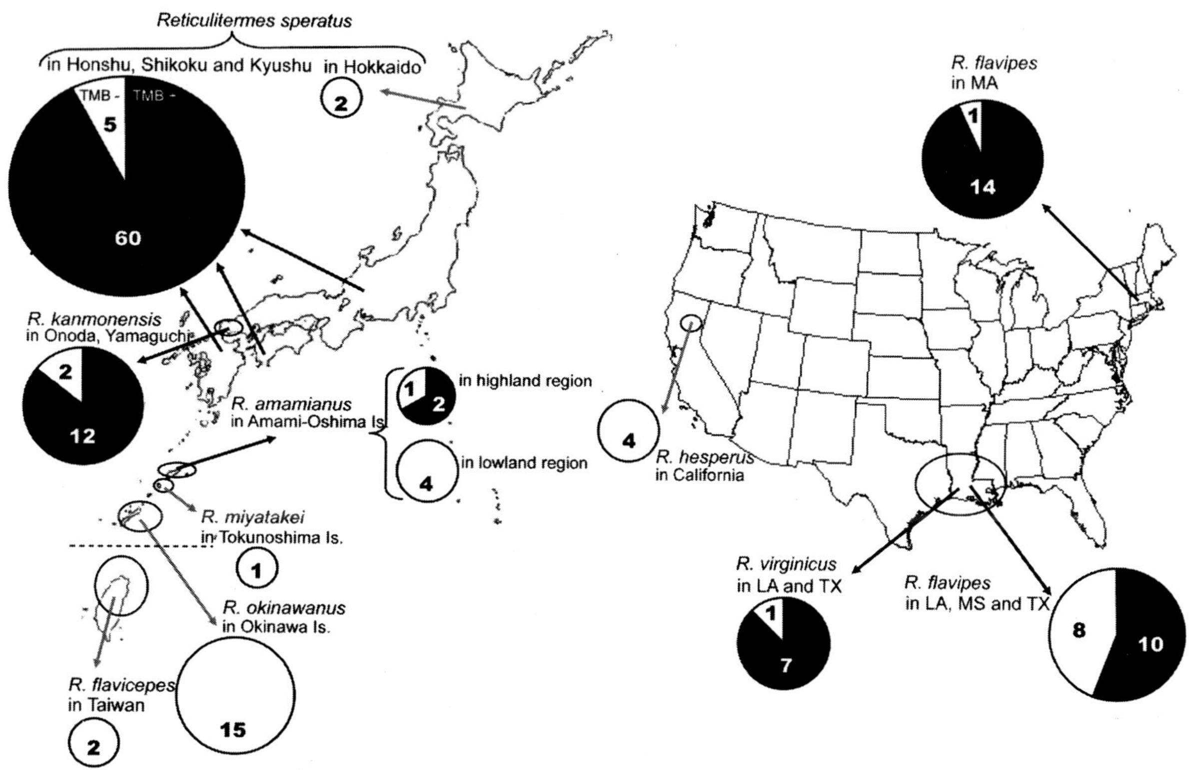

Fig. 2. Distribution of termite balls in *Reticulitermes* termites in Japan, Taiwan and in the United States. The pie charts show termite ball possession rates of each termite species (modified from Yashiro and Matsuura, 2007)

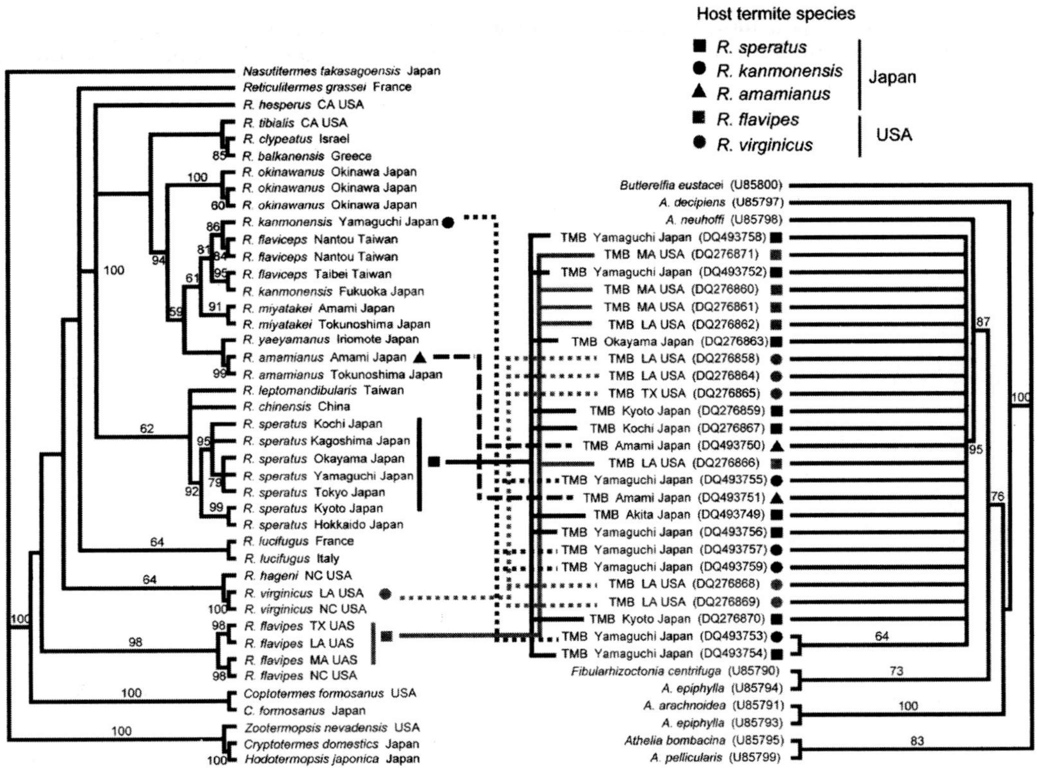

Fig. 3. Phylogenetic trees of termites based on mitochondrial sequences of the COII regions (left), and the termite-ball fungi (TMB) based on nucleotide sequences of the ITS regions (right). Bootstrap values are indicated for nodes having more than 50% support (500 replicates) (modified from Yashiro and Matsuura, 2007)

that was relatively more rotten had many termite balls in the egg piles. This was probably because fungistatic substances in fresh cedar wood can inhibit the growth of this fungus.

DNA sequence data have been used to estimate phylogenies of the host termites and the termite ball fungi isolated from the nests of the termites (Fig. 3). The phylogenetic tree of termite ball fungi based on nucleotide sequences of the ITS region showed that the termite ball fungi isolated from five different hosts (*R. speratus, R. kanmonensis, R. amamianus, R. flavipes and R. virginicus*) were very similar among host species and geographical locations without significant molecular differences (Fig. 3). Phylogenetic analysis of the fungal isolates based on nucleotide sequences of IGS region also showed no significant differences among host termite species (Yashiro et al. unpublished data), suggesting no host race formation.

Mechanism of termite-egg mimicry by termite balls

Egg-recognition bioassay. Why do termites harbor termite balls in their nursery cells together with eggs? There are two alternative hypotheses: (1) Termites recognize termite balls as their eggs, that is, termite balls mimic termite eggs. (2) Termites may harbor termite balls because of their potential benefits to the colony, although termites can distinguish termite balls from their eggs. To answer this question, it first needs to be determined how termites recognize their eggs. Dummy-egg bioassays were conducted using dummy eggs made of glass beads coated with egg-recognition chemicals to understand the egg recognition mechanism of termites (Matsuura et al., 2000; Matsuura, 2006). Termite eggs and dummy eggs were randomly arranged on moist unwoven cloth in a Petri dish, and workers were released in the dish. Dishes were maintained at 25°C in the dark. After 24 h, the acceptance rates were determined by counting the number of dummy eggs carried onto egg piles. Glass beads are piled with the true eggs only when workers recognized them as eggs. It was confirmed that this bioassay simulated termite behavior in the natural condition, where workers bring the dummy eggs into nursery cells from anywhere in their territory.

The size of termite eggs changes as the embryo develops (Matsuura and Kobayashi, 2007), and nutritional condition may also affect the egg size, while there is no significant difference among species in the genus *Reticulitermes*. As termite workers always grasp the short side of the oval eggs when they carry them, the short diameter of the eggs is an important physical cue in termite egg recognition. In *Reticulitermes* termites, the short diameters of eggs range from 0.26 mm to 0.44 mm (Matsuura, 2006). The 0.4 mm glass beads, whose diameter was similar to that of eggs, were carried by termites when they were coated with egg chemicals, whereas slightly larger (0.6 mm) and slightly smaller (0.2 mm) glass beads were rejected by termites, even when they were densely coated with egg chemicals. These results showed that the termites recognize their eggs by chemical cues (egg recognition pheromone) and physical cues, that is, by the spherical shape and the size.

Morphological mimicry. If termite balls mimic termite eggs, the sensory system used for egg recognition and the cognitive constraints of termites should affect the evolution of egg mimicry by the termiteball fungus. As termite workers live in the dark and do not have eyes, they cannot visually recognize eggs. Therefore, no selection acts on the color of termite balls, which is consistent with the observation of various colors of termite balls that are quite different from the color of termite eggs (Fig. 1A). This contrasts with egg mimicry in cuckoos, in which the color and pattern of the eggshell resembles that of the host bird eggs, because host birds recognize their eggs visually (Servedio and Lande, 2003; Brooke and Dacies, 1988).

To assess whether strict morphological mimicry of eggs is required for termite balls to be carried by termites, we compared the short diameters of termite eggs, termite balls produced by fungal isolates from termite nests and termite balls collected from egg piles, i.e., those accepted by termites (Matsuura, 2006). The diameter range of termite balls produced by the fungus did not exactly match the range of the short diameter of termite eggs. Nevertheless, there was no significant difference between the mean diameter of accepted termite balls and the mean short diameter of termite eggs. Termite balls <0.24 mm in diameter were never tended by termites, even when densely coated with egg chemicals. This result clearly shows that termites selectively carry termite balls whose diameters are similar to the eggs. In other words, the strong stabilizing selection by termites acts on the size of termite balls. This can reasonably explain why termite ball size is so stable to fit the short diameter of termite eggs in comparison with closely related sclerotium-forming fungi (Fig. 4).

Moreover, scanning electron microscopic observations revealed sophisticated mimicry of the smooth surface texture of eggs. Termite balls showed marked differences in surface texture from the sclerotia of the closely related fungus *A. epiphylla* (Fig. 4). The sclerotia of *A. epiphylla* were never tended by termites, even if they were densely coated with egg chemicals. These results provide clear evidence for morphological egg mimicry by the termite ball fungus.

Chemical mimicry. Egg recognition bioassay showed that termites recognize eggs based on a combination of the morphological and chemical cues. In addition to morphological mimicry, therefore, chemical mimicry should be required for termite balls to be recognized as eggs and tended by termites. Insect eggs are generally covered with cuticular waxes. The

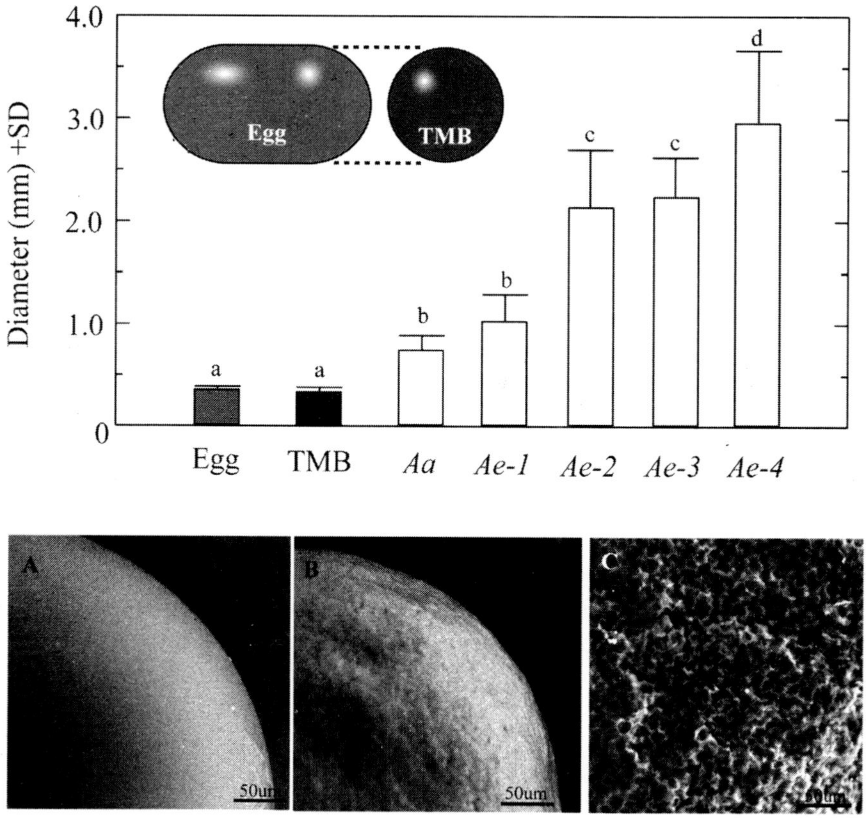

Fig. 4. Size comparison of termite eggs and termite balls and other closely related sclerotium-forming fungi. *Aa*—*Athelia archnoidea*, *Ae*—*Athelia epiphylla* (4 isolates). There were significant differences between different letters at the 0.05 level by Scheffe's test. Comparative observations of the micro-structure of the surfaces of an egg of *R. flavipes* (A), a termite ball (B) and a sclerotium of the closely related corticioid fungus *Athelia epiphylla* (C) using a variable-pressure scanning electron microscope. Note that this termite ball was formed on an agar plate without termites; therefore, the smooth surface structure is not the result of grooming by termites

primary function of cuticular lipids is to provide protection from desiccation (Lockey, 1988). In some ants, cuticular hydrocarbons on egg surfaces play an essential role as the egg recognition pheromone, allowing colony members to distinguish between worker- and queen-laid eggs (D'Ettorre et al., 2006). In termites, however, cuticular waxes extracted from egg surfaces with hexane showed no egg recognition activity (Matsuura, 2002). In contrast to ants, termites cannot discriminate eggs from nestmates and non-nestmates. Termite workers perform egg protection behavior even toward the eggs of different species indiscriminately, at least in the genus *Reticulitermes*, suggesting a broad cross-species activity of termite egg recognition pheromone (TERP).

Most recently, it was found by us that TERP is not a cuticular lipid but a protein. While the TERP protein has already been identified, it is not possible to open the unpublished information of the pheromone in this chapter. The identification of the TERP revealed an important fact that the TERP has another practical function and the chemical is abundant in termite nests. More than 90% of the termite balls collected from egg piles in termite nests were accepted as eggs by termites, while the termite balls obtained from a fungal colony cultured on potato dextrose agar (PDA) (Fig. 1D) were not recognized as eggs. Interestingly, termite balls obtained the egg recognition chemical when the extract of termite nest material was added to the PDA (Matsuura et al. unpublished data). This suggests that the fungus sequester the egg recognition chemical from termite nests rather than manufacturing their own egg mimicking chemical.

Interaction between termites and termite ball fungi: parasitism or mutualism?

By mimicking termite eggs, the fungus is protected and can be transported by the termites to the competitor-free habitat. Eggs and termite balls are groomed frequently by workers; their surfaces are coated with saliva, which contains antibiotic substances, thus providing protection from desiccation and pathogens (Matsuura et al., 2000). Worker grooming is indispensable for eggs and termite balls to survive since both are not dryness-resistant. On the other hand, the termite balls in the egg piles are inhibited from germination by worker grooming (Matsuura et al., 2000). If workers were removed experimentally, most of the termite balls germinated in the egg piles and grew by consuming termite eggs. Over the long term, termites do not completely inhibit sclerotial germination and fungal growth. The fungus sometimes consumed termite eggs, even when they were under the care of workers, although the occurrence of egg consumption by the fungus was rare (Matsuura, 2006). The termites remove from the egg piles sclerotia that have shrunk or deformed. These termite balls germinate and grow on termite excretions in the corner of the nest (Fig. 1E). The new termite balls produced in the nest are then carried by termites into the egg piles. Hence, the number of termite balls increases and sometimes exceeds the number of termite balls in the egg piles (Matsuura, 2005). Termites do not consume termite balls, and the termiteball fungus confers no nutritional benefit to termites (Matsuura, 2002). Workers need to groom tens of thousands of eggs and termite balls daily, thereby incurring substantial time and energetic costs. Therefore, the net outcome of the interaction seems most likely to be negative for termites.

Once it was speculated that the interaction between the termites and the termite ball fungus may include a symbiotic aspect for the termites because termite balls sometimes enhance egg survival under a certain experimental conditions (Matsuura et al., 2000; Matsuura, 2003). One possible explanation for the positive effect of termite balls on egg survival may be that the antibiotic substances produced by the termiteball fungus protect termite eggs from pathogens. However, comparison of egg survival rates with and without termite balls under semi-natural condition showed no significant effect (Matsuura, 2006). Under natural conditions, therefore, such a positive effect seems to be unlikely or a rare occurrence as a result of antagonistic interactions between the termiteball fungus and other microorganisms. The evidence obtained to date indicates that the interaction is parasitic, in that it is beneficial for the fungus but costly for the host termites at least in the short term. This explains why strict morphological and chemical mimicry of eggs is required for termite balls to be tended by termites.

Transmission of the termite ball fungi

As already mentioned, phylogenetic analyses of the termiteball fungi and their host termites showed no host race formation and no geographic variation of the termiteball fungi, even between Japan and the United States, suggesting long-distance gene flow through spore dispersion (Fig. 3). The transmission pathway of the termite ball fungus is still poorly understood. In *Reticulitermes* termites, new colonies are founded either by a pair of swarming winged reproductives (alates) and/or by secondary reproductives cut off from the main colony (budding). The termite ball fungus can certainly be transmitted to daughter colonies through budding, and colony fusion (Matsuura and Nishida, 2001) might also allow the fungus to be transferred to unrelated colonies. However, the high possession rate seems unlikely to be explained only by transmission through budding and colony fusion. In addition, alates do not carry fungal spores during their mating flight. Incipient colonies of *R. speratus*, *R. flavipes*, and *R. virginicus* founded by alates contained no termite balls, although even founding pairs in incipient colonies tended termite balls along with their eggs when they were experimentally provided with termite balls (Fig. 1F). These facts suggest that termite ball transmission in *Reticulitermes* termites is predominantly horizontal.

Most recently, a sclerotium-forming fungus was isolated from rotten leaves in the field where these fungi fruit spores. No significant difference was found in ITS sequences between this fungus and the termiteball fungus (Matsuura et al., unpublished data), suggesting that the termiteball fungus might fruit spores in nature and frequently transfer among termite nests. It is most likely that this fungus has a free-living sexual stage independent of termites.

However, making the fungus fruit spores in laboratory conditions has not as yet succeeded, despite attempts to induce spore formation using the methods described by Punja et al. (1982). It is generally difficult to induce basidiocarp formation in isolates of *Fibularhizoctonia* (teleomorph: *Athelia*) in culture because it is influenced by nutritional and environmental conditions (Punja et al., 1982). There may be any missing trigger for spore formation of termite ball fungi. More detailed knowledge on the reproductive modes of the termite ball fungus is essential to understand transmission pathways and the entire life cycle of the fungus.

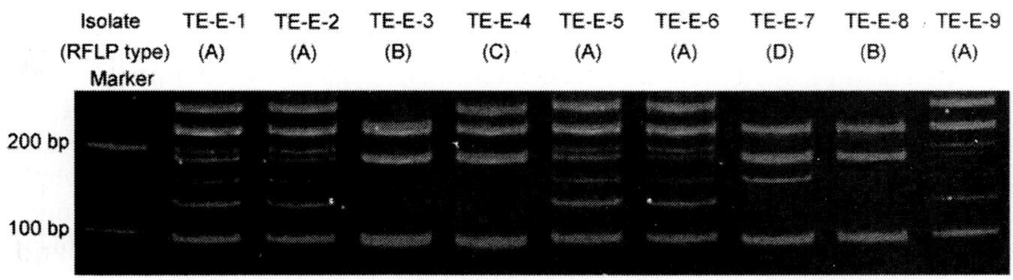

Fig. 5. Four RFLP types out of nine termite balls obtained from a single *R. speratus* colony

In the mutualistic interaction between fungus-growing termites and *Termitomyces*, only a single fungal clone is present per colony (Aanen et al., 2002). In contrast, multiple genotypes of the termite ball fungus per termite colony were found. For example, four different RFLP types out of nine termite balls obtained from a single *R. speratus* colony were found (Fig. 5). This high genetic variation of termite balls in each termite nest suggests that termites repeatedly bring the termite ball fungus into their nests from the surroundings.

Acknowledgements

We thank Drs. N.E. Pierce, E.O. Wilson, E. Vargo, S.-P. Quek, D.S. Hibbett, J.N. Thompson, T. Nishida, C. Tanaka, K. Fujisaki, T. Tamura and F. Nakasuji for the helpful advice and discussion.

References

Aanen, D.K. and Boomsma, J.J. 2005. Evolutionary dynamics of the mutualistic symbiosis between fungus-growing termites and *Termitomyces* fungi, pp. 191-210. In: Insect-Fungal Associations: Ecology and Evolution. F.E. Vega and M.Blackwell (eds.) Oxford University Press, New York, USA.

Aanen, D.K., Eggleton, P., Rouland-Lefevre, C., Guldberg-Froslev, T., Rosendahl, S. and Boomsma, J.J. 2002. The evolution of fungus-growing termites and their mutualistic fungal symbionts. Proceedings of the National Academy of Sciences of the United States of America, 99: 14887-14892.

Allan, R.A., Capon, R.J., Brown, W.V. and Elgar, M.A. 2002. Mimicry of host cuticular hydrocarbons by salticid spider *Cosmophasis bitaeniata* that preys on larvae of tree ants *Oecophylla smaragdina*. Journal of Chemical Ecology, 28: 835-848.

Brooke, M. de L. and Davies, N.B. 1988. Egg mimicry by cuckoos *Cuculus canorus* in relation to discrimination by hosts. Nature, 335: 630-632.

D'Ettorre, P., Tofilski, A., Heinze, J. and Ratnieks, F.L.W. 2006. Non-transferable signals on ant queen eggs. Naturwissenschaften, 93: 136-140.

Ito, F., Hashim, R., Huei, Y.S., Kaufmann, E., Akino, T. and Billen, J. 2004. Spectacular Batesian mimicry in ants. Naturwissenschaften, 91: 481-484.

James, T.Y., Kauff, F., Schoch, C., Matheny, B., Hofstetter, V., Cox, C., Celio, G., Gueidan, C., Fraker, E., Miadlikowska, J., Lumbsch, T., Rauhut, A., Reeb, V., Arnold, A.E., Amtoft, A., Stajich, J., Hosaka, K., Sung, G.H., Johnson, D., O'Rourke, B., Crockett, M., Binder, M., Curtis, J., Slot, J., Wang, Z., Wilson, A., Schusler, A., Longcore, J., O'Donnell, K., Mozley-Standridge, S., Porter, D., Letcher, P., Powell, M., Taylor, J., White, M., Griffith, G., Davies, D., Humber, R.A., Morton, J.B., Sugiyama, J., Rossman, A., Rogers, J., Pfister, D., Hewitt, D., Hansen, K., Hambleton, S., Shoemaker, R., Kohlmeyer, J., Volkmann-Kohlmeyer, B., Spotts, R., Serdani, M., Crous, P., Hughes, K., Matsuura, K., Langer, E., Langer, G., Unterreiner, W., Lucking, R., Budel, B., Geiser, D., Aptroot, A., Diederich, P., Schmitt, I., Schultz, M., Yahr, R., Hibbett, D., Lutzoni, F., McLaughlin, D., Spatafora, J., and Vilgalys, R. 2006. Reconstructing the early evolution of Fungi using a six-gene phylogeny. Nature, 443: 818-822.

Kistner, D.H. 2000. A new genus and species of a queen mimicking termitophile from Brazil (Coleoptera: Staphylinidae) Sociobiology, 35: 191-195.

Kitade, O., and Hayashi, Y. 2002. Localized distribution of an alien termite *Reticulitermes kanmonensis* (Isoptera: Rhinotermitidae). Entomological Science, 5: 197-201.

Lindstrom, L., Alatalo, R.V., Lyytinen, A. and Mappes, J. 2004. The effect of alternative prey on the dynamics of imperfect Batesian and Mullerian mimicries. Evolution, 58: 1294-1302.

Lockey, K.H. 1988. Lipids of the insect cuticle: origin, composition and function. Comparative biochemistry and physiology, B 89: 595-645.

Malcolm, S.B. 1990. Mimicry: status of a classical evolutionary paradigm. Trends in ecology and evolution, 5: 57-62.

Matsuura, K. 2002. Sociobiology of the Termite *Reticulitermes speratus* (in

Japanese). Doctoral thesis, Kyoto University, Kyoto, Japan.

Matsuura, K. 2003. Symbionts affecting termite behavior, pp. 131-143. In: Insect Symbiosis. T. Miller and K. Bourtzis (eds.), CRC Press, Boca Raton, Fl., USA.

Matsuura, K. 2005. Distribution of termite egg-mimicking fungi ("termite balls") in *Reticulitermes* spp. (Isoptera: Rhinotermitidae) nests in Japan and the United States. Applied Entomology and Zoology, 40: 53-61.

Matsuura, K. 2006. Termite-egg mimicry by a sclerotium-forming fungus. Proceedings of the Royal Society B: Biological Sciences, 273: 1203-1209.

Matsuura, K. and Nishida, T. 2001. Colony fusion in a termite: What makes the society "open" ?. Insectes Sociaux, 48: 378-383.

Matsuura, K. and Kobayashi, N. 2007. Size, hatching rate, and hatching period of sexually and asexually produced eggs in the facultatively parthenogenetic termite *Reticulitermes speratus* (Isoptera: Rhinotermitidae). Applied Entomology and Zoology, 42: 241-246.

Matsuura, K., Tanaka, C. and Nishida, T. 2000. Symbiosis of a termite and a sclerotium-forming fungus: Sclerotia mimic termite eggs. Ecological Research, 15: 405-414.

Puja, Z.K., Grogan, R.G. and Adams, Jr. G.C. 1982. Influence of nutrition, environment, and the isolate, on basidiocarp formation, development, and structure in *Athelia* (*Sclerotium*) *rolfsii*. Mycologia, 74: 917-926.

Rosegaus, R.B. and Traniello, J.F.A. 2001. Disease susceptibility and the adaptive nature of colony demography in the damp wood termite *Zootermopsis angusticollis*. Behavioral Ecology and Sociobiology, 50: 546-556.

Rothstein, S.I. 1990. A model system for coevolution: avian brood parasitism. Annual Review of Ecology and Systematics, 21: 481-508.

Servedio, M.R. and Lande, R. 2003. Coevolution of an avian host and its parasitic cuckoo. Evolution, 57: 1164-1175.

Singer, R.B., Flach, A., Koehler, S., Marsaioli, A.J. and Amaral, M.D.E. 2004. Sexual mimicry in *Mormolyca ringens* (Lindl.) Schltr. (Orchidaceae : Maxillariinae). Annals of Botany, 93(6): 755-762.

Wickler, W. 1968. Mimicry in Plants and Animals. McGraw-Hill, New York, USA.

Wilson, E.O. 1971. The Insect Societies. - Harvard Univ. Press, Cambridge, MA, USA.

Wood, T.G., and Thomas, R.J. 1989. The mutualistic association between Macrotermitinae and *Termitomyces*, pp. 69-92. In: Insect-Fungus Interactions. N.Wilding, N.M. Collins, P.M. Hammond and J.F. Webber (eds.), Academic Press, New York, USA.

Yashiro, T. and Matsuura, K. 2007. Distribution and phylogenetic analysis of the termite-egg mimicking fungi "termite balls" in *Reticulitermes* termites. Annals of the Entomological Society of America, 100: 532-538.

11

The Hallucinogenic Mushrooms: Diversity, Traditions, Use and Abuse with Special Reference to the Genus *Psilocybe*

Gastón Guzmán

Instituto de Ecologia, Km 2.5 carretera antigua a Coatepec No. 351
Congregación El Haya, Apartado postal 63, Xalapa,
Veracruz 91070, Mexico
E-mail: gaston.guzman@inecol.edu.mx

Abstract

The traditions, uses and abuses, and studies of hallucinogenic mushrooms, mostly species of *Psilocybe*, are reviewed and critically analyzed. *Amanita muscaria* seems to be the oldest hallucinogenic mushroom used by man, although the first hallucinogenic substance, LSD, was isolated from ergot, *Claviceps purpurea*. *Amanita muscaria* is still used in North Eastern Siberia and by some North American Indians. In the past, some Mexican Indians, as well as Guatemalan Indians possibly used *A. muscaria*. *Psilocybe* has more than 150 hallucinogenic species throughout the world, but they are used in traditional ways only in Mexico and New Guinea. Some evidence suggests that a primitive tribe in the Sahara used *Psilocybe* in religions ceremonies centuries before Christ. New ethnomycological observations in Mexico are also described.

INTRODUCTION

After hallucinogenic mushrooms were discovered in Mexico in 1956-1958 by Mr. and Mrs. Wasson and Heim (Heim, 1956; Heim and Wasson, 1958; Wasson, 1957; Wasson and Wasson, 1957) and Singer and Smith (1958), a lot of attention has been devoted to them, and many publications have flooded the literature (e.g. Singer, 1958a, b, 1978; Gray, 1973; Schultes, 1976; Oss and Oeric, 1976; Pollock, 1977; Ott and Bigwood, 1978; Wasson, 1980; Ammirati et al., 1985; Stamets, 1996). However, not all the fungi reported really have hallucinogenic properties, because several of them were listed by erroneous interpretation of information given by the ethnic groups originally interviewed or by the bibliography.

In spite of the above studies those of Heim (1956), Wasson and Wasson (1957), Schultes (1939) were the first to review the use of hallucinogenic mushrooms in Mexico. Schultes got two packages of sacred mushrooms from the Indians, and he also gathered more mushrooms in a third package. These three packages were deposited for identification in the Herbarium of Harvard University. Only specimens in the third package were identified as *Panaeolus campanulatus* var. *sphinctrinus* (today known as *P. sphinctrinus*). This identification was cited by Schultes in his first publication (Schultes, 1939), wherein he claimed that the Indians throughout Mexico used *P. sphinctrinus* as a sacred mushroom, which they identified as 'teonanácatl'. However Schultes made two mistakes, because *P. sphinctrinus* is not use as sacred mushroom in any part of the country, and the name 'teonanácatl' is an Aztec term known only from literature of the 16th century (e.g. Sahagún, 1569-1582). The true identity of the 'teonanácatl' of Sahagún was an enigma for centuries. Moreover, other specialists cited the paper of Schultes (1939) as evidence that all the species of *Panaeolus* (including *Panaeolina*) are narcotic (e.g. Singer, 1949; Gray, 1973; Schultes and Hofmann, 1973; Ott and Bigwood, 1978), despite Safford's (1915) claim that 'teonanácatl' is *Lophophora williansii*, a cactaceous plant with hallucinogenic properties used by the Indians of northern Mexico and the southern U.S.A. It is now known that Safford's interpretation was erroneous.

When Singer was at Harvard University in the 1940s, he identified the first mushroom package of Schultes obtained from the Indians. Singer identified it as *Psilocybe cubensis*. At the same time Singer checked the third package and agreed with the *Panaeolus sphinctrinus'* identification. This information was published in his monumental work 'The Agaricales' (Singer, 1949) but only as a little note in the chapter on *Psilocybe*: "At least one species is used as a drug in Mexico (causing a temporary narcotic state of hilarity)...." Singer (1949) also wrote in his chapter on *Panaeolus*, "*Panaeolus sphinctrinus* and *P. papilionaceus* are used as intoxicating drugs in Central America together with *Psilocybe cubensis*...." These two notes of Singer were the first indication that a species of *Psilocybe* is hallucinogenic. In 1970 Guzmán (1983) restudied specimens in the second Indian package of Schultes at the Harvard University Herbarium, previously identified by Singer as *Deconica* sp., Guzmán identified those mushrooms as *P. caerulescens*, another important sacred mushroom among the Indians, which Heim

(1956) considered as *P. mazatecorum* Heim, and then (in Heim and Wasson, 1958) as *P. caerulescens* var. *mazatecorum* (Heim) Heim. Guzmán (1959) had earlier found that the Indians in a locality of the state of Puebla called this mushroom 'teotlaquilnanácatl', but not 'teonanácatl' (both words are Aztec names; 'teotlaquilnanácatl' means 'sacred mushroom that describes or paints', and 'teonanácatl' means only 'sacred mushroom' (from nanácatl = mushroom, teo = sacred, and tlaquil = paint). Another confusion about the hallucinogenic mushrooms is whether or not the genus *Stropharia* has hallucinogenic species, because *Psilocybe cubensis* was first described as *Stropharia cubensis* from Cuba in the early 20th century; Heim used the name *Stropharia cubensis* in all his publications. However Singer (1949) for taxonomic reasons transferred the species to *Psilocybe*, a position followed since by all mycologists. No species of *Stropharia sensu stricto* have hallucinogenic properties. The other confusion on the hallucinogenic mushrooms will be discussed below.

In the present work the diversity of the hallucinogenic mushrooms in the world is described. This contribution is based on the researches of the author since 1957, mainly in the genus *Psilocybe* (e.g. Guzmán, 1959, 1983, 1995, 2005).

The known hallucinogenic fungi

More than 200 species of fungi have been reported with hallucinogenic properties (Table 1). They belong to more than 10 genera, both Basidiomycota (the majority) and Ascomycota. Table 1 shows the most important fungi reported in the literature as hallucinogenic, but *Psilocybe, Amanita muscaria* and *Claviceps purpurea* are the true hallucinogenics. For *Panaeolus* and *Panaeolina* see the discussion above. Species of *Conocybe, Copelandia, Gymnopilus, Inocybe* and *Pluteus* have hallucinogenic properties, because they have psilocybin (Schultes and Hofmann, 1979; Schultes, 1976; Stijve et al., 1985). *Conocybe siligineoides* was reported by Heim (Heim and Wasson, 1958; Wasson, 1957) only one time as an hallucinogenic mushroom by the Mazatec Indians of Mexico. It is a very rare fungus, because the author in his numerous explorations in the region, never found this mushroom or heard about it from the Indians. Species of the genera *Boletus, Cordyceps, Elaphomyces, Heimiella, Psathyrella, Russula, Lycoperdon* and *Vascellum* are not hallucinogenic mushrooms. The two puff-balls considered above, *Lycoperdon* and *Vascellum,* were reported by Heim and Wasson (Heim et al., 1967) as narcotic fungi in an Indian Mixtec town of Mexico. They reported *Lycoperdon mixtecorum* (actually *Vascellum qudenii*) and *Lycoperdon marginatum* (a synonym of *L. candidum*). Both are edible mushrooms, without any narcotic effect as Guzmán described in Ott et al. (1975). The report on *Cordyceps* (Table 1) is based on *C. capitata* and *C.*

Table 1. The hallucinogenic fungi known in the world (those with ? means doubtful hallucinogenic properties; those in boldface are used by ethnic groups in a traditional way; in some species the isolated hallucinogenic substance is presented)

Ascomycotina, Hypocreales

Claviceps purpurea (Fr.) Tul., 'ergot', LSD

? ***Cordyceps capitata* (Holms.: Fr.) Link**

? ***C. ophioglossoides* (Ehrh.) Link**

? ***Elaphomyces granulatus* Fr.**

Basidiomycotina, Agaricales

***Amanita muscaria* (L.: Fr.) Pers. ex Hooker,** ibotenic acid

? ***Boletus kumaeus* R. Heim**

? ***B. manicus* R. Heim**

Conocybe cyanopus (G.K. Atk.) Kühnar, psilocybin

***C. siligineoides* R. Heim,** psilocybin

C. smithii Watling, psilocybin

Copelandia cyanescens (Berk. & Broome) Singer, psilocybin

C. westii (Murrill) Singer & Weeks

Gymnopilus aeruginosus (Peck) Singer, psilocybin

G. braendlei (Peck) Hesler, psilocybin

G. luteoviridis Thiers, psilocybin

G. purpuratus (Cooke & Mass.) Singer, psilocybin

? *G. spectabilis* (Fr.) A.H. Smith

? *Heimiella anguiformis* (R. Heim)

Inocybe aeruginascens Babos

I. calamistrata (Fr.: Fr.) Gill.

? *Panaeolina foenisecii* (Pers.: Fr.) Maire

? *Panaeolus campanulatus* (L.: Fr.) Quél. s.l.

? *P. fimicola* (Pers.) Gill.

? *P. sphinctrinus* (Fr.) Quél.

P. subbalteatus (Berk. & Broome) Sacc., psilocybin

? *P. venenosus* Murrill

Pluteus salicinus (Pers.: Fr.) P. Kumm., psilocybin

P. nigroviridis Babos

? ***Psathyrella sepulchralis* Singer, A.H. Smith & Guzmán**

***Psilocybe* spp.** (see Tables 2 & 3), psilocybin, baeocystin and others

? ***Russula agglutinata* R. Heim**

? ***R. nondorhingi* Singer**

? ***R. wahgiensis* Singer**

Basidiomycotina, Gasteromycetes

? ***Lycoperdon marginatum* Vitt.**

? ***L. mixtecorum* R. Heim**

? ***Vascellum intermedium* A.H. Smith** (all these gasteromycetous fungi are edible)

ophioglossoides, two strong fungi related to 'ergot'. *Elaphomyces*, based on *E. granulatus* and other species, are edible hypogeous fungi used by Mexican Indians in ceremonies with sacred mushrooms with *Psilocybe muliercula*. The species of *Boletus* and *Heimiella* in the Boletaceae may be toxic but not narcotic. They were reported in a confused way as neurotropic from Papua New Guinea by Heim and Wasson (1965). The species of *Russula* were also reported by Heim and Wasson (1965) and Singer (1958a, b). The case of *Psathyrella sepulchralis* [= *P. asperospora* (Clel.) Guzmán, Bandala and Montoya] was reported (Singer et al., 1958) because the Zapotec Indians in a Mexican town confused it with *Psilocybe zapotecorum*.

Claviceps purpurea is the first hallucinogenic fungus known in the history of the neurotropic fungi. Known as 'ergot', it parasitizes the tassels of rye. However, this fungus has never been used as an hallucinogenic. 'Ergot' produces poisonous sclerotia (small, hard, black horns) on the tassels. When these sclerotia are accidently mixed in flour used to make bread, the bread becomes an important and often fatally poisonous and, when eaten, produces delirium, convulsions and hallucinations. This intoxication was common in Europe in the Middle Ages and it was known as 'Saint Antony's Fire'. In the 1940s Hofmann isolated from those sclerotia the first hallucinogenic substance known in a fungus, the lysergic acid of diethylamide, known as LSD, one of the numerous alkaloids which this fungus produces (Samorini, 2001; Schultes and Hofmann, 1973, 1979). LSD is an indole similar to the psilocybin of hallucinogenic species of *Psilocybe* (see below).

Amanita muscaria is the first narcotic fungus historically used by man. Primitive tribes of northeastern Siberia used and still use this mushroom in special ceremonies (Ford and Clark, 1914; Wasson and Wasson, 1957; Schultes, 1976; Schultes and Hofmann, 1979). Moreover, this mushroom is still used by North American Indians from Canada and the U.S.A. (Wasson, 1979) (this latter information strengthens the theory that man came from Asia across the Bering Strait). It is even very probable that Mexican and Guatemalan Indians used *A. muscaria* before they used species of *Psilocybe*, as discussed by Lowy (1974) and Guzmán (1990, 2001). Also, *A. muscaria*, which is common in Europe, has interesting relationships with the beginning of Christianity, as shown in a fresco found in an old church in France: a stylized *A. muscaria* is positioned between Adam and Eve as 'The Tree of Life' (Wasson and Wasson, 1957; Samorini, 2001; Gray, 1973). The hallucinogenic compound of *Amanita muscaria* is ibotenic acid, first erroneously reported as bufotenin, an indole isolated from a toad (Schultes and Hofmann, 1973). The ibotenic acid is an indole like LSD and psilocybin. However, the ingestion of *A. muscaria* produces first a gastrointestinal intoxication due to muscarine (a toxic glucoside) which this mushroom also contains. It is interesting to observe that one of first neurotropic effects of *A. muscaria* is to see all surrounding things as gigantic, as will be discussed below which concern Mexican Indians traditions. These enlarging effects relate directly to the history of the gnomos in Europe, which supposedly began with the ingestion of *A. muscaria*. Stijve (1995) reported ibotenic acid also in other species of *Amanita*, such as *A. regalis* (Fr.) Michael and *A. pantherina* (DC.: Krombh.).

The genus *Psilocybe*

The hallucinogenic species of *Psilocybe* are the most diverse and those with the widest geographical distribution, and the most important mushrooms in the hallucinogenic fungi. Moreover, the hallucinogenic species of *Psilocybe* are the most important in ethnomycology. They probably were used in many ethnic groups in the world, but at the present only certain groups of Mexican Indians (Wasson and Wasson, 1957; Guzmán, 2001) and the Kuma in Papua New Guinea (Heim and Wasson, 1965; Heim et al., 1967) considered these mushrooms as sacred. Some important hallucinogenic species of *Psilocybe* in the world are show in Figs. 1-14. See also Tables 2 & 3.

Guzmán (1983) reported 91 species of hallucinogenic species of *Psilocybe* in the world versus no more than 20 species reported by Heim and Wasson (1958) and Singer and Smith (1958). The new edition of 'The Genus *Psilocybe*' in preparation by the author includes more than 150 species of hallucinogenic species of *Psilocybe* throughout the world, of which more than 55 are in Mexico; 50 in Latin America including the Caribbean, but excluding Mexico; 22 in the U.S.A. and Canada; 16 in Europe, 15 in Asia; 15 in Australia and eastern islands, and only 4 in Africa (Guzmán, 2005) (Table 2) (Fig. 15). Several species are in more than one continent, for example *P. semilanceata,* which is common in Europe and the U.S.A., is also known from Chile and Tasmania. *Psilocybe cyanescens* is known from Europe and Canada. *Psilocybe yungensis* and *P. zapotecorum* are known from Mexico and South America. With the recent study of *P. fagicola*-complex (Guzmán et al., 2005) this complex is known from Mexico, Colombia and Indonesia. *Psilocybe cubensis* and *P. subcubensis* are pantropical species. Table 3 shows the most common species of *Psilocybe* in the world.

The bluing reaction of fresh fruit bodies of hallucinogenic species of *Psilocybe* is the best way to distinguish them from those that lack that property. *Psilocybe coprophila, P. montana* and *P. argentina,* among others, have a wide world distribution but do not have neurotropic properties. However, sometimes the bluing reaction is difficult to observe, depending on the developmental stage of the fruit body, as is the case with *P. semilanceata* and *P. mexicana*. It is important to note that the bluing reaction

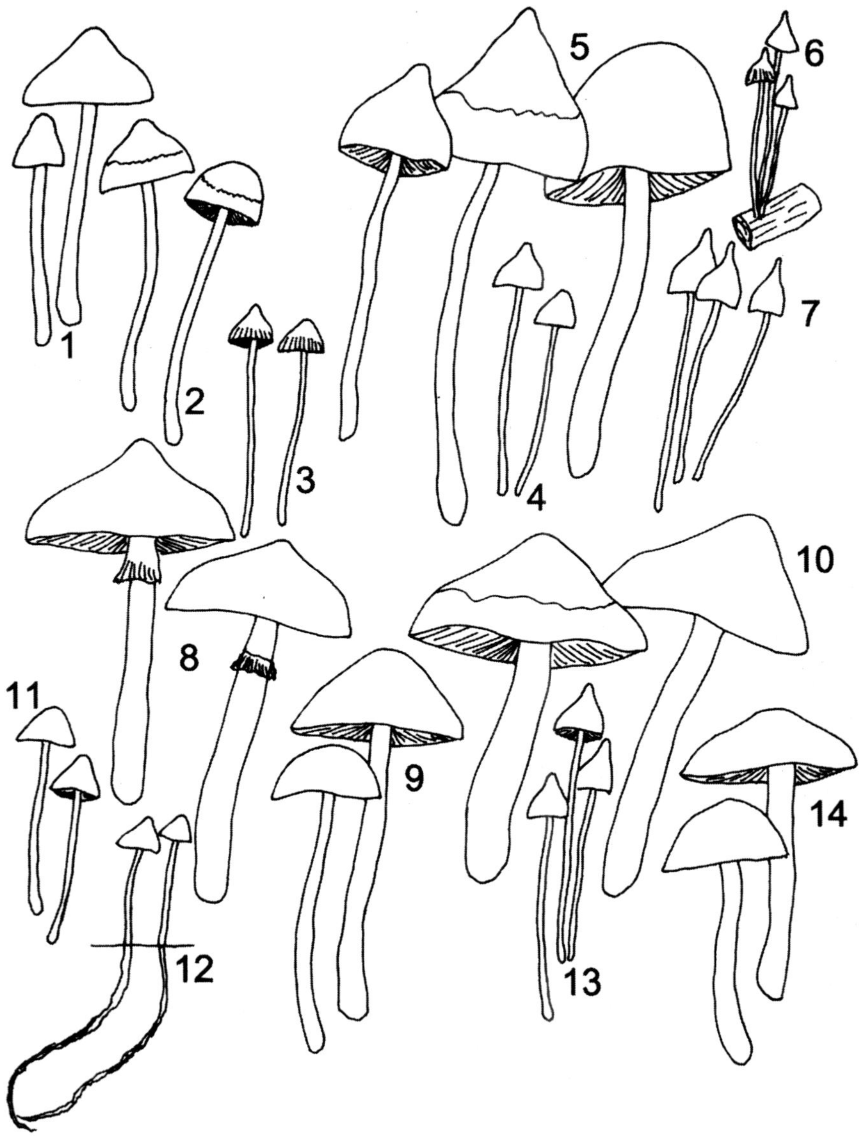

Figs. 1-14. Some species of *Psilocybe*. 1: *P. baeocystis*, 2: *P. aztecorum*, 3: *P. silvatica*, 4: *P. pelliculosa*, 5: *P. zapotecorum*, 6: *P. yungensis*, 7: *P. hoogshagenii*, 8: *P. cubensis*, 9: *P. caribaea*, 10: *P. caerulescens*, 11: *P. brasiliensis*, 12: *P. fagicola*, 13: *P. semilanceata*, 14: *P. cyanescens*

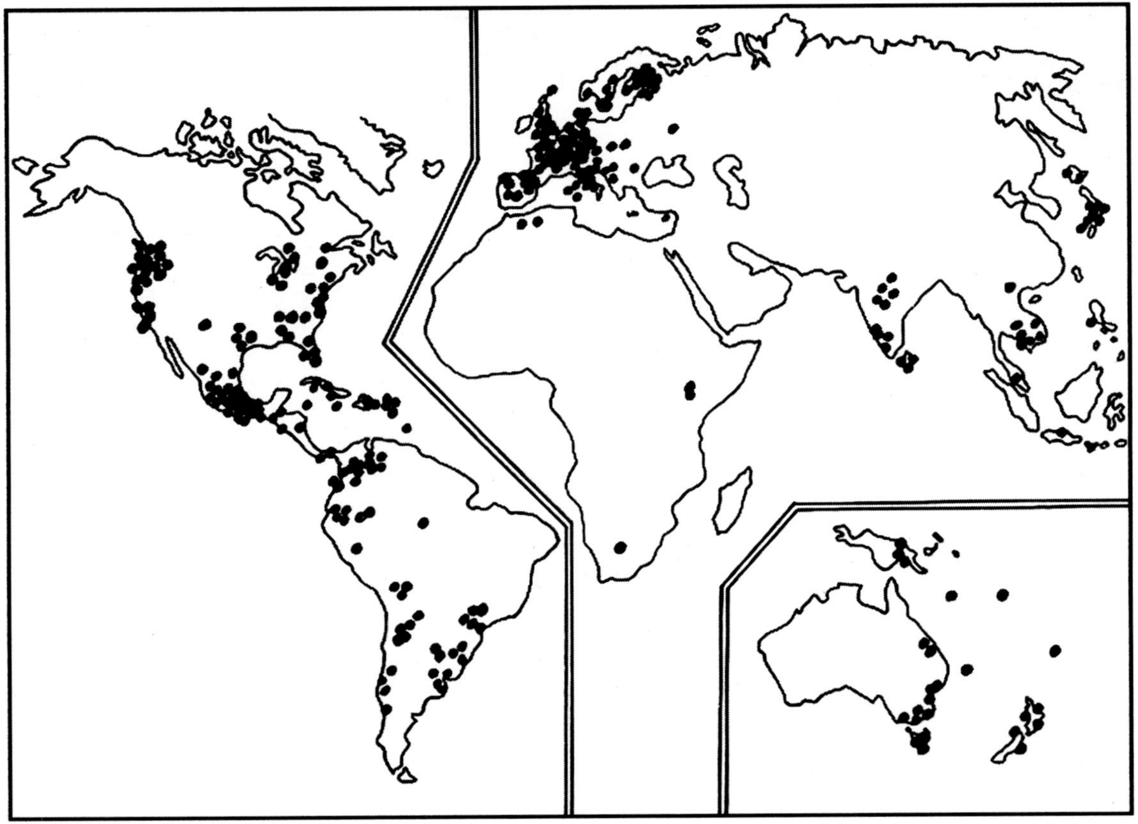

Fig. 15. World distribution of the hallucinogenic species of *Psilocybe* (each dot is one or several localities)

Table 2. Distribution of the known 150 hallucinogenic species of *Psilocybe* in the world

Mexico	55 species
Latin America including the Caribbean, but excluding Mexico	50 species
United States and Canada	22 species
Europe	16 species
Asia	15 species
Australia and eastern islands	15 species
Africa	4 species

The total is 177, because some species are common in several continents (e.g. *P. cubensis, P. cyanescens* and *P. semilanceata*).

is common in other mushrooms, edible or poisonous, for example in *Boletus satanas* (poisonous) and in *B. erythropus* (edible). Moreover some species of *Conocybe, Copelandia, Gymnopilus* and *Pluteus* turn blue, because they contain psilocybin, as described in Table 1.

Ethnomycological studies

The traditional use of hallucinogenic mushrooms was very important in the past, as discussed above with *Amanita muscaria* and with sacred species of *Psilocybe* in Mexico (Wasson and Wasson, 1957; Wasson, 1980). In Mexico, as stated in many documents, e.g. Sahagún (1569-1582), and some codexes, as the Magliabechiano Codex (Wasson and Wasson, 1957; Gray, 1973; Wasson, 1980) the Indians considered hallucinogenic mushrooms as sacred. They identified them with several common names, such as 'teonanácatl' (Sahagún, 1569-1582), 'teotlaquilnanácatl' (Guzmán, 1959), 'apipiltzin' (= little boy of the rain) or 'siwatsintli' (= little women) (Heim and Wasson, 1958), among others. Mazatec and Zapotec Indians in Oaxaca have around one hundred common names depending on the species and region. Guzmán (1983) presented a list of more than one hundred Indians names of the hallucinogenic mushrooms in Mexico, but later (Guzmán, 1997) described about one thousand names, both in Spanish and Indian languages, from Mexico.

When the Indians use hallucinogenic mushrooms in their ceremonies, they are always supervised by an old person or a shaman of the community. Shamans were and they are very common throughout the Mexican Indian world. These shamans are women or men and are specialists in

Table 3. Some common species of neurotropic species of *Psilocybe* in the world

P. argentipes Yokoyama. Known only from Japan.

P. aztecorum R. Heim emend. Guzmán. Known only from Mexico (Fig. 2).

P. baeocystis Singer & A.H. Smith emend. Guzmán. Known from the U.S.A. and Canada (Fig. 1).

P. barrerae Cifuentes & Guzmán emend. Guzmán. Known only from Mexico.

P. brasiliensis Guzmán. Known only from Brazil (Fig. 11).

P. caeruleoanulata Singer ex Guzmán. Known from Brazil and Uruguay.

P. caerulescens Murrill. Known from SE of the U.S.A., Mexico and South America (Fig. 10).

P. caribaea Guzmán, T.J. Baroni & Tapia. Known from the Caribbean and Mexico (Fig. 9).

P. columbiana Guzmán. Known only from Colombia.

* *P. cubensis* (Earle) Singer. Pantropical species (Fig. 8).

P. cyanescens Wakef. (a complex). Known from Europe, the U.S.A. and Canada (Fig. 14).

P. cyanofibrilosa Guzmán & Stamets. Known only from the U.S.A.

P. fagicola R. Heim & Cailleux emend. Guzmán. Known only from Mexico (Fig. 12).

P. fimetaria (P.D. Orton) Watling. Known in North America, Europe and Chile.

P. guilartensis Guzmán F., Tapia & Nieves-Riv. emend. Guzmán. Known from the Carribbea.

P. hoogshagenii R. Heim. Known from Mexico and South America (Fig. 7).

P. keralensis K.A. Thomas, Manim. & Guzmán. Known only from India.

P. kumaenorum R. Heim. Known only from Papua New Guinea.

P. laetissima Hauskn. & Singer. Known only from Europa.

P. liniformans Guzmán & Bas. Know only in Europe.

P. mammillata (Murrill) A.H. Smith. Known from the U.S.A., Jamaica, Bolivia and Mexico.

P. mexicana R. Heim. Known form Mexico and Guatemala.

P. pelliculosa (A.S. Smith) Singer & A.H. Smith. Known from North America and Europe (Fig. 4).

P. plutonia (Berk. & M.A. Curtis) Sacc. Known from Mexico, the Caribbean and South America.

* *P. semilanceata* (Fr.: Secr.) P. Kumm. Known from Europe, North America, Chile and Tasmania (Fig. 13).

P. silvatica (Peck) Singer & A.H. Smith. Known from North America and Europe (Fig. 3).

Contd.

Table 3 continued

P. stuntzii Guzmán & J. Ott. Known only from the U.S.A.
P. subcaerulipes Hongo. Known only from Japan.
P. subaeruginosa Clel. Known only from Australia.
P. subzapotecorum Guzmán. Known only from Mexico.
* *P. tampanensis* Guzmán & S.H. Pollock. Known only from Florida, U.S.A.
P. venenata (S. Imai) Imazeki & Hongo. Known only from Japan.
P. yungensis Singer & A.H. Smith. Known from Mexico and South America (Fig. 6).
P. zapotecorum R. Heim emend Guzmán. Known from Mexico and South America (Fig. 5).

sacred mushrooms, as well as other sacred plants. The Wassons (Wasson and Wasson, 1957; Heim and Wasson, 1958) found the shaman María Sabina in Huautla de Jiménez town; she showed them all the sacred traditions related to the mushroom ceremonies. María Sabina's knowledge of the mushrooms had such an impact on the Wassons, that they and their collaborators published a huge book with four long-playing records about it (Wasson et al., 1974). The book presented the shamanic ceremony performed in 1958 in Huautla de Jiménez, with canticles and prayers by María Sabina. After this publication on María Sabina, many books and articles were published about her, appointing her as the principal shaman on sacred mushrooms all over Mexico. However, it must be emphasized that at that time important Indian shamans with knowledge of the sacred mushrooms were common in Mexico, as the author saw in his field expeditions during that period (Fig. 17).

Unfortunately, due to the advance of modern civilization, the traditions of hallucinogenic mushrooms are being lost. Today it is difficult to find a good shaman specialist in sacred mushrooms, but it is easy to find 'meztizos' (crossbreeds between Indian and Spanish men) who trade in the Indian traditions and the mushrooms (Guzmán, 1990, 2001). Wasson (1980) and Lowy (1974), among others, discussed the probable but now lost use of sacred mushrooms among the people of Central America. Also, Lowy (1974) described the probable sacred use of *Amanita muscaria* among Mayan Indians of Mexico and Guatemala in relation to the Thunderbolt Legend and from the study of several codexes. Guzmán (1990) discussed the probable use of *A. muscaria* among the Purepechas of Mexico, in relation to a little stone figure found at an archaeological site: it seems to be the button of an *A. muscaria*. Also, Schultes and Hofmann (1979) presented a pottery figure which represents an Indian seated below an *A. muscaria*, which this author (Guzmán, 2001) has as an evidence of the use of *A.*

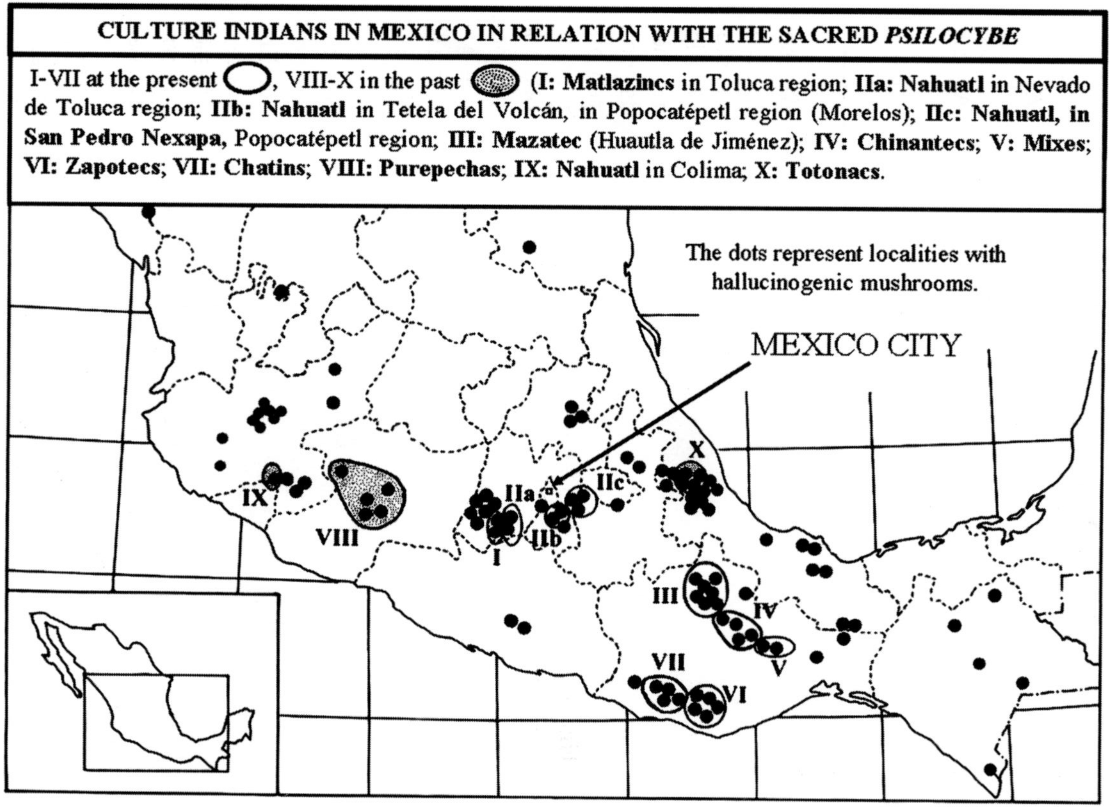

Fig. 16. Distribution of the hallucinogenic species of *Psilocybe* in Mexico, and the Indian culture related to them

Fig. 17. A typical Mexican Indian shaman in relationship with the sacred mushroom (photo by Guzmán in 1958)

Fig. 18. Pottery figure from Colima (Mexico) showing the gigantism effect after the ingestion of the hallucinogenic mushrooms. Observe the eyes and the snake on the hats of the Indians (from Schultes and Hofmann, 1979)

Fig. 19. A possible representation of Quetzalcóatl in a Colima pottery piece. Note the similarity of this with Fig. 18, where is a mushroom in the center (from del Villar, 2005)

muscaria among the Nahuatl Indians from Jalisco and Colima of Mexico in the past, as they also used the species of *Psilocybe* (see below).

At present in Mexico, six Indian cultures are related to hallucinogenic species of *Psilocybe*, as shown in Fig. 16. They are the Matlazincs and Nahuatls in central Mexico, and the Mazatecs, Chinantecs, Mixes, Zapotecs and Chatins in the State of Oaxaca. In the past it was probably the Colima Indians of Nevado de Colima, Purepechas of Michoacan and Totonacs of Veracruz who used these mushrooms. In regard to the group from Colima, the author found in Schultes and Hofmann (1979) and in del Villar (2005) and Díaz (2003) in the anthropological studies of the journal *Antropología Mexicana*, three interesting Indian pottery figures, which have a strong relationship to the traditional use of hallucinogenic species of *Psilocybe* and gigantism effects (Figs. 18-20). Fig. 18 shows four Indians surrounding a tall mushroom, which undoubtedly is a *Psilocybe* of the *Zapotecorum* group following Guzmán's classification (Guzmán, 1983). According to Schultes and Hofmann (1979), who first published the picture, these four Indians are dancing. However this interpretation is wrong, because an effect of the hallucinogenic mushrooms makes it difficult to stay standing, let alone

Fig. 20. An Indian woman with the effect of gigantism after eating hallucinogenic mushrooms. Note the gigantic mushroom in one of her hands and the eyes very big and out of orbit (from del Villar, 2005)

walking or dancing; the four Indians are really embracing, observing with admiration the gigantic mushroom which supposedly they ate. Other important features of this figure are the hats of the Indians, which are snakes, as are in their arms. Snakes were very important in the religion of several Indian cultures in Mexico, which considered these animals as sacred. To the Nahuatls snakes represented the god Quetzalcóatl, as was also true for the Teotihuacans, who represented this god in their pyramids. Even more interesting is Fig. 19, which resembles Fig. 18 except instead of the tall mushroom, the center contains another Indian with snakes on his head, neck and hands. This author concluded from these two figures in Colima that the cult of hallucinogenic mushrooms was related to the god Quetzalcóatl an important god among the Nahuatl Culture. Unfortunately neither an anthropological nor ethnomycological study has been devoted to these figures nor the relationships with the Quetzalcóatl god. Thus, the relationships of the sacred mushrooms with Quetzalcóatl is presented for the first time here.

Another pottery figure, also from Colima (Fig. 20) (del Villar, 2005; Díaz, 2003) shows a woman with a gigantic mushroom in her right hand, like that of Fig. 18. Thus we have here another representation of the gigantism symptomatic of hallucinogenic mushrooms. Note the eyes of the woman: they are very big, are out of orbit, probably due to the neurotropic action of the mushrooms and to see one of them as a gigantic one. This author experienced the gigantism effects when he ate *Psilocybe cubensis* in a traditional ceremony in 1958. At the beginning of his hallucinations, this author saw gigantic black men dancing around him, but those 'men' were actually one little dog which was chasing a cat around his bed.

Traditional use of sacred mushrooms in Indian ceremonies in Mexico

It is very important to know the procedures and care followed by Mexican Indians in their nocturnal ceremonies with hallucinogenic mushrooms. Such knowledge is useful for avoiding psychological problems or possible poisoning. Indians regard hallucinogenic mushrooms as sacred and use them with respect and care; accordingly, they never encounter health problems from using them. They eat the mushrooms to find health or to talk with God or dead relatives, although these are suggestive. They relate the ceremony to the Catholic religion, and the rite is performed in front of a little altar in the house where the ceremony takes place. They pray in a confused mix of their Indian language and in Spanish. The shaman or the director of the ceremony passes the mushroom through incense before they are eaten. Indians eat the mushrooms fresh or dry, alone or with water, but never with food. Some modern modifications introduce honey or chocolate, but this is rare. During ingestion of the mushrooms they follow simple rules to prevent health problems: 1) They eat the mushrooms only at night to avoid noise and distractions, in order to concentrate on the visions. 2) They do not take meals, alcohol, medicines or drugs. 3) The dosage is more than six but not more than 12 fresh fruit bodies, twice as many if the mushrooms are dry. 4) They always take the mushrooms under the supervision of an old person or shaman, never alone. 5) They avoid travel during the ensuing seven days, resting at home under proper care, because the nervous system remains very sensitive. Indians have followed these five rules for centuries and never had health problems. It is interesting to talk with shamans who took hallucinogenic mushrooms all their life, and to observe that they are normal persons in all respects. The important conclusion about the Indian use of the hallucinogenic mushrooms is that health problems are prevented by following these five rules.

Effects of the hallucinogenic mushrooms and their application in neuropsychiatry

Hallucinogenic mushrooms do not always produce hallucinations, but they always act on the central nervous system; accordingly, a more correct term for them is neurotropic, although there are several other names used in the bibliography, such as psychedelics, psychotropics, divines, healing, visionary and entheogens, among others (Guzmán, 1997, 2001). These mushrooms produce a dual perception between reality and the imagination. Those who eat the mushrooms do not forget who they are, but at the same time they accept another perception of themselves, and transform objects they see to visions remembered or imagined. The former are illusions and the latter are hallucinations, but in general both are initially gigantic figures with vivid colors. They hear voices from the objects they envision. All the effects vary according to the cultural background of the person, the quantity of mushrooms ingested and the surrounding circumstances. The effects of the neurotropic fungi begin approximately 30 minutes after ingestion and continue for 4-6 hours. When the effects subside, the person sleeps normally, and upon awakening is in a normal condition but remembers perfectly and will never forget the sensations experienced (Guzmán, 1990, 2001, 2003).

When this author ate neurotropic mushrooms as an experiment during a traditional Indian ceremony in an Indian house in 1958, he experienced illusions and hallucinations. He had decided to try these mushrooms, because he was finishing more than two years of field research on them. The author was gathering hallucinogenic mushrooms for a Swiss laboratory, and he doubted that the mushrooms were hallucinogenic. Following Indian counsel, he ate 12 fresh fruit bodies of *Psilocybe cubensis* gathered by him that morning. The ingestion was an interesting Indian nocturnal ceremony conducted by an old Mazatec woman, the mother of the owner of the Indian house, where this author was living, in a little Indian farm close to Huautla de Jiménez. After the ceremony this author remained talking for some minutes at a table with the only Indian there who spoke Spanish, the son of the woman who conducted the ceremony. This man had not consumed any mushrooms. The author's mind at that moment was clear, so he thought that the hallucinogenic effects were untrue. Then he decided to go to his straw sleeping mat, which was in a corner of the room where the ceremony had taken place.

While he was on his straw sleeping mat, he suddenly saw a caricature of a gigantic, colored castle with two human faces. The castle was his gasoline-lamp mushroom dryer at an opposite corner of the room. The castle smiled and said to him: "Come, come to me, don't be afraid". The author was greatly surprised and frightened, and he reached for his eyeglasses to see his mushroom dryer. Then the castle laughed loudly. He decided to turn his back on it and sleep, but sleep was impossible, because he began to see many attractive, bright colors wherever he turned, regardless of whether his eyes were open or closed. These colors gradually transformed into gigantic black men, who danced around him, singing. In that moment the author felt very comfortable and asked the castle to please keep silent, because he was busy and happy looking at the spectacle. As discussed above in the section *Ethnomycological Studies*, these gigantic men were really a little, dark dog chasing a cat. After the above visions, he saw many others, spectacularly colored things and persons, and experimented many other cases, but the castle was there all night, saying to him, "come, come to me....". Finally he fell into a normal sleep after almost 6 hours of neurotropic effects (Guzmán, 1990).

These experiences with the hallucinogenic mushrooms, practiced by Mexican Indians for centuries, attracted the attention of chemistry laboratories. Medical and psychotherapeutic researchers were very interested to study these mushrooms or the psilocybin isolated from them to better understand neurological disorders and schizophrenia. They conducted experiments in Europe and the U.S.A. Nevertheless, overdoses of either mushrooms or psilocybin can result in convulsions or death (McCawley et al., 1962). Psilocybin in the blood inhibits serotonin, the hormone that controls the central nervous system. The absence of serotonin causes a temporal, abnormal psychotic stage with illusions and hallucinations, that disappears when the serotonin is replaced by normal biochemical processes of the body (Heim and Wasson, 1958; Singer, 1958b, 1978). However, as it will be described below, all these researches were stopped because of abuses of the use of these mushrooms by young people.

It is important to note that the effects of neurotropic mushrooms or psilocybin are produced not only by their ingestion. The hallucinations will start also after inhaling the volatiles produced by the mushrooms, as this author determined some time ago in field work. He found that psilocybin is volatilized from fresh fruit bodies. This is why the Indians said that old, dry specimens of sacred mushrooms are not good; they keep dry mushrooms no more than six months. One night in 1958 this author was trying to sleep in an Indian house in a small, closed room containing plenty of fresh hallucinogenic mushrooms, which he had gathered for a Swiss laboratory. The author started to see hallucinations when he was on his camp bed. He saw great numbers of little, colored bubbles flying in the air of the room, but he observed that these colored bubbles appeared each time a little leak dripped on his bed (it was raining outside). Each drip exploded into many colored bubbles that flew through the air. He felt nervous because he thought that the colored hallucinations he saw months ago with ingestion of the mushrooms (see above) would start. He needed to go out of the room to breathe fresh air and to shower in the rain to stop the intoxication from the air of the room. This volatility of psilocybin does not appear to have been reported in the literature before now.

Diffusion of knowledge, abuses, trade and confiscation of hallucinogenic mushrooms

The recreational use of hallucinogenic mushrooms, now so common among young people, started after the diffusion of knowledge of these mushrooms in the 1960's as discussed in the Introduction. The increase of interest in eating hallucinogenic mushrooms forced the governments of many countries to introduce legislation aimed at preventing the use, trade, culture and distribution of these mushrooms (Guzmán, 2003). The problem was with people who carelessly consumed these mushrooms, ignoring the traditional use of the mushrooms by the Indians. Many Mexican places with Indian traditions of sacred mushrooms, such as Huautla de Jiménez in Oaxaca, were invaded by foreign young people, who profaned the Indian traditions.

However, the legislation against use of hallucinogenic mushrooms passed by several governments stopped scientific and medical research on these mushrooms, with serious consequences. Also, in some parts of Mexico the Indians were restricted in their traditional use of these mushrooms. Nevertheless, at present a trade exists in Indian ceremonies with these mushrooms flowers in Huautla de Jiménez, where abuses, confiscations or adulterations are common. Governments in North America, Europe and Japan vigorously enforce their laws to stop trade with these mushrooms (e.g. Canada in the case of Heim, 1966). Consequently, it is difficult or almost impossible today to mail dried scientific specimens of *Psilocybe* for research purposes such as taxonomic study, even non-hallucinogenic taxa, because they are confiscated by customs!

Corollary

As hallucinogenic mushrooms were discovered in modern science more than 50 years ago, it might be supposed that all research on them has been completed. However, many matters of ethnomycology, anthropology, taxonomy, distribution, chemistry, physiology and medicine are still poorly known. The scant information from Africa (Morocco, Algeria and Libia) on mushroom drawings in the Sahara Desert, in relationship with *Psilocybe mairei* or also on the use of *P. kumaenorum* in Papua New Guinea, are examples of the poor knowledge that we have on these mushrooms, as are those pottery Indian figures from Colima (Mexico) discussed earlier. Concerning the taxonomy of *Psilocybe,* there are several species not well known yet, for example the *P. cyanescens* complex, and many species remain to be described, mainly from the tropics. Mexico alone has more than 50 hallucinogenic *Psilocybe* species, but only 50, 22, 16, 15, 15 and 4 species are known, respectively, from South America, North America, Europe, Asia,

Australia and eastern islands, and Africa. Much more collection is needed in such under-explored regions. It is necessary also to restore chemical and medical research on hallucinogenic mushrooms, stopped in the 60's because they were erroneously listed as drugs. This situation stemmed from the abuse of these mushrooms by young people, followed by a substantial, clandestine trade in them.

Acknowledgments

This author thanks the members of his Institution, Manuel Hernández, Bertha Ulloa, Juan Lara, Etelvina Gándara, Florencia Ramírez-Guillén and Virginia Ramírez-Cruz for their help in computation, and the herbarium and laboratory. He also expresses his gratitude to Dr. James M. Trappe (University of Oregon, U.S.A.) for critically reviewing this chapter. A special acknowledgment is expressed to the editors of the Mexican journal Arqueología Mexicana for their authorization to publish the Figures 19 and 20 here.

References

Ammirati, J.F., Traquiar, J.A. and Horgen, P.A. 1985. Poisonous Mushrooms of the Northern United States and Canada. University of Minnesota Press, Minneapolis, USA.

del Villar, M. 2005. Los Tesoros de Colima. Arqueología Mexicana, (Consejo Nacional para la Cultura y las Artes), special number, Mexico City, Mexico.

Díaz, J.L. 2003. Las plantas mágicas y la conciencia visionaria. Arqueología Mexicana (Consejo Nacional para la Cultura y las Artes, Mexico), 59: 18-25.

Ford, W.W. and Clark, E.D. 1914. A consideration of the properties of poisonous fungi. Mycologia, 6: 167-191.

Gray, W.D. 1973. The Use of Fungi as Food and Food Processing, II. CRC Press, Cleveland, USA.

Guzmán, G. 1959. Nueva localidad de importancia etnomicológica de los hongos neurotrópicos mexicanos (Necaxa, Pue., México). Ciencia, (Mexico), 20: 85-87.

Guzmán, G. 1983. The Genus *Psilocybe*. Beih. Nova Hedwigia 74, Cramer, Vaduz, Liechtenstein.

Guzmán, G. 1990. Wasson and the Development of Mycology in Mexico. In: The Sacred Mushroom Seeker. Th. J. Riedlinger (ed.), Dioscorides Press, Portland, USA.

Guzmán, G. 1995. Supplement to the monograph of the genus *Psilocybe*. In: Taxonomic Monograph of Agaricales. O. Petrini, and E. Horak (eds.), Biblioth. Mycol. 159, Cramer, Berlin.

Guzmán, G. 1997. Los nombres de los hongos y lo relacionado con ellos en América Latina. Instituto de Ecología, Xalapa, Mexico.

Guzmán, G. 2001. Hallucinogenic, Medicinal, and Edible Mushrooms in Mexico and Guatemala: Traditions, Myths, and knowledge. International Journal of Medicinal Mushrooms, 3: 399-408.

Guzmán, G. 2003. Traditional Uses and Abuses of Hallucinogenic Fungi: Problems and Solutions. International Journal of Medicinal Mushrooms, 5: 57-59.

Guzmán, G. 2005. Species Diversity of the Genus *Psilocybe* in the World Mycobiota, with Special Attention to Hallucinogenic Properties. International Journal of Medicinal Mushrooms, 7: 305-331.

Guzmán, G., Jacobs, J.Q., Ramírez-Guillén, F., Murrieta, D. and Gándara, E. 2005. The taxonomy of *Psilocybe fagicola*-complex. J. Microbiol. (Korea), 43: 158-165.

Heim, R. 1956. Les champignons divinatoires utilisés dans les rites des Indiens Mazatèques, recuillis au cour de leur premier voyage au Mexique, en 1953, par Mme Valentina Pavlovna Wasson et M.R. Gordon Wasson. Comp. Pend. Sean. Acad. Sc. 242: 965-968 (reprinted in 1957 in Rev. Mycol, 22: 58-62).

Heim, R. 1966. Botanical and Chemical Characterisation of a Forensic Mushroom Specimens of the Genus *Psilocybe*. Journal of the Forensic Science Society, 6: 192-201.

Heim, R., and Wasson, R.G. 1958. Les champignons hallucinogènes du Mexique (with the collaboration of A. Hofmann, R. Cailleux, A. Cerletti, A. Brack, H. Kobel, J. Delay, P. Pichot, Th. Lemperière and J. Nicolas-Charles). Arch. Mus. Nat. d'Hist. Nat., sér. 7, VI, Paris, France (reprinted in 1958 by Mus. Nat. d'Hist. Nat., Paris).

Heim, R. and Wasson, R.G. 1965. The "mushroom madness" of the Kuma. Bot. Mus. Leaflets Harvard University, 21: 1-36.

Heim, R., Cailleux, R., Wasson, R.G. and Thévenard, P. 1967. Nouvelles Investigtions sur les Champignons Hallucinogènes. Arch. Mus. Nat. Hist. Nat., sér. 7, IX, Paris. Archives du Muséum National d'Histoire. Naturelle, Paris, France.

Lowy, B. 1974. *Amanita muscaria* and the Thunderbolt Legend in Guatemala and Mexico. Mycologia, 66: 188-190.

McCawley, E.L., Brummet, R.E. and Dana, G.W. 1962. Convulsions from *Psilocybe* mushroom poisoning. Proceedings of the Western Pharmacology Society, 5: 27-33.

Oss, O.T. and Oeric, O.N. 1976. Psilocybin: Magic Mushroom Grower's Guide. And/Or Press, Berkeley, USA.

Ott, J. and Bigwood, J. 1978. Teonanácatl, Hallucinogenic Mushrooms of North America. Madrona, Seattle, USA.

Ott, J., Guzmán, G., Romano, J. and Díaz, J.L. 1975. Nuevos datos sobre los supuestos licoperdáceos psicotrópicos y dos casos de intoxicación provocados por hongos del género *Scleroderma* en México. Bol. Soc. Mex. Mic, 9: 67-76.

Pollock, S.H. 1977. Magic Mushroom Cultivation. Herbal Medicine Research Foundation, San Antonio, USA.

Safford, W.E. 1915. An Aztec Narcotic. Journal of Heredity, 6: 291-311.

Sahagún Fray B. de. 1569-1582. Historia de las cosas de la Nueva España, Mexico City (with several reprints in Spanish, English and French; e.g. 1950-1969:

Dibble & Anderson, Univ. Uthah Press, Salt Lake, U.S.A.).

Samorini, G. 2001. Fungi Allucinogenic. Studi Etnomicologici. Telesterion, Dozza, Italy.

Schultes, R.E. 1939. Plantae Mexicanae II. Identification of Teonanácatl, a Narcotic Basidiomycete of the Aztecs. Botanical Museum Leaflets Harvard University, 7: 37-56.

Schultes, R.E. 1976. Hallucinogenic Plants. Golden Press, New York, USA.

Schultes, R.E. and Hofmann. 1973. The Botany and Chemistry of Hallucinogens. Charles C. Thomas Publ., Spring field, USA.

Schultes, R.E. and Hofmann. 1979. Plants of the Gods: Origins of Hallucinogenic Use. McGraw-Hill, New York, USA.

Singer, R. 1949. The Agaricales (Mushrooms) in Modern Taxonomy. Lilloa, 22: 5-832.

Singer, R. 1958a. A *Russula* provoking hysteria in New Grinea. In: Observations on Agarics Causing Cerebral Mycetisms. Mycopathologia et Mycologia Applicata, 9: 261-284.

Singer, R. 1958b. Pilze, die Zerebralmyzetismen verusachen. Bulletin Suisse de Mycologie, 36: 81-89.

Singer, R. 1978. Hallucinogenic mushrooms. In: Mushroom poisoning: Diagnosis and Treatment. B.H. Rumack and E. Salzman (eds.), CRC Press, West Palm Beach, U.S.A.

Singer, R. and Smith, A.H. 1958. Mycological investigations on Teonanácatl, the Mexican hallucinogenic mushroom. II. A taxonomic monograph of *Psilocybe*, section *Caerulescentes*. Mycologia, 50: 262-303.

Singer, R., Smith, A.H. and Guzmán, G. 1958. A new species of *Psathyrella*. Lloydia, 21: 26-28.

Stamets, P. 1996. Psilocybin Mushrooms of the World. Ten Speed Press, Berkeley, U.S.A.

Stijve, T. (1995). Worldwide occurrence of psychoactive mushrooms–an update. Czech Mycology, 48: 11-19.

Stijve, T., Klán, J. and Kuyper, Th.W. 1985. Occurrence of psilocybin and baeocystin in the genus *Inocybe* (Fr.) Fr. Persoonia, 12: 469-473.

Wasson, R.G. 1957. Seeking the Magic Mushroom. Life, May 13, New York, USA (translated to several language editions).

Wasson, R.G. 1979. Traditional use in North America of *Amanita muscaria* for Divinatory Purpose. Journal of Psychedelic Drugs, 11: 25-27.

Wasson, R.G. 1980. The Wondrous Mushroom. Mycolatry in Mesoamerica. McGraw-Hill, New York, USA.

Wasson, V.P. and Wasson, R.G. 1957. Mushrooms, Russia and History. Pantheon Books, New York, USA.

Wasson, R.G., Cowan, G., Cowan, F. and Rhodes, W. 1974. María Sabina and her Mazatec Mushroom Velada. Harcourt Brace Jovanovich, New York, USA.

12

Environmental Impacts on Fatty Acid Composition of Fungal Membranes

Cene Gostinčar, Martina Turk and Nina Gunde-Cimerman*

University of Ljubljana, Biotechnical Faculty, Department of Biology, Večna pot 111, SI-1000 Ljubljana, Slovenia

*Corresponding author: E-mail: nina.gunde-cimerman@bf.uni-lj.si

Abbreviations

ACC – acetyl-CoA carboxylase
ACP – acyl carrier protein
CoA – coenzyme A
ELO – elongase
FAA1 – long chain fatty acyl-CoA synthetase with a preference for C12:0-C16:0 fatty acids, involved in the activation of imported fatty acids
FAD – fatty acid desaturase
FAS – fatty acid synthase
GNS1 – ELO2 (*S. cerevisiae*)
OLE1 – Δ9 desaturase
PUFA – polyunsaturated fatty acids
SUR4 – ELO3 (*S. cerevisiae*)

Abstract

The chapter reviews the studies carried out to assess the responses of changing environmental stresses on the cell membranes, particularly the fatty acid composition, metabolism and emphasizes the need for further research.

Any shifts in physicochemical parameters of the surrounding environment disturb the proper functioning of the cell. One of the major changes is the altered fluidity of the membranes, which is compensated by adjustments in membrane lipid composition. It appears that the change in membrane physical state actually triggers at least part of the response mechanisms. The differences in membrane fatty acid composition and enzymes involved in their metabolism in different fungi may partially explain their different tolerance to stresses. Although the basic enzymology involved in fatty acid biosynthesis is well worked out, many questions related to the regulation of this process as well as understanding of fatty acid response to environmental changes are still unresolved.

Although mesophilic *Saccharomyces cerevisiae* is an intensively studied model organism, data on other, extremotolerant organisms would contribute to our knowledge. In contrast with *S. cerevisiae,* most other fungi have at least one more, in many cases (especially filamentous fungi) even several other enzymes involved in fatty acid metabolism with different specificity, roles in general metabolism and presumably also regulatory mechanisms. These fungi are more tolerant to different extreme environmental conditions and have different lipid responses to stress. However, more data is needed regarding the expression responses of genes encoding fatty acid modifying enzymes to different environmental conditions. High throughput techniques, such as microarrays and real-time PCR, could provide valuable information, especially in poorly investigated topics such as responses to different pH or hydrostatic pressure. Finally, the identification of novel fatty acid modifying enzymes increases the possibility of success in genetically modifying industrially important fungi for lipid production.

In almost all experiments, where organisms were subjected to sudden shock because of a shift in temperature, salinity, etc., it has been observed that transcriptional response following a shock is, to a very large extent, transient and follows a distinct temporal pattern. Authors' findings confirm these observations in case of fatty acid modifying genes. They also show that life at a constant value of a specific physicochemical parameter, no matter how stressful it is, represents to cells a fundamentally different challenge than a sudden shift to the same conditions. Therefore, more experiments with fungi in different constant environmental conditions (not just their reaction to shock) would significantly improve our understanding of adaptations to extreme environments. Fungi are increasingly recognized as convenient research models in eukaryotic lipidomics. Understanding the role of fatty acids and the enzymes involved in their metabolism will not only shed light on the ability of life in different (even extreme) environments, but could also have an industrial potential.

INTRODUCTION

Membranes form a framework, which compartmentalizes the different biochemical processes of the cell, controlling movement of substrates and products and providing the anchorage for enzymes and carriers. Besides, they form a continuous dynamic system, the composition of which differs in individual organelles and regions of the cell (Lösel, 1990).

To function properly, membranes have to preserve a suitable dynamic state of the bilayer even in changing environments, which alters their fluidity. This is achieved by active restructuring of membrane lipids composition.

The choice of different lipid species is very large. No other class of biological molecule displays a greater diversity of structure over such a narrow range of physical and chemical parameters, so close to physiological conditions (Mariani et al., 1990). Consequently, environmentally induced perturbations in membrane structure may result in significant disruption of its physiological function (Hazel and Williams, 1990).

Organisms exposed to changes in the environmental conditions, such as temperature, water activity, or oxygen availability adjust their membrane lipid composition. By synthesizing a proper combination of acyl chain and polar head group structures, the organisms maintain membrane lipids in a lamellar liquid crystalline phase, and avoid the formation of a lamellar gel phase (Rilfors and Lindblom, 2002).

It is well documented that all kinds of organisms can adapt their membrane lipid composition according to the prevailing environmental and physiological conditions. Four strategies seem to be utilized: (i) changes in the acyl chain structure; (ii) changes in the polar head group structure; (iii) changes in the amount of sterols, and (iv) reshuffling of acyl chains to form new lipid molecular species without changing the average acyl chain composition (Russell, 1989b; Hazel and Williams, 1990; Suutari and Laakso, 1994; Rilfors and Lindblom, 2002).

The aim of this chapter is to give an overview of membranes fatty acid composition and metabolism in fungi in the light of adjustments necessary to respond to fluctuating environmental conditions.

1. Lipid composition of fungal membranes

Phospholipids (glycerophospholipids and sphingolipids) are the major class of membrane lipids, together with glycolipids, sterols and according to some reports other neutral lipids.

The amount of individual membrane lipids can vary greatly due to the developmental stage, experimental conditions and methodology used.

Additionally, generalizations concerning fungal membrane lipid composition are based on few selected species investigated, mainly different mesophilic yeasts cells and relatively few filamentous fungi. Most data derive from entire membrane systems or even total lipids, rather than the separate analysis of individual membrane fractions. This more precise approach has been adopted for a few fungi, e.g. *Aspergillus niger* (Letoublon et al., 1982), *Neurospora crassa* (Aaronson et al., 1982; Bowman et al., 1987), and *Saccharomyces cerevisiae* (Zinser et al., 1991; van der Rest et al., 1995; Daum et al., 1998; Schneiter et al., 1999; Tuller et al., 1999).

The analysis of membrane fractions of *N. crassa* by Bowman et al. has shown that the plasma membranes, endoplasmic reticulum, vacuolar membranes and mitochondrial membranes differed markedly in their phospholipids and sterols, but showed a generally similar fatty acid composition. An exception represented mitochondrial membranes, which had a higher content of C18:2 and C18:3 unsaturated fatty acids (Bowman et al., 1987).

The plasma membrane of *S. cerevisiae*, one of the most investigated fungi, is highly enriched in sterols and sphingolipids, saturated species of phosphatidylserine (PS) and phosphatidylethanolamine (PE), and deviates most from all other membranes in this yeast (Schneiter et al., 1999).

1.1. Sterols

Sterols are essential lipid components of eukaryotic membranes. They are responsible for a number of important physical characteristic of the membranes, such as regulation of membrane permeability, fluidity and stability. Sterol molecules are of a suitable shape to lock across the fatty acid chains of the polar lipids, thus increasing the chain order in liquid crystalline state (condensing effect) and decreasing it in the gel state (liquefying effect). A decrease in the ratio of sterols to phospholipids indicates an increase in membrane fluidity, provided that compensatory changes in other membrane components do not occur.

The fungal sterol, ergosterol, differs from the animal sterol, cholesterol, by the presence of unsaturations at C-7,8 in the ring structure and at C-22 in the side chain and by the presence of a methyl group at C-24 on the side chain (Demel and De Kruyff, 1976; Rodriguez et al., 1985; Weete, 1989; Weete and Gandhi, 1996).

1.2. Phospholipids

The arrangement of phospholipid molecules in membranes results from the polar head group being orientated towards the cytoplasm and the hydrophobic fatty acid chains directed inwards, towards the fatty acid chains of the opposite lipid layer. The several types of phospholipids present in membranes differ in their polar groups, and the fatty acids within the individual phospholipid classes vary in both chain length and degree of unsaturation. Through the introduction of one or more double bonds, the configuration of the fatty acid chains is altered, affecting the fluidity of the membrane (Lösel, 1990). Structural roles of the major phospholipids in membranes of fungal cells correspond to those established for other eukaryotes. In fluctuating environmental conditions, the remodelling of the membrane phospholipids by changes in chain length, saturation or positions of fatty acids can facilitate the maintenance of membrane structure and permeability (Lösel, 1988; Lösel, 1990).

1.2.1. Glycerophospholipids

Glycerophospholipids are regarded as a primary structural element of most biological membranes and consist of a glycerol backbone esterified with fatty acids in the *sn*-1 and *sn*-2 positions, and a phosphate group in the *sn*-3 position. One hydroxyl group of the phosphate is linked to a polar head group, which is relevant for the physical properties of these molecules (and hence of biological membranes) and also for the classification of the molecule. In *S. cerevisiae* the fatty acids in the *sn*-1 position are mostly saturated whereas those in the *sn*-2 position are unsaturated (Daum et al., 1998).

In fungal membranes, the major glycerophospholipids by order of decreasing occurrence are phosphatidylcholine and phosphatidylethanolamine followed by phosphatidylinositole and phosphatidylserine. Inner mitochondrial membrane differs from all other membrane fractions in its substantial amount of cardiolipin, while phosphatidylserine is present at low concentrations in most membranes except in plasma membrane. Phosphaditylglycerol appears to be rare. Although phospholipids are indispensable as bulk membrane components, it is not clear whether or not all individual classes of phospholipids are essential, with the exception of phosphatidylinositole (Rattray et al., 1975; Kaneko et al., 1976; Lösel, 1988; Lösel, 1990; Zinser et al., 1991; Daum et al., 1998).

1.2.2. Sphingolipids

The understanding of sphingolipids' function in fungi is still poor in comparison to our knowledge about their glycerophospholipids. Sphingolipids include (i) ceramides, in which long-chain sphingosine-type bases (C_{14} to C_{22}) are amide-linked to 2-hydroxy fatty acids or normal fatty acids, and (ii) glycosyl ceramides (cerebrosides) or glycophosphosphingolipids, in which sugars and oligosaccharides are attached to ceramides via a phosphoinositol bridge (Lösel, 1988; Lösel, 1990; Daum et al., 1998). The majority of sphingolipids in *S. cerevisiae* are localized in the plasma membrane, a lesser part in the Golgi and in the vacuole, with trace amounts in the mitochondria. Besides being a constitutive element of fungal membranes, ceramides also serve as an anchor for some membrane proteins, they have a role in cell wall synthesis, and are necessary for survival at 37°C, low pH and osmotic stress (van der Rest et al., 1995; Daum et al., 1998; Cowart and Obeid, 2007).

1.3. *Fatty acids*

Intrinsic properties of fatty acids are used in multiple ways. First, the hydrophobic nature of acyl chains creates the force to establish membrane bilayer structures that are the basis of subcellular compartimentalisation. Second, fatty acids provide an ideal storage form of metabolic energy. The energy contained in their C–C bonds can be efficiently released by β-oxidation, a reaction formally equivalent to the reverse of fatty acid biosynthesis. Acetyl-CoA generated by fatty acid degradation is utilized as an anabolic building block, or further catabolized by the TCA cycle. Third, specific fatty acids serve as precursors for biologically more active compounds, and may thus harbor signaling functions, e.g., arachidonic acid in mammalian cells (Tehlivets et al., 2007). Accordingly, ample evidence exists that links fatty acid synthesis to organelle structure, function and inheritance (Schneiter and Kohlwein, 1997). They are considered to have a key role in the structure and functioning of the membranes.

Fatty acids normally occur in a bound form, as components of glycerophospholipids, sphingolipids, and glycolipids. Reports of free fatty acids from the membranes are regarded as indicators of phospholipase activity during extraction.

The spectrum of fatty acids in fungal membranes is rather simple, consisting mostly of C_{16} and C_{18} fatty acids, with varying degrees of unsaturation. In studied fungi from *Chytridiomycota* the major fatty acids present in the glycerophospholipids of membranes are palmitic (C16:0) and oleic (C18:1) fatty acids with γ-linolenic (C18:3), linolenic (C18:2), stearic (C18:0), and palmitoleic (C16:1) fatty acids in decreasing order. In *Mucoromycotina* (formerly *Zygomycota*) the composition is similar with the addition of myristic (C14:0) fatty acid among the major fatty acids in the cells (Lösel, 1988).

Among *Ascomycota* subphylum *Pezizomycotina* the linolenic (C18:2) fatty acid prevails, followed by α-linolenic (C18:3) (instead of γ-linolenic (C18:3) fatty acid), palmitic (C16:0), oleic (C18:1) and stearic (C18:0) fatty acids, with palmitoleic (C16:1) fatty acid in smaller amounts. In subphylum *Saccharomycotina* the principal fatty acids are palmitoleic (C16:1), palmitic (C16:0), stearic (C18:0), oleic (C18:1), and linolenic (C18:2) with trace amounts of γ-linolenic (C18:3) fatty acid (Kaneko et al., 1976; Lösel, 1988). In *S. cerevisiae* no linolenic (C18:2) fatty acid is found, while fatty acids like myristic (C14:0) fatty acid and C-26 fatty acids are detected in minor amounts. They play essential functions in protein modification or as components of sphingolipids and GPI-anchors, respectively (van der Rest et al., 1995; Daum et al., 1998; Tehlivets et al., 2007). Typically, 80% of *S. cerevisiae* fatty acids are monounsaturated via a reaction catalyzed by the ER-resident and essential Δ9 desaturase, Ole1 (Martin et al., 2007). Tuller et al. (Tuller et al., 1999) analyzed the fatty acid composition of subcellular fractions of *S. cerevisiae* and reported that the mitochondria, microsomes, and vacuoles had similar fatty acid compositions, in which palmitoleic acid (C16:1) is the major fatty acyl constituent followed by palmitic acid (C16:0), oleic acid (C18:1), stearic acid (C18:0), with minor amounts of myristic acid (C14:0). In contrast, the fatty acid composition of the plasma membrane was found to contain higher levels of palmitic and oleic acid and significantly lower amounts of palmitoleic acid.

In *Schizosaccharomycetes* (subphylum *Taphrynomycotina, Ascomycota*) the oleic (C18:1) acid comprised more than 50% of all fatty acids, followed by palmitic (C16:0) and stearic (C18:0) fatty acids with trace amounts of linolenic (C18:2), palmitoleic (C16:1), and γ-linolenic (C18:3) fatty acids (Jeffery et al., 1997).

Basidiomycetous fungi from the phylum *Agaricomycotina* show the prevailence of palmitic (C16:0), linolenic (C18:2), oleic (C18:1), and stearic (C18:0) fatty acids, in subphylum *Ustilaginomycotina* the dominant fatty acid is palmitic (C16:0) together with oleic (C18:1) fatty acid, while in *Pucciniomycotina* almost 70% of all fatty acids is comprised by α-linolenic (C18:3) fatty acids with palmitic (C16:0) and linolenic (C18:2) fatty acids in minor amounts (Lösel, 1988).

Yeast and filamentous fungi fatty acids that are generally associated with sphingolipids of the plasma membrane, are commonly 2-hydroxy fatty acids of chain-lengths up to C24:0 as well as normal fatty acids of composition similar to glycerophospholipids (Lösel, 1988; Lösel, 1990).

Differences in fatty acid composition according to the optimal temperature and solute concentration were observed in fungi. The polar lipids in thermophilic fungi are generally more saturated than the polar lipids of mesophilic species. Mesophiles contain more C16:1, C18:2, and C18:3 than thermophiles (Lösel, 1988; Lösel, 1990; van der Rest et al., 1995). Similarly, psychrophiles have more unsaturated fatty acyl chains than mesophiles (van der Rest et al., 1995). In ascomycetous halophilic/halotolerant black yeasts from order *Dothideales* (*Pezizomycotina, Ascomycota*) the phospholipids have more unsaturated fatty acids than salt-sensitive yeasts like *S. cerevisiae* (Table 1, Turk et al., 2004).

Table 1. The unsaturation index and the average number of C atoms in fatty acid´s chain from different fungal groups

Phylum	*Subphylum*	*Order*	*Unsaturation index (UI)*		*Average C atoms in fatty acid´s chain (C_a)*	
			Total FA	*FA from PL*	*Total FA*	*FA from PL*
Chytridiomycota			0.79	1.11	17.13	17.32
Mucoromycotina			1.07	0.85	17.28	17.03
Ascomycota	*Pezizomycotina*		1.13	1.65	17.48	17.69
		Dothideales	1.27	1.06	17.67	17.56
	Saccharomycotina		1.01	n.d.	17.24	n.d.
	Taphrynomycotina	*Shizosaccharomycetales*	0.78	0.76	17.69	17.53
Basidiomycota	*Agaricomycotina*		1.33	0.73	17.61	17.19
	Ustilaginomycotina		1.26	0.35	17.47	17.08
	Pucciniomycotina		0.8	2.31	17.45	17.70

Adapted from Kaneko (Kaneko et al., 1976), Lösel (Lösel, 1988), van der Rest et al. (van der Rest et al., 1995), Stahl and Klug (Stahl and Klug, 1996), Jeffery et al. (Jeffery et al., 1997), Turk et al. (Turk et al., 2004).

Unsaturation index (UI) = $(\Sigma\%C_x{:}1+2 \times \Sigma\%C_x{:}2+3 \times \Sigma\%C_x{:}3)/(\Sigma\%C_x{:}0+ \Sigma\%C_x{:}1+ \Sigma\%C_x{:}2+ \Sigma\%C_x{:}3)$; average C atoms ($C_a$) = $(I \times \Sigma\%C_I+(I+1) \times \Sigma\% C_{(I+1)}+.../(\Sigma\%C_I+ \Sigma\%C_{(I+1)}+...)$ where I is the number of C atoms.

FA—fatty acids; PL—phospholipids; n.d.—not determined.

2. Enzyme systems involved in fatty acid metabolism

2.1. Fatty acid synthesis

In eukaryotes, fatty acid synthesis proceeds by the addition of two carbon units from malonyl-CoA. This reaction sequence requires seven reaction centres in a multifunctional protein, fatty acid synthase (FAS) and the carrier of the acyl chain intermediates, acyl carrier protein (ACP). In *S. cerevisiae* the final reaction is catalyzed by a reaction centre with transferase activity, which substitutes coenzyme (CoA) for ACP and allows the release of a long chain acyl-CoA, generally palmitoyl-CoA (C16:0). This is the substrate for enzymes that catalyze subsequent desaturation and elongation reactions required to generate monounsaturated and very long chain fatty acids (reviewed in Black and DiRusso, 2007).

In *S. cerevisiae* the main genes involved in fatty acid biosynthesis are *FAS1* and *FAS2*. They encode the β and α subunits of the fatty acid synthase complex (Chirala et al., 1987; Mohamed et al., 1988), organized as a hexameric α6β6 complex and *ACC1*, which encodes acetyl-CoA carboxylase (Al-Feel et al., 1992; Hasslacher et al., 1993).

Proper function of these genes is essential for the survival of *S. cerevisiae* in normal media. In organisms other than green plants, the FAS responsible for bulk fatty acid synthesis is a soluble cytoplasmic enzyme (Schweizer and Hofmann, 2004). Fungal mitochondria contain their own FAS, which is structurally and functionally independent of the cytoplasmic system (Mikolajczyk and Brody, 1990; Brody et al., 1997). In mitochondria (as well as in chloroplasts and in most bacteria) the reactions are catalyzed by dissociated, individual gene products (type II FAS systems), similarly to the initial ACC reaction. Nevertheless, the organellar system is unable to compensate for the loss of cytoplasmic FAS in fas1 or fas2 yeast mutants (Schweizer and Hofmann, 2004).

The chain length of fatty acids produced by FAS types I and II varies quite significantly, however, the molecular basis for chain-length determination is still not clear. In yeasts, cytosolic FAS is capable of *in vitro* synthesizing fatty acids up to C_{20}; *in vivo*, however, the interaction with cytosolic factors, such as acyl-CoA binding protein (Acb1), and the availability of malonyl-CoA provided by Acc1 (see above) clearly determine acyl-chain length distribution (Tehlivets et al., 2007). Interestingly, FAS was suggested to interact at least temporarily with another set of elongation (Elo) enzymes, catalyzing acyl-chain elongation up to C_{26} carbon atoms in the endoplasmic reticulum (Rossler et al., 2003; Tehlivets et al., 2007).

Fungi differ from other organisms in releasing the synthesized acyl chain. Most bacteria catalyze direct acyl-transfer from acyl-ACP to lipids, and mammalian cells release free fatty acids from FAS by a thioesterase activity. In *S. cerevisiae* cytosolic FAS acyl-residues are transferred from ACP by an intrinsic acyltransferase activity to coenzyme A, yielding long-chain acyl-CoAs (Tehlivets et al., 2007).

The amount of FAS holoenzyme is determined by the level of Fas1 protein and is established at both the level of gene expression and protein stability. Changes in expression of *FAS1* trigger similar changes in expression of *FAS2*, while *FAS1* expression is independent of the level of *FAS2* (or *ACC1*) expression (Wenz et al., 2001). Interestingly, despite the presence of Fas1 and Fas2 subunits in equimolar amounts in the FAS complex, expression of both genes shares only limited similarity, which is also reflected in rather unrelated promoter sequences (Mohamed et al., 1988). A common feature of *FAS1* and *FAS2* genes and other genes involved in lipid synthesis, is coordinated control by lipid precursors, inositol and choline (Henry and Patton-Vogt, 1998) and repression in the presence of long-chain fatty acids (Chirala, 1992).

FAS1 and *FAS2* genes are subject to activation by general transcription factors, such as Gcr1 and Reb1, and harbor the respective binding sites in their promoter regions (Schuller et al., 1994; Greenberg and Lopes, 1996). *FAS1* belongs to a small group of genes regulated by both the SAGA and TFIID complexes, as determined by microarray analysis. The *FAS1* promoter activity is sensitive to alterations in histone acetylation (Huisinga and Pugh, 2004).

In addition to the primary fatty acid synthesis system, other FAS systems occur in some fungi, serving a variety of specialized functions (Schweizer and Hofmann, 2004). *Aspergillus nidulans,* for example, contains an additional sFAS, required for secondary metabolism (sFAS). While FAS mutants require long chain fatty acids for growth, sFAS mutants grow normally but cannot synthesize sterigmatocystin (ST), a carcinogenic secondary metabolite structurally and biosynthetically related to aflatoxin (Brown et al., 1996).

In *S. cerevisiae* acetyl-CoA carboxylase protein is encoded by *ACC1* (or *FAS3*) (Al-Feel et al., 1992; Hasslacher et al., 1993). Its regulation occurs at multiple levels and is described in more detail in (Cho et al., 1998; Carman and Henry, 1999; Shirra et al., 2001; Huisinga and Pugh, 2004; Tehlivets et al., 2007). Regulation of Acc1 also exists on the level of protein phosphorylation, biotinylation and oligomerization, as reviewed by (Tehlivets et al., 2007).

2.2. *Fatty acid degradation*

To maintain proper fatty acid composition of the cell, fatty acid biosynthetic pathways have to be carefully balanced with fatty acid degradation.

S. cerevisiae is able to degrade both saturated and unsaturated fatty acids (Dommes et al., 1981). In contrast to higher eukaryotes, in which β-oxidation occurs in both the mitochondria and the peroxisome, *S. cerevisiae* lacks the enzymes required for mitochondrial β-oxidation (Kunau et al., 1988; Kunau et al., 1995). β-oxidation of fatty acids is restricted to peroxisomes and proceeds via the same set of enzymatic reactions as in mitochondria and peroxisomes from higher eukaryotes (Kunau et al., 1995; van Roermund et al., 2003).

Peroxisomes contain a series of enzymes needed for the main reactions as well as reoxidation the NADH produced during fatty acid β-oxidation, reduction of NADP+ produced in the 2,4-dienoyl-CoA reductase reaction, take up of fatty acids from the cytosol and export of acetyl-CoA units to mitochondria for utilization in the citric acid cycle (reviewed in Trotter, 2001; van Roermund et al., 2003).

The first step in the β-oxidation of activated fatty acids involves the introduction of a double bond between the α and β carbon atoms. The following reactions are catalyzed by a multifunctional enzyme, with both 2-enoyl-CoA hydratase activity and 3-hydroxyacyl-CoA dehydrogenase activity, which result in the generation of 3-ketoacyl-CoA. The final reaction of the peroxisomal β-oxidation pathway is catalyzed by a single 3-ketoacyl-CoA thiolase, which cleaves 3-ketoacyl-CoA esters into a C2-shortened acyl-CoA and acetyl-CoA (reviewed in Trotter, 2001; van Roermund et al., 2003).

Peroxisome biogenesis and activity is induced by exposure to fatty acids. In *S. cerevisiae*, the expression of peroxisomal β-oxidation enzymes is subject to several types of regulation, primarily at the level of transcription (Trotter, 2001), for example by the transcription factors Pip2p and Oaf1p (van Roermund et al., 2003).

In most organisms, the degradation of fatty acids also provides a major source of metabolic energy. In yeast, there are at least two sources of fatty acids for energy production: triglycerides stored in the lipid body and fatty acids imported from the environment (Black and DiRusso, 2007).

Recent evidence suggest that breakdown of fatty acids may be important in the metabolism, development, and pathogenicity of many fungi (Hynes et al., 2006).

2.3. *Fatty acid modifications*

Generation of a diversity of fatty acid phenotypes, which represent evolutionary advantages, requires elaborate metabolic pathways (Laoteng et al., 2005). Palmitate, often the major product of the fatty acid synthetase pathway, is the precursor of longer chain saturated and unsaturated fatty acids through the actions of elongases and desaturases.

2.3.1. *Elongation*

In contrast to the structurally related FAS of mycobacteria, which synthesizes both C16 to C18 and C24 to C26 fatty acids, such bimodality of products has not been observed for *S. cerevisiae* FAS *in vitro*. It may nevertheless be speculated that *in vivo*, a fraction of cellular FAS is associated with the microsomal membrane and thereby engages in VLCFA synthesis (Schweizer and Hofmann, 2004). The general potential of yeast FAS to synthesize VLCFAs has been demonstrated in *Schizosaccharomyces pombe* (Yokoyama et al., 2001).

However, in *S. cerevisiae* enzyme systems, fully dedicated to elongating fatty acyl-CoAs up to 26 carbons are known. These require malonyl-CoA as the two-carbon donor, NADPH, and a medium- or long-chain fatty acyl-CoA primer with greater than 10 carbons (Dittrich et al., 1998; Trotter, 2001). Elongases are not soluble cytoplasmic enzymes but are localized in the microsomal membrane fraction, in contrast to enzymes involved in *de novo* fatty acid synthesis. The reason for this localization is presumably the hydrophobic nature of the VLCFA, which can be directly inserted into ceramide by ceramide synthase, at their site of production in the ER membrane (Schweizer and Hofmann, 2004; Tehlivets et al., 2007).

In *S. cerevisiae,* at least three different elongase systems have been identified, which differ according to their primer usage and product specificities. Elo1 catalyses elongation of medium and long-chain fatty acids (Toke and Martin, 1996; Dittrich et al., 1998) and is also able to elongate monounsaturated fatty acids (Schneiter et al., 2000); Elo2 (elongation up to C_{22}) and Elo3 (elongation up to C_{26}) are required for very long chain fatty acid synthesis (Oh et al., 1997; Rossler et al., 2003).

Hydrophobicity analyses indicate that Elo2p and Elo3p are intrinsic membrane proteins with multiple membrane-spanning regions. The lethality caused by the simultaneous disruption of the *ELO2* and *ELO3* genes indicates that their products have a high degree of overlapping functions (Oh et al., 1997).

Mutations in any of the genes involved in fatty acid elongation are associated with pleiotropic deficiencies as a result of altered sphingolipid metabolism, protein trafficking, cell wall structure and function. Thus, elongation-defective mutants have been identified in different screens that were designed to identify factors involved in the synthesis and secretion of membrane or cell wall proteins, drug or ion uptake and detoxification. *ELO1* mutants for example are unable to grow on myristic acid, in the absence of a functional fatty acid synthase (Toke and Martin, 1996; Dittrich et al., 1998; Tehlivets et al., 2007).

Knowledge about the transcriptional regulation of elongation enzymes is scarce, and is mainly based on large-scale transcriptome profiling experiments (Tehlivets et al., 2007). Expression of the *ELO1* gene is upregulated 3- to 5-fold by myristic acid supplementation (Toke and Martin, 1996) and repressed by palmitic acid. Expression is induced in the late G1 phase of the cell cycle (Cho et al., 1998) and repressed in stationary phase (Gasch et al., 2000). Similarly, *ELO2/FEN1* and *ELO3/SUR4* gene expression is down-regulated in the stationary phase, and under nitrogen depletion conditions (Gasch et al., 2000). Data suggest growth rate or starvation-dependent regulation of fatty acid elongation at the level of transcription of the condensing enzymes (Tehlivets et al., 2007).

Genes encoding elongases have been found in other fungi as well, for example two elongases from *Hansenula polymorpha*, *HpELO1* and *HpELO2* (Prasitchoke et al., 2007a; Prasitchoke et al., 2007b), *ELO2* from *Pichia stipitis* (Jeffries et al., 2007), *FEN1* from *Candida albicans* (Jones et al., 2004), *GNS1* from *Aspergillus fumigatus* (Nierman et al., 2005), *ELO* gene from *Cryptococcus neoformans* (Loftus et al., 2005), a PUFA-specific elongation enzyme from *Mortierella alpina*, *GLELOp* (Das et al., 2000; Parker-Barnes et al., 2000). Other putative fungal elongases can be found in a search through the GenBank databases.

2.3.2. Desaturation

Increases in the degree of fatty acid unsaturation in biological membranes can have profound effects on membrane order and stability. The introduction of a *cis* C–C double bond and a 308° bend into a fatty acyl chain results in the fatty acid occupying an expanded conformation, packing less compactly in a bilayer, and having a lower melting point. Thus, UFAs are key molecules in the regulation of cellular membrane fluidity. Fungi and most other organisms (with the exception of some bacteria) possess a fatty acid synthetase capable of producing only SFA. Therefore, UFA must be produced by desaturases, which convert a single bond between two carbon atoms to a double bond in a fatty acyl chain (Hazel and Williams, 1990; Los and Murata, 1998; Aguilar and de Mendoza, 2006) typically in the more effective *cis* configuration (Los and Murata, 1998).

Acyl-CoA desaturases are present in animal and fungal cells, and they introduce unsaturated bonds into fatty acids that are bound to CoA (Los and Murata, 1998). Most consist of 300-350 amino acid residues (Nakashima et al., 1996; Meesters et al., 1997; Los and Murata, 1998), they are hydrophobic and span the lipid bilayer of membranes (Shanklin et al., 1994). They accept electrons from an electron-transport system that is composed of cytochrome b5 and NADH-dependent cytochrome b5 reductase (Dailey and Strittmatter, 1980; Mitchell and Martin, 1995).

Depending on the enzyme studied, introduction of a double bond can occur counting carbons from the carboxyl end (Δ-desaturases) or from the methyl terminus (ω-desaturases) of the acyl chain. Other enzymes may use a pre-existing double bond as a reference point for subsequent desaturation

(Heinz, 1993). Changes to an enzyme's regiospecificity typically require between two to six specific alterations at key locations along the amino acid chain (Broadwater et al., 2002; Guy et al., 2007). A recent study on two desaturases from *A. nidulans* discovered two domains near the active site or directly involved in forming the active site, which are apparently involved in desaturase specificity (Hoffmann et al., 2007).

The most common and widespread are the Δ9 desaturases. They can be divided into three well-known groups: (1) soluble 18:0-ACP desaturases in higher plants, (2) membrane-bound 18:0-lipid desaturases in cyano-bacteria, and (3) membrane-bound 18:0-CoA desaturases in mammals and fungi (Pereira et al., 2003).

The best characterized representative in fungi is the enzyme from *S. cerevisiae*. It primarily synthesizes palmitoleic (16:1) and oleic (18:1) acids and is encoded by the *OLE1* gene (Stukey et al., 1990; Mitchell and Martin, 1995). Elo1 consists of 510 amino acid residues (Stukey et al., 1989), which is an unusually large number for an acyl-CoA desaturase. The sequence of the carboxy-terminal portion of the protein is homologous to that of cytochrome b5. This domain takes care of electron transport and can function as an electron donor in the process of desaturation. Such chimeric enzyme appears to have evolved through an event in which the NH2-terminal protein coding regions of an ancestral cytochrome *b*5 gene were fused to the carboxyl terminus of coding sequences within an independent desaturase gene. This could potentially speed up the electron transfer by eliminating the need for diffusion and reorientation of the reduced cytochrome *b*5 (Mitchell and Martin, 1995).

Ole1 localizes into the endoplasmic reticulum, where most of the lipid biosynthetic machinery resides (Aguilar and de Mendoza, 2006). It can catalyse the desaturation of a wide range of substrates, ranging from C12 to C19 carbon saturated fatty acids.

As unsaturated fatty acids comprise most of the mass of membrane lipids, and exert a strong influence on bilayer fluid properties, Ole1 is a highly regulated key enzyme of lipid metabolism (Martin et al., 2002). Transcription of *OLE1* is weakly induced by saturated fatty acids (1.6-fold) and severely repressed (up to 60-fold) by unsaturated fatty acids (McDonough et al., 1992; Choi et al., 1996), which is also reflected in desatu-rase enzyme activity (Bossie and Martin, 1989). Besides transcription, *OLE1* mRNA level in cells is also regulated with alteration in the mRNA's stability. In cells grown without unsaturated fatty acids, *OLE1* mRNA is defined as a moderately stable species. In the presence of unsaturated fatty acids, *OLE1* mRNA becomes one of the most unstable transcripts reported to date (Gonzalez and Martin, 1996). *OLE1* mRNA stability is affected by the Mga2: it can either stabilize the transcript or confer its regulated instability (Gonzalez and Martin, 1996; Vemula et al., 2003; Kandasamy et al., 2004). Another factor involved in the regulation of fatty acid desaturation by Ole1 is the RSP5 gene encoding a ubiquitin ligase (Kaliszewski et al., 2006). Mechanisms regulating *OLE1* gene expression described in more detail can be found in (Kandasamy et al., 2004) or (Kaliszewski et al., 2006).

Microarray studies have showed that *OLE1* gene is induced in the early G1 phase of the cell cycle (Cho et al., 1998).

It seems that desaturation is co-ordinated by feedback regulation of transcription in organisms ranging from bacteria to human. Recent advances in the study of the regulation of fatty acyl desaturases from different organisms have revealed a conceptual convergence: (i) the changed physical properties of the membrane cause conformational change in sensor proteins and activate a signal transduction pathway that controls the expression of desaturase genes and (ii) UFAs, the product of desaturases, act as negative signalling molecules that turn the pathway off (Aguilar and de Mendoza, 2006).

Δ9 desaturase exists in almost all organisms as the most important desaturase in unsaturated fatty acid biosynthesis (Abe et al., 2006). Often unstable mutants in Δ9 desaturase genes exhibit a requirement for unsaturated fatty acids for growth, a defective development and reduced respiration rates (Certik et al., 1998). Δ9 desaturase has already been identified in different fungi: *Cryptococcus curvatus* (Ykema et al., 1990; Meesters et al., 1997), *Sc. pombe* (McDonough and Roth, 2004), *A. niger* (Chattopadhyay et al., 1985), *N. crassa* (Goodrich-Tanrikulu et al., 1994), *Lentinula edodes* (Sakai and Kajiwara, 2003), *Rhizopus arrhizus* (Wei et al., 2007), three desaturases in *Mo. alpina* (Wongwathanarat et al., 1999) and several others.

S. cerevisiae and *Sc. pombe* unusually form only monounsaturated fatty acids. Most other fungi also express membrane bound Δ12 and Δ15 desaturases (Martin et al., 2007).

The existence of Δ12 desaturase has been reported in *A. nidulans* (Calvo et al., 2001), *Mo. alpina* (Huang et al., 1999; Sakuradani et al., 1999a), *Mucor rouxii* (Passorn et al., 1999), *Kluyveromyces lactis* (Kainou et al., 2006), *Saccharomyces kluyveri* (Oura and Kajiwara, 2004; Watanabe et al., 2004), *Le. edodes* (Sakai and Kajiwara, 2005), *Rhi. arrhizus* (Wei et al., 2004), *Pichia pastoris* (Wei et al., 2006a), *Coprinus cinereus* (Zhang et al., 2007), *Aspergillus parasiticus* (Wilson et al., 2004), *A. flavus* (AY280867), *A. nidulans* (AF528822) and others.

Desaturases with different specificities have been discovered especially in filamentous fungi. *M. alpina* for example has five types of desaturases (Δ9, Δ12, Δ6, Δ5, and ω3) in the PUFA biosynthetic pathway (Sakuradani et al., 1999a; Sakuradani et al., 1999c; Sakuradani et al., 1999b; Sakuradani et al., 2005). Two Δ6 desaturases were reported for *Mucor circinelloides* (Michinaka et al., 2003), one for *Rhi. arrhizus* (Zhang et al., 2004), a PUFA-specific ω3 desaturase gene from *S. kluyveri* (Oura and Kajiwara, 2004), etc.

More unpublished desaturase genes can be found in the GenBank database.

3. Influence of different environmental factors on fatty acid composition

The composition of lipid acyl chains determines largely the physical properties of biological membranes including membrane thickness, intrinsic curvature, and fluidity, which affect membrane barrier function, the activity of membrane-associated enzymes, and other dynamic processes like membrane fusion and fission. Compositions differ between various membranes and membrane leaflets, and are subject to adaptation in response to the changing environment (de Kroon, 2007). The ability of an organism to manipulate the fatty acid composition of its membranes is dependent on many aspects of metabolism including synthesis, degradation, and incorporation of exogenous fatty acids. Organisms alter their membrane fatty acid composition utilizing strategies like changes in the acyl chain structure (saturation, length, cyclization, branching) and reshuffling of acyl chains to form new lipid molecular species without changing the average acyl chain composition (Russell, 1989b; Hazel and Williams, 1990; Suutari and Laakso, 1994; Rilfors and Lindblom, 2002). There are previous reports that suggest that membrane lipid composition specially fatty acids is correlated with tolerance to different stresses, including heat shock (Steels et al., 1994), heavy metals (Howlett and Avery, 1997), and exposure to the herbicide 2,4-dichlorophenoxyacetic acid (Viegas et al., 2005). The finding that biosynthesis of diunsaturated fatty acids increased tolerance to salt stress and freezing was not completely unexpected. Increasing the unsaturation index of yeast lipids fluidized the yeast membrane and altered the stress response of yeast cells. Unsaturation of the acyl chains produces a dramatic reduction in the lamellar to hexagonal-II phase transition temperature of phosphatidylethanolamine, effectively lowering it from nonphysiological to physiological temperatures for the majority of microorganisms (Russell, 1989a). The expression of genes for desaturase is very important since it provides the molecular basis for the acclimation of organisms to changing environmental temperatures (Murata and Wada, 1995; Los and Murata, 1998; Rodriguez-Vargas et al., 2007). The same protective mechanism may also operate in cells exposed to hyperosmolarity, since osmotic stress reduces cell membrane fluidity

(Laroche et al., 2001), which influences membrane permeabilization and cell death (Poirier et al., 1999). They found that fluidization of the yeast membrane produced a moderate increase in Na^+ tolerance but had no effects on resistance to pure osmotic stress.

3.1. *Temperature*

The major response of membrane composition to a change in temperature is an alteration in the fatty acid component of lipids (Russell, 1989a). Microorganisms counteract the propensity for membranes to rigidify at lower temperature by adapting to the conditions in order to maintain a more-or-less constant degree of membrane fluidity (homoeoviscous adaptation) (Quinn, 1981; Hazel and Williams, 1990). Thus, a low temperature stress initiates a co-ordinated response by fatty acid desaturase and dehydrases induction which in turn increases the ratio of polyunsaturated to saturated fatty acids and/or (but less frequently) actually decreases the length of the fatty acid chains in the membrane (Al-Fageeh and Smales, 2006).

3.1.1. *Low temperatures*

When poikilothermic organisms such as bacteria, plants and fish are exposed to suboptimal growth temperatures, their membrane lipids become more rigid, leading to subnormal functioning of cellular activities (Phadtare, 2004; Mansilla and de Mendoza, 2005; Al-Fageeh and Smales, 2006). Adaptation to such conditions involves an increase in the proportion of unsaturated fatty acids (UFA) and/or to a lesser extent shortening of fatty acids. The resulting increase in UFA content causes membrane lipid fluidity to return to its original state, or close to it, with concurrent restoration of normal cellular activity at the lower temperature (Suutari et al., 1990; Suutari and Laakso, 1994; Mansilla et al., 2004; Phadtare, 2004). Thus, cold tolerant individuals have a higher proportion of unsaturated lipids incorporated into membrane lipids, related to decreased physical order of the resulting bilayer. This effect more or less overcomes the ordered, rigidifying effects of cold and has been invoked as an explanation of both capacity and resistance cold adaptation at the level of the whole organism (Cossins et al., 2002).

In fungi three groups of responses to decreased temperature on the fatty acid level can be observed. The first consists of changes in polyunsaturated linoleic (C18:2) and/or linolenic (C18:3) fatty acid amount by at least three different modes when the temperature decreases: (i) the amount of less unsaturated acids decreases more than that of the most unsaturated acids, which leads a decrease in fatty acid content and increase of unsaturation; this mode can be observed in *C. utilis, Lipomyces starkeyi,* and

Rhodosporidium toruloides; (ii) the cellular content of the most unsaturated fatty acid (linoleic (C18:2) or linolenic (C18:3) acid) increases, which results in an increase of fatty acid unsaturation; this mode is employed by *Candida oleophila* and *A. niger*; (iii) the cellular content of polyunsaturated fatty acids decreases and consequently also fatty acid unsaturation (Suutari et al., 1990; Suutari and Laakso, 1994). Second, increasing the biosynthesis of monounsaturated fatty acids like palmitoleic (C16:1) acid in *S. cerevisiae*, shortens the average acyl chains and increases the unsaturation (Suutari et al., 1990; Suutari and Laakso, 1994; Los and Murata, 1998). In *Sc. pombe* cold stress increases fatty acid unsaturation by increasing oleic (C18:1) acid content (McDonough and Roth, 2004). The third group of responses is adaptation by induction of the elongation and desaturation and resulting production of long-chain polyunsaturated fatty acids like α-linolenic (C18:3) and stearidonic (C18:4) acids in *Mortierella antarctica* (Suutari et al., 1990; Suutari and Laakso, 1994; Zlatanov et al., 2001).

There are only a few studies dealing with the adaptation of psychrophilic/psychrotolerant fungi at low temperatures. These studies show that individual species employ different strategies for the maintenance of membrane fluidity by fatty acid composition at low temperatures. Some Antarctic fungi such as *Mo. antarctica* and *Cadophora fastigiata* show increased amounts of linoleic (C18:2) and arachidonic (C20:4) acid respectively, at low temperatures. In psychrotolerant yeasts such as *Cryptococcus albidus, Cr. laurentii, Rhodotorula mucilaginosa*, unsaturated fatty acids predominated in the membranes at lower temperatures (Zlatanov et al., 2001). With decreasing temperature the relative amount of palmitic (C16:0) fatty acid decreased while the amount of linolenic acid (C18:2) increased in *Aspergillus alliaceus, Cyathus stercoreus, Phanerochaete chrysosporium, Phialophora malorum, Poria placenta*, and *Syncephalastrum racemosum* (Stahl and Klug, 1996). Fatty acid unsaturation increased in *A. niger, Penicillium chrysogenum*, and *Trichoderma reesei* when the temperature was lowered. In *A. niger* and *T. reesei* this was due to the increase in linolenic acid (C18:3) content. In *P. chrysogenum*, the linolenic acid (C18:3) content increased concomitantly with a more pronounced decrease in the less-unsaturated fatty acid, oleic acid (C18:1), and in palmitic (C16:0) and linoleic acids (C18:2); consequently, the fatty acid content decreased. In *N. crassa*, fatty acid unsaturation was nearly temperature-independent (Suutari, 1995).

3.1.2. High temperatures

In living organisms, including fungi, higher environmental temperatures generally produce an increase in the degree of saturation in membrane lipids (Suutari et al., 1990). In *S. cerevisiae*, sensitivity to heat stress depends on membrane lipid composition. Aerobic cells with membranes enriched in palmitoleic (C16:1) and oleic acids (C18:1) evidenced the highest degrees of resistance to heat stress (52°C, 5 min) and oxidative stress (Steels et al., 1994; Swan and Watson, 1997). In response to heat shock, *S. cerevisiae* slightly decreases the degree of acyl chain unsaturation (Swan and Watson, 1997). Kim et al. observed with regard to cellular fatty acid composition that levels of unsaturated fatty acids (UFA) were increased significantly at high temperatures (43°C), and this was particularly true of oleic acid (C18:1) (Kim et al., 2006).

In yeast species like *Candida utilis, C. oleophila, Li. starkeyi, Rho. toruloides* and *S. cerevisiae*, there were nc temperature related changes in the composition of major fatty acids of any of them, but the absolute amounts and relative compositions of the fatty acids did alter. In *S. cerevisiae* temperature-induced changes in the mean fatty acid chain length were observed, whereas in *Rho. toruloides* there were changes in the degree of unsaturation. *C. oleophila, C. utilis* and *Li. starkeyi* showed the changes in the degree of unsaturation at temperatures above 20-26°C (Suutari et al., 1990).

S. cerevisiae cells are less sensitive to heat stress when they contain high levels of unsaturated fatty acids (UFA) in their membranes, since high percentage of UFA during heat shock can desensitize the heat-shock response (HSR) pathway (Chatterjee et al., 1997). The other major heat-sensing system, the general stress response (GSR) pathway also triggers a decreased heat sensitivity under the same conditions. Heat-stress-acclimated cells contained 9% more unsaturated fatty acids than cells grown at 25°C and required a 4°C higher temperature to induce the stress response pathways. Furthermore, remarkably similar changes occurred in cells that had been acclimated to salt stress. There was a 9% increase in the amount of unsaturated fatty acids (as had been the case with heat-stressed cells) and an identical decrease in the sensitivity of GSR induction by either salt or heat. It is suggested that heat stress is detected by membrane-linked »thermostat(s)« the activation of which depends not only on temperature but also on the type and percentage of unsaturated fatty acids present in the cell. The corollary to this hypothesis is that stress-inducible genes protect living cells during acute stress but that changes in cellular lipids are a major factor by which cells acclimate to long-term heat (or salt) stress (Chatterjee et al., 2000).

3.2. *pH*

Almost nothing has been published in relation to the effect of external pH on lipid composition in different fungi. The only publication reporting results on the adaptation to different pH values *in* fungi is about the marine yeast *Debaryomyces hansenii.* It shows that growth at low pH caused a decrease in fatty acids unsaturation, with the increase in palmitic acid (C16:0) and the decrease in oleic acid (C18:1) (Turk et al., 2007b).

3.3. *Water activity*

Integrity of biological membranes requires the presence of water, and changes in cellular water activity can have a profound influence on membrane stability (Hazel and Williams, 1990). Adaptation mechanisms, however, differ according to the solute, that lowered the water activity.

3.3.1. *Sugars*

Not much is known about the effect of high sugar concentration on fatty acyl composition in fungi. In *Penicillium expansum* and *Aspergillus chevalieri* at 50% sucrose the unsaturation index of membranes decreased by 20-25%, indicating that the plasma membrane is less fluid at this concentration. At 80% sucrose a similar trend was observed for *P. expansum*, but for *A. chevalieri* the unsaturation index was changed only little compared to the control (Hefnawy and Abou-Zeid, 2003).

3.3.2. *Salt*

Increased NaCl concentration stimulate osmotic responses in essentially the same way as sugars or sugar alcohols at a similar water activity. Na^+, however, is toxic, because it replaces K^+ in biomolecules. Therefore, Na+ stimulates additional detoxification responses (Hohmann, 2002).

In contrast to the response of membrane composition to changes in temperature, the major change in response to salt is the head-group of the lipids. The most common alteration is an increased proportion of anionic phospholipids and/or glycolipids. However, salt and other solutes may also influence the fatty acid composition of the membrane lipids (Russell, 1989a; Russell et al., 1995).

The effect of salt stress on lipid composition and membrane fluidity has been investigated primarily in a restricted group of salt-sensitive fungi, including *S. cerevisiae*. In these fungi salt stress caused saturation of phospholipid-esterified fatty acids (Tunblad-Johansson and Adler, 1987; Sharma et al., 1996). In *S. cerevisiae* salt stress induced a depletion of oleic acid (C18:1) and a concurrent enrichment in palmitoleic acid (C16:1) (Turk et al., 2004). Membrane composition and properties were also examined in the halotolerant yeasts *Zygosaccharomyces rouxii, D. hansenii, Candida membranefaciens* and *Yarrowia lipolytica*. These fungi showed different responses under salt stress. *Z. rouxii* showed decreased fatty acid unsaturation (Hosono, 1992; Yoshikawa et al., 1995), while high salinity did not induce significant changes in the unsaturation of fatty acids in *Y. lipolytica* (Andreishcheva et al., 1999). In contrast, *C. membranefaciens* grown at high NaCl concentration exhibited increases in fatty acid unsaturation (Khaware et al., 1995), while in *D. hansenii*, salt stress caused a relatively small increase in the fatty acid unsaturation and an increase in length of fatty acyl residues of the phospholipids (Tunblad-Johansson et al., 1987; Turk et al., 2007a).

Due to the lack of knowledge on the biology of halophilic eukaryotes, only recently studies on the adaptation of membranes at high salinity have been performed on the halophilic black yeasts (*Dothideales, Ascomycota*). The study on the influence of increased NaCl concentration on membrane lipid composition and fluidity in the halophilic *Hortaea werneckii* and *Phaeotheca triangularis* and the halotolerant *Aureobasidium pullulans* showed an increase in the phospholipid-esterified fatty acid unsaturation. The most abundant fatty acids in the phospholipids were C_{16} and C_{18} fatty acids with a high percentage of linolenic (C18:2) fatty acid. In *H. werneckii*, increased salinity was accompanied by a decrease in palmitic fatty acid (C16:0) together with an increase in linoleic acid (C18:2). More stenotolerant *P. triangularis* showed a relative decrease of linoleic acid (C18:2) at higher salinities. In *Au. pullulans* linoleic acid (C18:2) predominated among all fatty acids and increased with raised salinity, whereas stearic (C18:0) and oleic acid (C18:1) decreased (Turk et al., 2004).

In addition to effects on specific membrane-dependent functions, large changes in membrane lipid order (resulting from changes in temperature or fatty acid unsaturation) have been correlated with increased passive permeability of membranes to ions such as K^+ and Na^+ (Murata, 1989; Hazel and Williams, 1990).

3.4. High hydrostatic pressure

Hydrostatic pressure interferes with cellular membrane structure increasing the order of lipid molecules, especially in the vicinity of proteins. This phenomenon is driven by the smaller volume associated with a more ordered, tighter packing. The consequence is a decrease in cell membrane fluidity followed by an increase in thickness (Braganza and Worcester, 1986; Reyes Mateo et al., 1993). Increased levels of fatty acid chain unsaturation may reduce the sensitivity of bilayer order to variations in pressure (Skanes et al., 2006).

The ratio of unsaturated fatty acids to saturated fatty acids in a barotolerant mutant of *S. cerevisiae* was high during all phases of growth, in contrast to the parent strain (Fujii and Fulco, 1977).

3.5. *Oxidative stress*

Aerobic organisms produce reactive oxygen intermediates as a result of incomplete reduction of oxygen during respiration, *β*-oxidation of fatty acids, exposure to radiation, and the action of some drugs (Storz et al., 1990). Some heavy metals and various genotoxic agents, including UV irradiation and alkylating agents, also generate oxidative stress (Ikner and Shiozaki, 2005). These harmful reactive oxygen intermediates interfere with cellular processes due to their ability to damage DNA, lipid membranes, and proteins. Unsaturated fatty acyl groups in membranes are a major target for the hydroxyl radical and the protonated superoxide anion and this attack initiates autocatalytic lipid peroxidation resulting in the formation of reactive lipid radicals and lipid hydroperoxides (Wiseman and Halliwell, 1996). Breakdown of lipid hydroperoxides leads to the generation of a great diversity of highly reactive aldehydes (Levine, 2002).

A *Hevea brasiliensis* Δ12 fatty acid desaturase gene was cloned and functionally expressed in *S. cerevisiae* wild-type strains. Cells producing linolenic (C18:2) and 9, 12-hexadecadienoic (C16:2) polyunsaturated fatty acids become more sensitive to oxidative challenge by treatment with H_2O_2 or other compounds inducing oxidative stress, such as tBH or paraquat. Desaturase expression does not seem to globally increase cellular sensitivity, as for instance transient exposure to 2 M NaCl did not affect growth (Cipak et al., 2006). In *S. cerevisiae* exposed to H_2O_2 treatment the content of unsaturated fatty acids decreased (Swan and Watson, 1997). The most oxidative stress-resistant *S. cerevisiae* cells in anaerobiosis have membranes enriched in saturated fatty acids while in aerobiosis their membranes were enriched in palmitoleic (C16:1) and oleic (C18:1) fatty acids (Steels et al., 1994).

3.6. *Ethanol stress*

Ethanol is well known as an inhibitor of microbial growth. Many studies have documented the alteration of cellular lipid composition in response to ethanol exposure. Although conflicting reports on this issue can be found in the literature, most authors find that the ethanol tolerance of *S. cerevisiae* correlates with an increased degree of fatty acid unsaturation and that the presence of ethanol causes an increase in the amount of monounsaturated fatty acids in cellular lipids and a decrease in the proportion of saturated fatty acids (Šajbidor and Grego, 1992; Alexandre et al., 1994; Chi and Arneborg, 1999; Swan and Watson, 1999). Moreover, it was found that the more ethanol tolerant strain incorporates more long-chain fatty acids, i.e. stearic (C18:0) and oleic (C18:1) acid into the membrane phospholipids at the expense of shorter chain fatty acids, i.e. palmitic (C16:0) and palmitoleic (C16:1) fatty acid, compared with the less ethanol-tolerant strain (Chi and Arneborg, 1999). You et al. report that UFA composition is a significant determinant of ethanol tolerance in *S. cerevisiae* and that oleic (C18:1) fatty acid is the most efficacious UFA in overcoming the toxic effects of ethanol in growing yeast cells. Ethanol tolerance in yeast thus results from the incorporation of oleic acid in lipid membranes, effecting a compensatory decrease in membrane fluidity that counteracts the fluidising effects of ethanol (You et al., 2003). Also in *S. pombe* the ethanol tolerance is connected with a high content of oleic acid (C18:1) and thus with increased unsaturation of fatty acids (Koukou et al., 1990), while in *K. lactis* and *Zygosaccharomyces bailii* ethanol decreased the unsaturation of membrane fatty acids and increased the mean acyl chain length (Baleiras Couto and Huis in't Veld, 1995; Heipieper et al., 2000). A different response to ethanol stress was seen in *Mu. rouxii* where Jeennor et al. identified some unusual fatty acids including odd-numbered fatty acids and 7-hydroxy dodecanoic acid (7-OH C12:0) in addition to the more common fatty acids (Jeennor et al., 2006).

4. Genetic basis of changes in fatty acid composition

The expression of genes involved in fatty acid metabolism is very important since it provides the molecular basis for the acclimation of organisms to changing environmental conditions. Even FAS expression that occurs mostly at an intermediate, constitutive level, may be modulated by distinct metabolic conditions, as shown in animals (Semenkovich, 1997) and bacteria (Marrakchi et al., 2002).

It seems likely that cells perceive these alterations via sensory proteins embedded in their membranes. These proteins transfer the signals from the environment to networks of signal-transduction pathways, with the resultant regulation of gene expression (Murata and Los, 1997; Los and Murata, 2000, 2004). It has been postulated that a change in membrane fluidity might be the primary signal in the perception of cold stress and, possibly, of osmotic stress. It is very likely that the rigidification of membrane lipids at low temperatures and under hyperosmotic stress is the primary trigger for the corresponding acclimatory responses in cells. It is possible that some common sensor(s) recognizes the rigidification of membrane lipids irrespective of the nature of the stimulus (for example, cold stress, hyperosmotic stress, or salt stress) (Los and Murata, 2004).

The regulation of expression of enzymes involved in fatty acid metabolism is complex (as was demonstrated in the case of *S. cerevisiae OLE1* gene) and still very poorly understood. The existence of several different enzymes, catalyzing different or even the same reaction that contribute to membrane fluidity further complicate the issue. As mentioned before, *M.*

alpina for example has five types of desaturases (Δ9, Δ12, Δ6, Δ5, and ω3) (Sakuradani et al., 1999a; Sakuradani et al., 1999c; Sakuradani et al., 1999b; Sakuradani et al., 2005). In addition to this Wongwathanarat et al. (1999) reported multiple Δ9 desaturases in some *M. alpina* strains and suggested that the expression of their genes may be strain-specific or induced under certain physiological conditions (Wongwathanarat et al., 1999).The same was observed in the case of two Δ6 desaturases from a similar fungus (Abe et al., 2006).

Some expression responses of different genes involved in fatty acid metabolism, however, are known and will be described in the following paragraphs.

4.1. *Adaptations to different temperatures*

4.1.1. *Low temperatures*

Elevated activities of desaturase enzymes appear to be a widespread adaptation for life at cold temperatures and may be responsible for increased levels of unsaturated fatty acids (Hazel and Landrey, 1988; Hazel and Williams, 1990). Fatty acid desaturases are involved in cold acclimation in bacteria (Kaan et al., 2002), animals (Murray et al., 2007), plants (Khodakovskaya et al., 2006; Wei et al., 2006b) and also fungi (Rodriguez-Vargas et al., 2007). Transcriptional expression of fatty acid desaturase genes of several fungi is induced by low-temperature stress (Laoteng et al., 1999; Calvo et al., 2001; Nakagawa et al., 2002; Sakai and Kajiwara, 2003).

The gene for the best studied fungal desaturase, *S. cerevisiae OLE1*, belongs to a group of early cold response genes (Schade et al., 2004). It is induced by low temperature through ubiquitin/proteasome-dependent processing of membrane bound transcription factors (Zhang et al., 1999; Hoppe et al., 2000; Nakagawa et al., 2002) and was induced together with components involved in its regulation, including the ER-membrane-bound transcription factor Mga2 (Hoppe et al., 2000; Schade et al., 2004).

Another example of induction by low temperature are desaturases from *S. kluyveri*. *Sk-FAD3* (encoding ω3 desaturase), *Sk-OLE1* (encoding Δ9 desaturase) and *Sk-FAD2* (encoding Δ12 desaturase) were all induced after a temperature downshift to 10°C. The induction of *Sk-FAD3* was relatively slow and permanent, whereas the induction of both *Sk-OLE1* and *Sk-FAD2* was much faster, but only transient (Oura and Kajiwara, 2004).

The levels of the *Le. edodes Le-FAD2* transcript encoding a Δ12 desaturase following a shift from 25 to 18°C did not show any changes (Sakai and Kajiwara, 2005). On the other hand, an increase in the

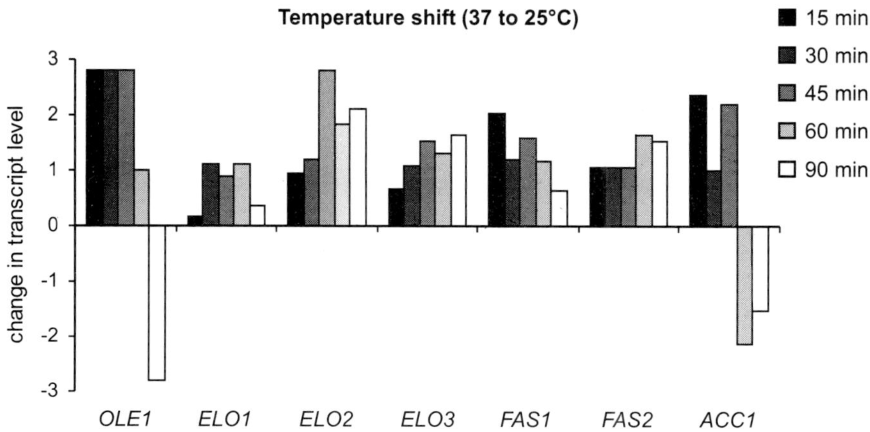

Fig. 1. Changes in expression level of major genes involved in fatty acid metabolism in *S. cerevisiae* following a temperature shift from 37°C to 25°C. *OLE1*—gene encoding Δ9 desaturase; *ELO1, ELO2, ELO3*—genes encoding desaturases; *FAS1, FAS2*—genes encoding two subunits of fatty acid synthase, *ACC1*—gene encoding acetyl-CoA carboxylase. The experiment was performed using microarrays. Values exceeding 2.8-fold induction/ repression are truncated at 2.8. Adapted from (Gasch et al., 2000)

expression level of a gene for Δ9 fatty acid desaturase *Le-FAD1* has been observed after the same temperature change (Sakai and Kajiwara, 2003).

Microarray experiments with *S. cerevisiae* (Gasch et al., 2000) have shown a strong transient induction of OLE1 gene after a shift from 37 to 25°C and an induction of all three elongase genes. The same has been observed for both FAS and ACC1 gene (Figure 1).

Although the effects of low temperature on expression of enzymes involved in fatty acid metabolism are the best studied among responses of these enzymes to different environmental changes, a lot of work still remains. The elongase expression profiles should be investigated in detail. The selective incorporation of primarily unsaturated acyl-CoAs at low temperature (or saturated acyl-CoA at high temperature) by the enzymes of complex lipid biosynthesis may also be important to the temperature-induced remodelling of membrane lipid composition (Hazel and Williams, 1990).

4.1.2. High temperatures

Less data is available regarding adaptations to high environmental temperatures. A substantial increase of expression of *FAA1*, the gene encoding a long chain fatty acyl-CoA synthetase involved in the activation of imported

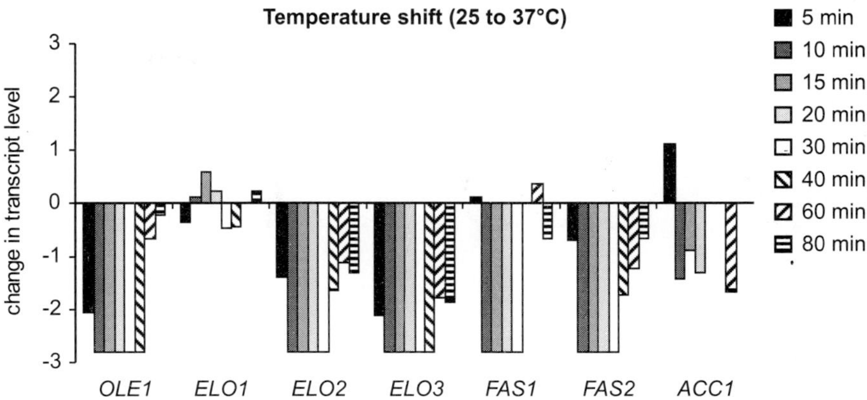

Fig. 2. Changes in expression level of major genes involved in fatty acid metabolism in *S. cerevisiae* following a temperature shift from 25°C to 37°C. *OLE1*—gene encoding Δ9 desaturase; *ELO1, ELO2, ELO3*—genes encoding desaturases; *FAS1, FAS2*—genes encoding two subunits of fatty acid synthase, *ACC1*—gene encoding acetyl-CoA carboxylase. The experiment was performed using microarrays. Values exceeding 2.8-fold induction/ repression are truncated at 2.8. Adapted from (Gasch et al., 2000)

fatty acids has been observed during heat shock and upon entry into stationary phase (Martinez et al., 2004). The relationship between FAA1 and heat stress response is probably due to a requirement for sphingolipid synthesis to modify the cell membranes and resist temperature elevation (Cowart and Hannun, 2005). *FAA3* and *FAA4* in contrast have reduced expression during heat shock (Gasch et al., 2000; Gasch and Werner-Washburne, 2002).

Microarray data showed that a shift from 25 to 37°C results in strong repression of *OLE1*, *ELO2* and *ELO3*, the expression of *ELO1* however, stays basically the same (Figure 2). A repression was also observed in case of *FAS1*, *FAS2* and *ACC1* genes (Gasch et al., 2000). A different study (Causton et al., 2001) reported similar responses with *FAS1*, *FAS2* and *ACC1* gene repression being more short-lived and *ELO1* gene being moderately induced.

4.2. *Acidic and alkaline conditions*

Virtually nothing is known about detailed expression responses of genes involved in fatty acid metabolism to different environmental pH values from targeted studies. All available data come from microarray experiments.

Lowering of pH from pH 6.0 to pH 4.0 led to moderate repression of *OLE1*, *ELO2* and *ELO3* and moderate induction of *ELO1* (Figure 3). *OLE1* was moderately repressed immediately after the shift, but substantially

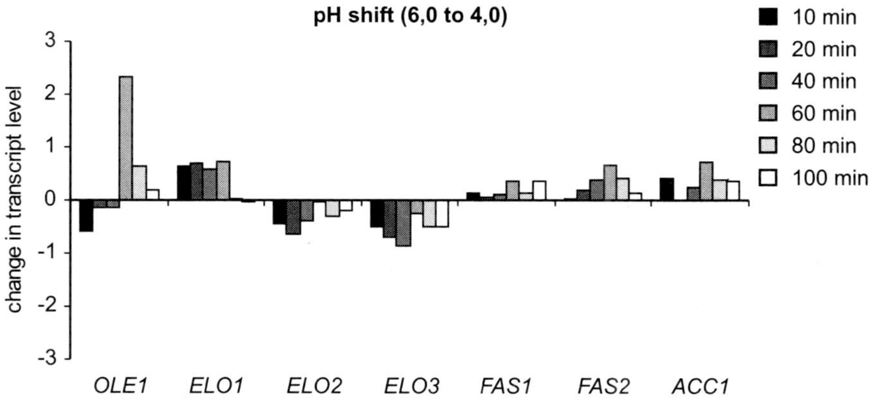

Fig. 3. Changes in expression level of major genes involved in fatty acid metabolism in *S. cerevisiae* following a pH shift from pH 6.0 to pH 4.0. *OLE1*—gene encoding Δ9 desaturase; *ELO1, ELO2, ELO3*—genes encoding desaturases; *FAS1, FAS2*—genes encoding two subunits of fatty acid synthase, *ACC1*—gene encoding acetyl-CoA carboxylase. The experiment was performed using microarrays. Values exceeding 2.8-fold induction/ repression are truncated at 2.8. Adapted from (Causton et al., 2001)

induced one hour later. Moderate induction of *FAS2* and *ACC1* was observed only at one time point.

Shift to alkaline conditions from pH 6.0 to pH 7.9 caused a severe repression of *ELO2, ELO3* and *FAS1,* moderate repression of *OLE1, FAS2* and *ACC1,* while the expression of *ELO1* remained unchanged (Figure 4).

4.3. Lowered water activities

Studies in *S. cerevisiae* indicate that osmotic stress reduces cell-membrane fluidity (Laroche et al., 2001). Lipid saturation seems to be important for salt tolerance in *Z. rouxii* cells under high-NaCl conditions (Watanabe and Takakuwa, 1987; Hosono, 1992). Therefore, one would expect that fatty acid metabolism enzymes are involved in adaptations to lowered water activities.

Hyperosmotic shock, caused by the addition of sorbitol, envokes immediate strong—although very short-lived—repression of *OLE1, ELO2, ELO3* and *FAS2* and more moderate repression of *FAS1* gene in salt-sensitive *S. cerevisiae.* The response of *ELO1* gene is rather weak and somewhat ambiguous (Gasch et al., 2000; Causton et al., 2001). For more details see also Figure 5. Hypoosmotic shock from 1 M sorbitol to medium without sorbitol resulted in the opposite (although weaker) responses (Figure 6).

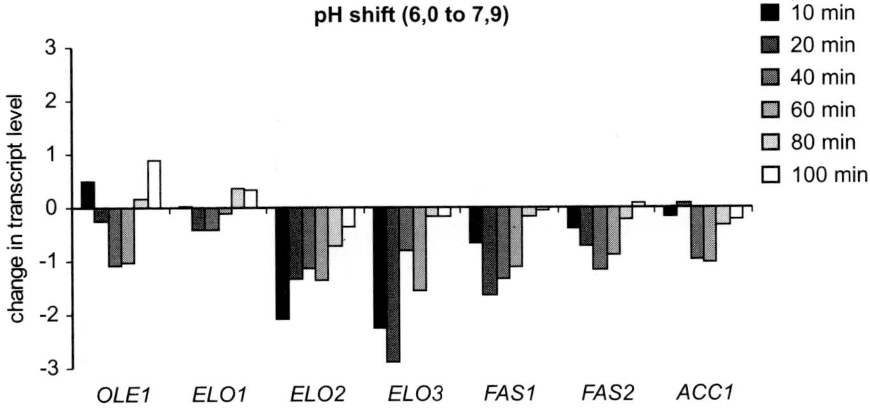

Fig. 4. Changes in expression level of major genes involved in fatty acid metabolism in *S. cerevisiae* following a pH shift from pH 6.0 to pH 7.9. *OLE1*—gene encoding Δ9 desaturase; *ELO1, ELO2, ELO3*—genes encoding desaturases; *FAS1, FAS2*—genes encoding two subunits of fatty acid synthase, *ACC1*—gene encoding acetyl-CoA carboxylase. The experiment was performed using microarrays. Values exceeding 2.8-fold induction/repression are truncated at 2.8. Adapted from (Causton et al., 2001)

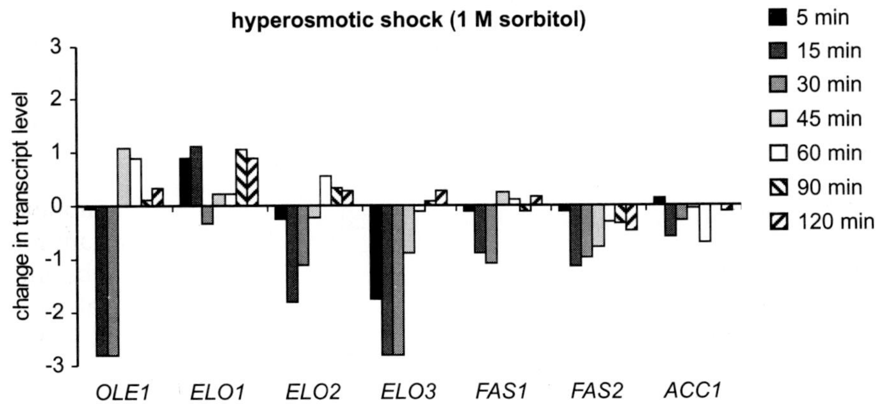

Fig. 5. Changes in expression level of major genes involved in fatty acid metabolism in *S. cerevisiae* following a hyperosmotic shock with 1 M sorbitol. *OLE1*—gene encoding Δ9 desaturase; *ELO1, ELO2, ELO3*—genes encoding desaturases; *FAS1, FAS2*—genes encoding two subunits of fatty acid synthase, *ACC1*—gene encoding acetyl-CoA carboxylase. The experiment was performed using microarrays. Values exceeding 2.8-fold induction/repression are truncated at 2.8. Adapted from (Gasch et al., 2000)

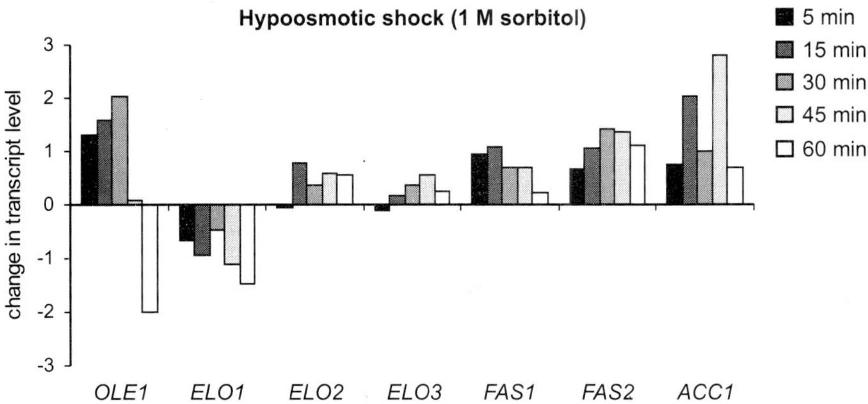

Fig. 6. Changes in expression level of major genes involved in fatty acid metabolism in *S. cerevisiae* following a shift from 1 M sorbitol to medium without sorbitol. *OLE1*—gene encoding Δ9 desaturase; *ELO1, ELO2, ELO3*—genes encoding desaturases; *FAS1, FAS2*—genes encoding two subunits of fatty acid synthase, *ACC1*—gene encoding acetyl-CoA carboxylase. The experiment was performed using microarrays. Values exceeding 2.8-fold induction/repression are truncated at 2.8. Adapted from (Gasch et al., 2000)

In *S. cerevisiae* the results of microarray studies investigating expression following hyperosmotic shock due to 1 M sodium chloride, revealed similar, although more pronounced responses to sorbitol (Figure 7). In case of *OLE1* gene the response is surprising, since the shock is followed by repression, instead of expected induction, which would compensate for higher rigidity of the membranes. However, such a response may be only temporary. In *S. cerevisiae* ORF's encoding many known yeast salinity stress response proteins are unaffected or down-regulated immediately following the up-shift and gradually induced only later (Yale and Bohnert, 2001). According to another microarray study, expression of *OLE1* diminished 45 minutes after a shift to high osmolarity (Rep et al., 2000) and increased again after 90 min (Yale and Bohnert, 2001).

Studies by Gostinčar and associates indicate a similar response in halotolerant black yeast *Au. pullulans*. Hyperosmotic shock caused by 1.71 M NaCl resulted in a repression of genes for elongase, Δ9 and Δ12 desaturases, while hypoosmotic shock had just the opposite effect. Nevertheless, elevated expression of all three genes was observed at high salinities, which was consistent with increased desaturation of fatty acids in membranes (Gostinčar and Gunde-Cimerman, 2007). The expression of homologues of these genes in extremely halotolerant black yeast *H. werneckii* also showed the same pattern after hyper- and hypoosmotic shocks caused

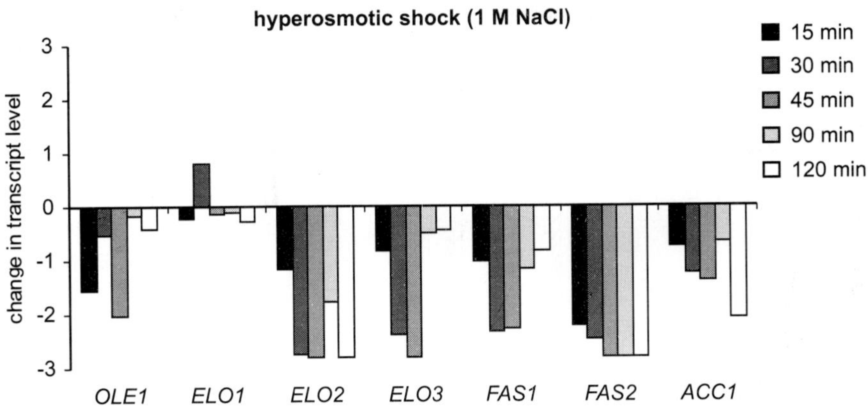

Fig. 7. Changes in expression level of major genes involved in fatty acid metabolism in *S. cerevisiae* following a hyperosmotic shock with 1 M NaCl. *OLE1*—gene encoding Δ9 desaturase; *ELO1, ELO2, ELO3*—genes encoding desaturases; *FAS1, FAS2*—genes encoding two subunits of fatty acid synthase, *ACC1*—gene encoding acetyl-CoA carboxylase. The experiment was performed using microarrays. Values exceeding 2.8-fold induction/repression are truncated at 2.8. Adapted from (Gasch et al., 2000)

by 2.9 M NaCl. The expression was lowest in cells grown in the medium with NaCl concentrations around 2 M and higher towards both salinity extremes (medium without salt and 4.27 M NaCl) (Gostinčar et al., 2005).

The listed differences between expression profiles in cells grown at constant salinity and cells exposed to sudden shock caused by the medium with the same salt concentration indicate that these two types of stress represent a fundamentally different challenge to cells. Therefore, data gained from the up-shift and down-shift salinity studies have to be interpreted with caution.

4.4. *High hydrostatic pressure*

Whole genome microarray hybridization study of *S. cerevisiae* cells subjected to a hydrostatic pressure of 200 MPa for 30 min detected a 2,4-fold induction of *OLE1* gene (Fernandes et al., 2004).

4.5. *Oxidative stress*

According to microarray studies, oxidative stress induced with 0.4 mM (Causton et al., 2001) or 0.32 mM (Gasch et al., 2000) hydrogen peroxide leads to transient repression of *OLE1*, followed by later induction, repression of *ELO2, ELO3* and *FAS2*, while the results for *ELO1, FAS1* and

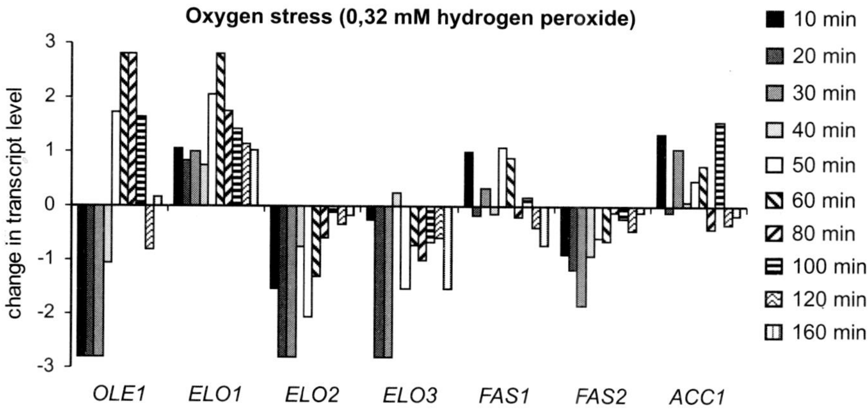

Fig. 8. Changes in expression level of major genes involved in fatty acid metabolism in *S. cerevisiae* following an oxidative shock with 0.32 mM H_2O_2. *OLE1*—gene encoding Δ9 desaturase; *ELO1, ELO2, ELO3*—genes encoding desaturases; *FAS1, FAS2*—genes encoding two subunits of fatty acid synthase, *ACC1*—gene encoding acetyl-CoA carboxylase. The experiment was performed using microarrays. Values exceeding 2.8-fold induction/repression are truncated at 2.8. Adapted from (Gasch et al., 2000)

ACC1 are contradictory (see Figure 8 for data from Gasch et al., 2000, data from Causton et al., 2001 not shown).

Oxydative stress does not effect only membrane compounds, but also the enzymes themselves. A subunit of cytosolic fatty acid synthase was shown to be one of the major targets of oxidative damage, its activity level diminished at around 60% (Cabiscol et al., 2000).

A normal amount of oxygen, however, is essential for complete fatty acid metabolism, since desaturation requires oxygen. Interestingly, sake yeasts, that live in an environment with very limited oxygen availability express unusually high amounts of *OLE1* mRNA, which may enable them to synthesize unsaturated fatty acids using the little oxygen available (Yamada et al., 2005).

4.6. High ethanol concentrations

Reports suggest a relationship between membrane fluidity and ethanol tolerance (Kajiwara et al., 1996).

S. cerevisiae cells overexpressing their own *OLE1* gene or the *FAD2* gene from *Arabidopsis thaliana* exhibited greater resistance to high concentrations of ethanol than did the control cells (Kajiwara et al., 1996; Kajiwara et al., 2000).

5. General stress response

It has been postulated that a change in membrane fluidity might be the primary signal in the perception of cold stress and, possibly, of osmotic stress. It is very likely that the rigidification of membrane lipids at low temperatures and under hyperosmotic stress is the primary trigger for the corresponding acclimatory responses in cells. The existence of sensors that perceive changes in the physical state of the membrane, disregarding the nature of the stress, is postulated (Los and Murata, 2004). This would make membranes not just a target of adaptations to environmental conditions, but also a crucial part of the mechanism, involved in triggering the adaptations. Membrane fluidity as an indicator of cell environment would also explain, why many stress responses are common regardless of the kinds of stress. As proposed by Panadero and co-workers, the fluidity state of the cell membrane might be a key factor to integrate the sensing mechanism of cold and hyperosmolarity (Panadero et al., 2006).

The so-called general stress response seems to aim mainly at the production of cellular protectants and the adjustment of energy metabolism. There are also genes whose expression appears to be affected only by a subset of stress conditions, such as osmotic and oxidative stress, but those stress-specific gene sets are surprisingly small (Causton et al., 2001; Hohmann, 2002). This may also explain the cross-tolerance to different environmental conditions. The pattern of gene expression induced by a mild temperature allows the cells to achieve tolerance to high temperature, osmotic pressure, dehydration and cryotreatment (Varela and Mager, 1996). Barotolerant mutants generated by random mutagenesis showed higher thermotolerance and greater tolerance against oxidative stress than their parent strain (Iwahashi et al., 1993).

This phenomenon of cross-protection is evidence of the existence of a general stress response. Nevertheless, this is not true for all situations, as cells exposed to high osmolarity do not acquire resistance against heat stress. Furthermore, the protection induced by a different stress form is usually not as efficient as a mild pretreatment with the same kind of stress. This indicates that, despite the general response, each stress situation leads to a particular gene expression profile (Hohmann and Mager, 1997; Fernandes et al., 2004). Results of *S. cerevisiae* microarray analysis performed under various stress conditions showed that even though there are common genes that respond to a variety of stresses, there is not a uniform response for all kinds of stress situations (Gasch et al., 2000).

The fluidity of the membranes may represent a base for the general stress response, which is complemented by additional sensory pathways, which fine tune the adaptations in dependence of the nature of the environmental insult.

6. Possible applications

As seen above, cell membranes play an important role in response to stressful environmental conditions. The sustenance of proper membrane lipid composition is one of the major adaptations to stress in which the diversity of fatty acids and their physical properties play a major role. Furthermore, the physical state of membranes may be involved in triggering of cell responses to stress. Therefore, enzymes involved in fatty acid modifications may be important in the efforts to engineer industrially important fungi (or even plants) that will be able to endure stressful conditions encountered in biotechnological processes.

Lipid composition may also be important from the medical point of view. Close interaction between membrane lipids and the drug sensitivities of yeast cells has been observed in *S. cerevisiae ERG* mutants. It could be attributed to membrane permeability changes, which may involve changes in passive diffusion across the membrane or in the active transport of these drugs (Mukhopadhyay et al., 2002).

Polyunsaturated fatty acids have received a lot of attention recently as pharmaceutical and nutraceutical compounds. Biotechnological strategies represent a tool for the creation of novel fungal strains synthesizing economically valuable lipid metabolites (Certik et al., 1998). The possible application of new knowledge on fungal fatty acid metabolism is described in several publications (Certik et al., 1998; Broun et al., 1999; Beaudoin et al., 2000; Dyer et al., 2002).

CONCLUSIONS

Cell membranes are in direct contact with the surrounding environment and as such immediately subjected to altered environmental conditions. The fluidity of membranes, which is crucial for proper functioning of the cell, is significantly affected by many physicochemical factors, such as temperature, salinity, etc. Many studies have shown that environmental insults trigger complex adjustments in membrane lipid composition, among which the adjustment of ratios between different fatty acids is one of the most important changes. Furthermore, it appears that the change in membrane physical state actually triggers at least part of the response mechanisms.

The differences in membrane fatty acid composition and enzymes involved in their metabolism in different fungi may partially explain their different tolerance to stress. Understanding of the phenomenon may be important for engineering fungi (or even plants) with elevated stress tolerance, for successful fighting against quickly emerging fungal pathogens and for biotechnological exploitation of fungi for production/modification of lipids.

However, although the basic enzymology involved in fatty acid biosynthesis has been well worked out, many questions related to the regulation of this process are still unresolved, as pointed out by Tehlivets et al., 2007. The understanding of fatty acid response to environmental changes contains many of these questions.

Most of the work on this topic was performed on *S. cerevisiae*. While bakers yeast as an intensively studied model organism with known genome sequence, commercially available molecular biology tools etc. is a sensible choice, data on other organisms would contribute to our knowledge for several reasons. *S. cerevisiae* has only one desaturase enzyme, while most other fungi have at least one more, in many cases (especially filamentous fungi) even several other enzymes with different specificity, roles in metabolism and presumably also regulatory mechanisms. Other fungi, more tolerant to different extreme environmental conditions, have different lipid responses to stress as was observed by our group in case of two black yeasts, halotolerant *Au. pullulans* and extremely halotolerant *H. werneckii*. Finally, the more fatty acid modifying enzymes are known from different organisms, the greater is the possibility of success in genetically modifying industrially important fungi for lipid production.

More data is needed on expression responses of genes encoding fatty acid modifying enzymes to different environmental conditions. The majority of the now existing data originate from large scale microarray experiments, the results of which are sometimes ambiguous or even contradictory. Other techniques, such as real-time PCR, would provide new information, especially in poorly investigated topics such as responses to different pH or hydrostatic pressure.

Almost all experiments, observing response to different environmental conditions, were to study organisms subjected to shock because of sudden shift in temperature, salinity or similar conditions. However, it was already observed some years ago, that transcriptional response following a shock is, to a very large extent, transient and follows a distinct temporal pattern. For many genes the peak of expression can be so sharp, that it is easily missed if measured as little as 15 minutes after the shock (Gasch et al., 2000; Rep et al., 2000; Hohmann, 2002). Our findings confirm these observations in case of fatty acid modifying genes and show us that life at a constant value of a specific environmental parameter, no matter how stressful it is, represents to cells a fundamentally different challenge than a sudden shift to the same environment (Gostinčar et al., 2005; Gostinčar and Gunde-Cimerman, 2007). Therefore, if we want to understand, how organisms adapt to certain environments, not just their reaction to shock, more experiments with fungi in different constant environmental conditions are needed.

Fungi represent convenient research models for increasing our knowledge on eukaryotic lipidomics. They are easier to handle in a laboratory than most other eukaryotes and at the same time have the complexity that prokaryotic model organisms lack. However, a lot of research has still to be done in the future. More knowledge about the role of fatty acids and the enzymes involved in their metabolism in adaptation to different environments would contribute not only to our understanding of life in different (even extreme) environments but could also have an industrial potential.

References

Aaronson, L.R., Johnston, A.M. and Martin, C.E. 1982. The effects of temperature acclimation on membrane sterols and phospholipids of *Neurospora crassa*. *Biochim* Biophys Acta, 713: 456-462.

Abe, T., Sakuradani, E., Asano, T., Kanamaru, H. and Shimizu, S. 2006. Functional characterization of Delta9 and omega9 desaturase genes in *Mortierella alpina* 1S-4 and its derivative mutants. Appl Microbiol Biotechnol, 70: 711-719.

Aguilar, P.S. and de Mendoza, D. 2006. Control of fatty acid desaturation: a mechanism conserved from bacteria to humans. Mol Microbiol, 62: 1507-1514.

Al-Fageeh, M.B. and Smales, C.M. 2006. Control and regulation of the cellular responses to cold shock: the responses in yeast and mammalian systems. Biochem Journal, 397: 247-259.

Al-Feel, W., Chirala, S.S. and Wakil, S.J. 1992. Cloning of the yeast *FAS3* gene and primary structure of yeast acetyl-CoA carboxylase. Proc Natl Acad Sci, USA, 89: 4534-4538.

Alexandre, H., Rousseaux, I. and Charpentier, C. 1994. Relationship between ethanol tolerance, lipid composition and plasma membrane fluidity in Saccharomyces cerevisiae and Kloeckera apiculata. FEMS Microbiol Lett, 124: 17-22.

Andreishcheva, E.N., Isakova, E.P., Sidorov, N.N., Abramova, N.B., Ushakova, N.A., Shaposhnikov, G.L., Soares, M.I.M. and Zvyagilskaya, R.A. 1999. Adaptation to salt stress in a salt-tolerant strain of the yeast *Yarrowia lipolytica*. Biochemistry (Moscow), 64: 1061-1067.

Baleiras Couto, M.M. and Huis in't Veld, J.H. 1995. Influence of ethanol and temperature on the cellular fatty acid composition of Zygosaccharomyces bailii spoilage yeasts. Journal of Appl Bacteriol, 78: 327-334.

Beaudoin, F., Michaelson, L.V., Hey, S.J., Lewis, M.J., Shewry, P.R., Sayanova, O. and Napier, J.A. 2000. Heterologous reconstitution in yeast of the polyunsaturated fatty acid biosynthetic pathway. Proc Natl Acad Sci, USA, 97: 6421-6426.

Black, P.N. and DiRusso, C.C. 2007. Yeast acyl-CoA synthetases at the crossroads of fatty acid metabolism and regulation. Biochim Biophys Acta, 1771: 286-298.

Bossie, M.A. and Martin, C.E. 1989. Nutritional regulation of yeast delta-9 fatty acid desaturase activity. Journal of Bacteriol, 171: 6409-6413.

Bowman, B.J., Borgeson, C.E. and Bowman, E.J. 1987. Composition of *Neurospora crassa* vacuolar membranes and comparison to endoplasmic reticulum, plasmalemmas and mitochondrial membranes. Exp Mycol, 11: 197-205.

Braganza, L.F. and Worcester, D.L. 1986. Structural changes in lipid bilayers and biological membranes caused hydrostatic pressure. Biochemistry (Mosc), 25: 7484-7488.

Broadwater, J.A., Whittle, E. and Shanklin, J. 2002. Desaturation and hydroxylation. Residues 148 and 324 of Arabidopsis FAD2, in addition to substrate chain length, exert a major influence in partitioning of catalytic specificity. Journal of Biol Chem, 277: 15613-15620.

Brody, S., Oh, C., Hoja, U. and Schweizer, E. 1997. Mitochondrial acyl carrier protein is involved in lipoic acid synthesis in *Saccharomyces cerevisiae*. FEBS Lett, 408: 217-220.

Broun, P., Gettner, S. and Somerville, C. 1999. Genetic engineering of plant lipids. Annu Rev Nutr, 19: 197-216.

Brown, D.W., Adams, T.H. and Keller, N.P. 1996. *Aspergillus* has distinct fatty acid synthases for primary and secondary metabolism. Proc Natl Acad Sci, USA, 93: 14873-14877.

Cabiscol, E., Piulats, E., Echave, P., Herrero, E. and Ros, J. 2000. Oxidative stress promotes specific protein damage in *Saccharomyces cerevisiae*. Journal of Biol Chem, 275: 27393-27398.

Calvo, A.M., Gardner, H.W. and Keller, N.P. 2001. Genetic connection between fatty acid metabolism and sporulation in *Aspergillus nidulans*. Journal of Biol Chem, 276: 25766-25774.

Carman, G.M. and Henry, S.A. 1999. Phospholipid biosynthesis in the yeast *Saccharomyces cerevisiae* and interrelationship with other metabolic processes. Prog Lipid Res, 38: 361-399.

Causton, H.C., Ren, B., Koh, S.S., Harbison, C.T., Kanin, E., Jennings, E.G., Lee, T.I., True, H.Ll., Lander, E.S. and Young, R.A. 2001. Remodeling of yeast genome expression in response to environmental changes. Mol Biol Cell, 12: 323-337.

Certik, M., Sakuradani, E. and Shimizu, H. 1998. Desaturase-defective fungal mutants: useful tools for the regulation and overproduction of polyunsaturated fatty acids. Trends Biotechnol, 16: 500-505.

Chatterjee, M.T., Khalawan, S.A. and Curran, B.P. 1997. Alterations in cellular lipids may be responsible for the transient nature of the yeast heat shock response. Microbiol, 143 (Pt 9): 3063-3068.

Chatterjee, M.T., Khalawan, S.A. and Curran, B.P. 2000. Cellular lipid composition influences stress activation of the yeast general stress response element (STRE). Microbiol, 146 (Pt 4): 877-884.

Chattopadhyay, P., Banerjee, S.K., Sen, K. and Chakrabarti, P. 1985. An unsaturated fatty acid mutant of *Aspergillus niger* with partially defective delta 9-

desaturase. Can J Microbiol, 31: 346-351.

Chi, Z. and Arneborg, N. 1999. Relationship between lipid composition, frequency of ethanol-induced respiratory deficient mutants, and ethanol tolerance in *Saccharomyces cerevisiae*. Journal of Appl Microbiol, 86: 1047-1052.

Chirala, S.S. 1992. Coordinated regulation and inositol-mediated and fatty acid-mediated repression of fatty acid synthase genes in *Saccharomyces cerevisiae*. Proc Natl Acad Sci, USA, 89: 10232-10236.

Chirala, S.S., Kuziora, M.A., Spector, D.M. and Wakil, S.J. 1987. Complementation of mutations and nucleotide sequence of FAS1 gene encoding beta subunit of yeast fatty acid synthase. Journal of Biol Chem, 262: 4231-4240.

Cho, R.J., Campbell, M.J., Winzeler, E.A. et al. 1998. A genome-wide transcriptional analysis of the mitotic cell cycle. Mol Cell, 2: 65-73.

Choi, J.Y., Stukey, J., Hwang, S.Y. and Martin, C.E. 1996. Regulatory elements that control transcription activation and unsaturated fatty acid-mediated repression of the *Saccharomyces cerevisiae OLE1* gene. Journal of Biol Chem, 271: 3581-3589.

Cipak, A., Hasslacher, M., Tehlivets, O. et al. 2006. *Saccharomyces cerevisiae* strain expressing a plant fatty acid desaturase produces polyunsaturated fatty acids and is susceptible to oxidative stress induced by lipid peroxidation. Free Radic Biol Med, 40: 897-906.

Cossins, A.R., Murray, P.A., Gracey, A.Y., Logue, J., Polley, S., Caddick, M., Brooks, S., Postle, T. and Maclean, N. 2002. The role of desaturases in cold-induced lipid restructuring. Biochem Soc Trans, 30: 1082-1086.

Cowart, L.A. and Hannun, Y.A. 2005. Using genomic and lipidomic strategies to investigate sphingolipid function in the yeast heat-stress response. Biochem Soc Trans, 33: 1166-1169.

Cowart, L.A. and Obeid, L.M. 2007. Yeast sphingolipids: recent developments in understanding biosynthesis, regulation, and function. Biochim Biophys Acta, 1771: 421-431.

Dailey, H.A. Strittmatter, P. 1980. Characterization of the interaction of amphipathic cytochrome b5 with stearyl coenzyme A desaturase and NADPH: cytochrome P-450 reductase. Journal of Biol Chem, 255: 5184-5189.

Das, T., Thurmond, J.M., Bobik, E., Leonard, A.E., Parker-Barnes, J.M., Huang, Y.S. and Mukerji, P. 2000. Polyunsaturated fatty acid-specific elongation enzymes. Biochem Soc Trans, 28: 658-660.

Daum, G., Lees, N.D., Bard, M. and Dickson, R. 1998. Biochemistry, cell biology and molecular biology of lipids of *Saccharomyces cerevisiae*. Yeast, 14: 1471-1510.

de Kroon, A.I. 2007. Metabolism of phosphatidylcholine and its implications for lipid acyl chain composition in *Saccharomyces cerevisiae*. Biochim Biophys Acta, 1771: 343-352.

Demel, R.A. and De Kruyff, B. 1976. The function of sterols in membranes. Biochim Biophys Acta, 457: 109-132.

Dittrich, F., Zajonc, D., Huhne, K. et al. 1998. Fatty acid elongation in yeast—biochemical characteristics of the enzyme system and isolation of elongation-defective mutants. Eur J Biochem, 252: 477-485.

Dommes, V., Baumgart, C. and Kunau, W.H. 1981. Degradation of unsaturated fatty acids in peroxisomes. Existence of a 2,4-dienoyl-CoA reductase pathway. Journal of Biol Chem, 256: 8259-8262.

Dyer, J.M., Chapital, D.C., Kuan, J.W., Mullen, R.T. and Pepperman, A.B. 2002. Metabolic engineering of *Saccharomyces cerevisiae* for production of novel lipid compounds. Appl Microbiol Biotechnol, 59: 224-230.

Fernandes, P.M., Domitrovic, T., Kao, C.M. and Kurtenbach, E. 2004. Genomic expression pattern in *Saccharomyces cerevisiae* cells in response to high hydrostatic pressure. FEBS Lett, 556: 153-160.

Fujii, D.K. and Fulco, A.J. 1977. Biosynthesis of unsaturated fatty acids by bacilli. Hyperinduction and modulation of desaturase synthesis. Journal Biol Chem, 252: 3660-3670.

Gasch, A.P. and Werner-Washburne, M. 2002. The genomics of yeast responses to environmental stress and starvation. Funct Integr Genomics, 2: 181-192.

Gasch, A.P., Spellman, P.T., Kao, C.M., Carmel-Harel, O., Eisen, M.B., Storz, G., Botstein, D. and Brown, P.O. 2000. Genomic expression programs in the response of yeast cells to environmental changes. Mol Biol Cell, 11: 4241-4257.

Gonzalez, C.I. and Martin, C.E. 1996. Fatty acid-responsive control of mRNA stability. Unsaturated fatty acid-induced degradation of the *Saccharomyces* OLE1 transcript. Journal of Biol Chem, 271: 25801-25809.

Goodrich-Tanrikulu, M., Stafford, A.E., Lin, J.T., Makapugay, M.I., Fuller, G. and McKeon, T.A. 1994. Fatty acid biosynthesis in novel ufa mutants of *Neurospora crassa*. Microbiol, 140 (Pt 10): 2683-2690.

Gostinčar, C. and Gunde-Cimerman, N. 2007. Expression of fatty acid modifying enzymes in halotolerant black yeast *Aureobasidium pullulans* subjected to salt stress. In: ISHAM Black Yeast Workshop: Black Yeasts between extremotolerance and human pathology Centraalbureau voor Schimmelcultures, Utrecht, The Netherlands.

Gostinčar, C., Turk, M., and Gunde-Cimerman, N. 2005. The effect of salt stress on the expression of elongase, Δ9-desaturase, and Δ12-desaturase genes in the halophilic black yeast *Hortaea werneckii*. In: 7th yeast lipid conference Swansea: University of Wales, School of medicine. Wales, UK.

Greenberg, M.L. and Lopes, J.M. 1996. Genetic regulation of phospholipid biosynthesis in *Saccharomyces cerevisiae*. Microbiol Rev, 60: 1-20.

Guy, J.E., Whittle, E., Kumaran, D., Lindqvist, Y. and Shanklin, J. 2007. The crystal structure of the ivy Delta4-16:0-ACP desaturase reveals structural details of the oxidized active site and potential determinants of regioselectivity. Journal of Biol Chem, 282: 19863-19871.

Hasslacher, M., Ivessa, A.S., Paltauf, F. and Kohlwein, S.D. 1993. Acetyl-CoA carboxylase from yeast is an essential enzyme and is regulated by factors that control phospholipid metabolism. Journal of Biol Chem, 268: 10946-10952.

Hazel, J.R. and Landrey, S.R. 1988. Time course of thermal adaptation in plasma membranes of trout kidney. II. Molecular species composition. Am J Physiol, 255: R628-634.

Hazel, J.R. and Williams, E.E. 1990. The role of alterations in membrane lipid composition in enabling physiological adaptation of organisms to their physical environment. Prog Lipid Res, 29: 167-227.

Hefnawy, M.A. and Abou-Zeid, A.M. 2003. Differential adaptation of membranes of two osmotolerant fungi, *Aspergillus chevalieri* and *Penicillium expansum* to high sucrose concentrations. Acta Microbiol Pol, 52: 53-64.

Heinz, E. 1993. Lipid metabolism in plants. CRC Press, London, UK.

Heipieper, H.J., Isken, S. and Saliola, M. 2000. Ethanol tolerance and membrane fatty acid adaptation in adh multiple and null mutants of Kluyveromyces lactis. Res Microbiol, 151: 777-784.

Henry, S.A. and Patton-Vogt, J.L. 1998. Genetic regulation of phospholipid metabolism: yeast as a model eukaryote. Prog Nucleic Acid Res Mol Biol, 61: 133-179.

Hoffmann, M., Hornung, E., Busch, S., Kassner, N., Ternes, P., Braus, G.H. Feussner, I. 2007. A small membrane-peripheral region close to the active center determines regioselectivity of membrane bound fatty acid desaturases from *Aspergillus nidulans*. Journal of Biol Chem in press.

Hohmann, S. 2002. Osmotic stress signaling and osmoadaptation in yeast. Microbiol Mol Biol Rev, 66: 300-372.

Hohmann, S. and Mager, W.H. 1997. Yeast stress responses. R.G. Landes (ed.), Chapman and Hall, New York, London, Austin.

Hoppe, T., Matuschewski, K., Rape, M., Schlenker, S., Ulrich, H.D. and Jentsch, S. 2000. Activation of a membrane-bound transcription factor by regulated ubiquitin/proteasome-dependent processing. Cell, 102: 577-586.

Hosono, K. 1992. Effect of salt stress on lipid composition and membrane fluidity of the salt-tolerant yeast *Zygosaccharomyces rouxii*. Journal of Gen Microbiol, 138: 91-96.

Howlett, N.G. and Avery, S.V. 1997. Induction of lipid peroxidation during heavy metal stress in *Saccharomyces cerevisiae* and influence of plasma membrane fatty acid unsaturation. Appl Environ Microbiol, 63: 2971-2976.

Huang, Y.S., Chaudhary, S., Thurmond, J.M., Bobik, E.G., Jr., Yuan, L., Chan, G.M., Kirchner, S.J., Mukerji, P. and Knutzon, D.S. 1999. Cloning of delta12- and delta6-desaturases from *Mortierella alpina* and recombinant production of gamma-linolenic acid in *Saccharomyces cerevisiae*. Lipids, 34: 649-659.

Huisinga, K.L. and Pugh, B.F. 2004. A genome-wide housekeeping role for TFIID and a highly regulated stress-related role for SAGA in *Saccharomyces cerevisiae*. Mol Cell, 13: 573-585.

Hynes, M.J., Murray, S.L., Duncan, A., Khew, G.S. and Davis, M.A. 2006. Regulatory genes controlling fatty acid catabolism and peroxisomal functions in the filamentous fungus *Aspergillus nidulans*. Eukaryot Cell, 5: 794-805.

Ikner, A. and Shiozaki, K. 2005. Yeast signaling pathways in the oxidative stress response. Mutat Res, 569: 13-27.

Iwahashi, H., Fujii, S., Obuchi, K., Kaul, S.C., Sato, A. and Komatsu, Y. 1993. Hydrostatic pressure is like high temperature and oxidative stress in the damage it causes to yeast. FEMS Microbiol Lett, 108: 53-57.

Jeennor, S., Laoteng, K., Tanticharoen, M. and Cheevadhanarak, S. 2006. Comparative fatty acid profiling of Mucor rouxii under different stress conditions. FEMS Microbiol Lett, 259: 60-66.

Jeffery, J., Kock, J.L., Botha, A., Coetzee, D.J. and Botes, P.J. 1997. The value of lipid composition in the taxonomy of the *Schizosaccharomycetales*. Antonie Van Leeuwenhoek, 72: 327-335.

Jeffries, T.W., Grigoriev, I.V., Grimwood, J. et al. 2007. Genome sequence of the lignocellulose-bioconverting and xylose-fermenting yeast *Pichia stipitis*. Nat Biotechnol, 25: 319-326.

Jones, T., Federspiel, N.A., Chibana, H. et al. 2004. The diploid genome sequence of *Candida albicans*. Proc Natl Acad Sci, USA, 101: 7329-7334.

Kaan, T., Homuth, G., Mader, U., Bandow, J. and Schweder, T. 2002. Genome-wide transcriptional profiling of the *Bacillus subtilis* cold-shock response. Microbiol, 148: 3441-3455.

Kainou, K., Kamisaka, Y., Kimura, K. and Uemura, H. 2006. Isolation of Delta12 and omega3-fatty acid desaturase genes from the yeast *Kluyveromyces lactis* and their heterologous expression to produce linoleic and alpha-linolenic acids in *Saccharomyces cerevisiae*. Yeast, 23: 605-612.

Kajiwara, S., Shirai, A., Fujii, T., Toguri, T., Nakamura, K. and Ohtaguchi, K. 1996. Polyunsaturated fatty acid biosynthesis in *Saccharomyces cerevisiae*: expression of ethanol tolerance and the *FAD2* gene from *Arabidopsis thaliana*. Appl Environ Microbiol, 62: 4309-4313.

Kajiwara, S., Aritomi, T., Suga, K., Ohtaguchi, K. and Kobayashi, O. 2000. Over-expression of the OLE1 gene enhances ethanol fermentation by *Saccharomyces cerevisiae*. Appl Microbiol Biotechnol, 53: 568-574.

Kaliszewski, P., Ferreira, T., Gajewska, B., Szkopinska, A., Berges, T. and Zoladek, T. 2006. Enhanced levels of Pis1p (phosphatidylinositol synthase) improve the growth of *Saccharomyces cerevisiae* cells deficient in Rsp5 ubiquitin ligase. Biochem Journal, 395: 173-181.

Kandasamy, P., Vemula, M., Oh, C.S., Chellappa, R. and Martin, C.E. 2004. Regulation of unsaturated fatty acid biosynthesis in *Saccharomyces*: the endoplasmic reticulum membrane protein, Mga2p, a transcription activator of the *OLE1* gene, regulates the stability of the *OLE1* mRNA through exosome-mediated mechanisms. Journal of Biol Chem, 279: 36586-36592.

Kaneko, H., Hosohara, M., Tanaka, M. and Itoh, T. 1976. Lipid composition of 30 species of yeast. Lipids, 11: 837-844.

Khaware, R.K., Koul, A. and Prasad, R. 1995. High membrane fluidity is related to NaCl stress in *Candida membranefaciens*. Biochem Mol Biol Int, 35: 875-880.

Khodakovskaya, M., McAvoy, R., Peters, J., Wu, H. and Li, Y. 2006. Enhanced cold tolerance in transgenic tobacco expressing a chloroplast omega-3 fatty acid desaturase gene under the control of a cold-inducible promoter. Planta, 223: 1090-1100.

Kim. I.S., Moon, H.Y., Yun, H.S. and Jin, I. 2006. Heat shock causes oxidative stress and induces a variety of cell rescue proteins in *Saccharomyces cerevisiae* KNU5377. Journal of Microbiol, 44: 492-501.

Koukou, A.I., Tsoukatos, D. and Drainas, C. 1990. Effect of ethanol on the phospholipid and fatty acid content of Schizosaccharomyces pombe membranes. Journal of Gen Microbiol, 136: 1271-1277.

Kunau, W.H., Buhne, S., de la Garza, M., Kionka, C., Mateblowski, M., Schultz-Borchard, U. and Thieringer, R. 1988. Comparative enzymology of beta-

oxidation. Biochem Soc Trans, 16: 418-420.
Kunau, W.H., Dommes, V. and Schulz, H. 1995. beta-oxidation of fatty acids in mitochondria, peroxisomes, and bacteria: a century of continued progress. Prog Lipid Res, 34: 267-342.
Laoteng, K., Anjard, C., Rachadawong, S., Tanticharoen, M., Maresca, B. and Cheevadhanarak S. 1999. Mucor rouxii delta9-desaturase gene is transcriptionally regulated during cell growth and by low temperature. Mol Cell Biol Res Commun, 1: 36-43.
Laoteng, K., Ruenwai, R., Tanticharoen, M. and Cheevadhanarak, S. 2005. Genetic modification of essential fatty acids biosynthesis in *Hansenula polymorpha*. FEMS Microbiol Lett, 245: 169-178.
Laroche, C., Beney, L., Marechal, P.A. and Gervais, P. 2001. The effect of osmotic pressure on the membrane fluidity of *Saccharomyces cerevisiae* at different physiological temperatures. Appl Microbiol Biotechnol, 56: 249-254.
Letoublon, R., Mayet, B., Frot-Coutaz, J., Nicolau, C. and Got, R. 1982. Subcellular distribution of phospholipids and of polyprenol phosphate in *Aspergillus niger* van Tieghem. Biochim Biophys Acta, 711: 509-514.
Levine, R.L. 2002. Carbonyl modified proteins in cellular regulation, aging, and disease. Free Radic Biol Med, 32: 790-796.
Loftus, B.J., Fung, E., Roncaglia, P. et al. 2005. The genome of the basidiomycetous yeast and human pathogen *Cryptococcus neoformans*. Science, 307: 1321-1324.
Los, D.A. and Murata, N. 1998. Structure and expression of fatty acid desaturases. Biochim Biophys Acta, 1394: 3-15.
Los, D.A. and Murata, N. 2000. Regulation of enzymatic activity and gene expression by membrane fluidity. Sci STKE 2000: PE1.
Los, D.A. and Murata, N. 2004. Membrane fluidity and its roles in the perception of environmental signals. Biochim Biophys Acta, 1666: 142-157.
Lösel, D.M. 1988. Fungal lipids, pp. 699-805. In: Microbial lipids. Vol. 1. C. Ratledge and S.G. Wilkinson (eds.). Academic Press.
Lösel, D.M. 1990. Lipids in the structure and function of fungal membranes, pp. 119-133. In: Biochemistry of the Cell Walls and Membranes in Fungi. P.J. Kuhn, A.P.J. Trinci, M.J. Jung, M.W. Goosey, and L.G. Copping (eds.). Springer-Verlag, Berlin, Heidelberg, New York, London, Paris, Tokyo, Hong Kong.
Mansilla, M.C., Cybulski, L.E., Albanesi, D. and de Mendoza, D. 2004. Control of membrane lipid fluidity by molecular thermosensors. Journal of Bacteriol, 186: 6681-6688.
Mansilla, M.C. and de Mendoza, D. 2005. The *Bacillus subtilis* desaturase: a model to understand phospholipid modification and temperature sensing. Arch Microbiol, 183: 229-235.
Mariani, P., Rivas, E., Luzzati, V. and Delacroix, H. 1990. Polymorphism of a lipid extract from *Pseudomonas fluorescens*: structure analysis of a hexagonal phase and of a novel cubic phase of extinction symbol Fd. Biochemistry (Mosc), 29: 6799-6810.
Marrakchi, H., Zhang, Y.M. and Rock, C.O. 2002. Mechanistic diversity and regulation of Type II fatty acid synthesis. Biochem Soc Trans, 30: 1050-1055.
Martin, C.E., Oh, C.S., Kandasamy, P., Chellapa, R. and Vemula, M. 2002. Yeast

desaturases. Biochem Soc Trans, 30: 1080-1082.

Martin, C.E., Oh, C.S. and Jiang, Y. 2007. Regulation of long chain unsaturated fatty acid synthesis in yeast. Biochim Biophys Acta, 1771: 271-285.

Martinez, M.J., Roy, S., Archuletta, A.B., Wentzell, P.D., Anna-Arriola, S.S., Rodriguez, A.L., Aragon, A.D., Quinones, G.A., Allen, C. and Werner-Washburne, M. 2004. Genomic analysis of stationary-phase and exit in *Saccharomyces cerevisiae*: gene expression and identification of novel essential genes. Mol Biol Cell, 15: 5295-5305.

McDonough, V.M. and Roth, T.M. 2004. Growth temperature affects accumulation of exogenous fatty acids and fatty acid composition in *Schizosaccharomyces pombe*. Antonie Van Leeuwenhoek, 86: 349-354.

McDonough, V.M., Stukey, J.E. and Martin, C.E. 1992. Specificity of unsaturated fatty acid-regulated expression of the *Saccharomyces cerevisiae OLE1* gene. Journal Biol Chem, 267: 5931-5936.

Meesters, P.A., Springer, J. and Eggink, G. 1997. Cloning and expression of the delta 9 fatty acid desaturase gene from *Cryptococcus curvatus* ATCC 20509 containing histidine boxes and a cytochrome b5 domain. Appl Microbiol Biotechnol, 47: 663-667.

Michinaka, Y., Aki, T., Shimauchi, T., Nakajima, T., Kawamoto, S., Shigeta, S., Suzuki, O. Ono, K. 2003. Differential response to low temperature of two Delta6 fatty acid desaturases from *Mucor circinelloides*. Appl Microbiol Biotechnol, 62: 362-368.

Mikolajczyk, S. and Brody, S. 1990. De novo fatty acid synthesis mediated by acyl-carrier protein in *Neurospora crassa* mitochondria. Eur J Biochem, 187: 431-437.

Mitchell, A.G. and Martin, C.E. 1995. A novel cytochrome b5-like domain is linked to the carboxyl terminus of the *Saccharomyces cerevisiae* delta-9 fatty acid desaturase. Journal of Biol Chem, 270: 29766-29772.

Mohamed, A.H., Chirala, S.S., Mody, N.H., Huang, W.Y. and Wakil, S.J. 1988. Primary structure of the multifunctional alpha subunit protein of yeast fatty acid synthase derived from FAS2 gene sequence. Journal of Biol Chem, 263: 12315-12325.

Mukhopadhyay, K., Kohli, A. and Prasad, R. 2002. Drug susceptibilities of yeast cells are affected by membrane lipid composition. Antimicrob Agents Chemother, 46: 3695-3705.

Murata, N. 1989. Low-temperature effects on cyanobacterial membranes. J Bioenerg Biomembr, 21: 61-75.

Murata, N. and Wada, H. 1995. Acyl-lipid desaturases and their importance in the tolerance and acclimatization to cold of cyanobacteria. Biochem Journal, 308 (Pt 1): 1-8.

Murata, N. and Los, D.A. 1997. Membrane Fluidity and Temperature Perception. Plant Physiol, 115: 875-879.

Murray, P., Hayward, S.A., Govan, G.G., Gracey, A.Y. and Cossins, A.R. 2007. An explicit test of the phospholipid saturation hypothesis of acquired cold tolerance in *Caenorhabditis elegans*. Proc Natl Acad Sci, USA, 104: 5489-5494.

Nakagawa, Y., Sakumoto, N., Kaneko, Y. and Harashima, S. 2002. Mga2p is a putative sensor for low temperature and oxygen to induce *OLE1* transcription in *Saccharomyces cerevisiae*. Biochem Biophys Res Commun, 291: 707-713.

Nakashima, S., Zhao, Y. and Nozawa, Y. 1996. Molecular cloning of delta 9 fatty acid desaturase from the protozoan Tetrahymena thermophila and its mRNA expression during thermal membrane adaptation. Biochem Journal, 317 (Pt 1): 29-34.

Nierman, W.C., Pain, A., Anderson, M.J. et al. 2005. Genomic sequence of the pathogenic and allergenic filamentous fungus *Aspergillus fumigatus*. Nature, 438: 1151-1156.

Oh, C.S., Toke, D.A., Mandala, S. and Martin, C.E. 1997. *ELO2* and *ELO3*, homologues of the *Saccharomyces cerevisiae ELO1* gene, function in fatty acid elongation and are required for sphingolipid formation. Journal of Biol Chem, 272: 17376-17384.

Oura, T. and Kajiwara, S. 2004. *Saccharomyces kluyveri* FAD3 encodes an omega3 fatty acid desaturase. Microbiol, 150: 1983-1990.

Panadero, J., Pallotti, C., Rodriguez-Vargas, S., Randez-Gil, F. and Prieto, J.A. 2006. A downshift in temperature activates the high osmolarity glycerol (HOG) pathway, which determines freeze tolerance in *Saccharomyces cerevisiae*. Journal of Biol Chem, 281: 4638-4645.

Parker-Barnes, J.M., Das, T., Bobik, E., Leonard, A.E., Thurmond, J.M., Chaung, L.T., Huang, Y.S. and Mukerji, P. 2000. Identification and characterization of an enzyme involved in the elongation of n-6 and n-3 polyunsaturated fatty acids. Proc Natl Acad Sci, USA, 97: 8284-8289.

Passorn, S., Laoteng, K., Rachadawong, S., Tanticharoen, M. and Cheeva-dhanarak, S. 1999. Heterologous expression of *Mucor rouxii* delta(12)-desaturase gene in *Saccharomyces cerevisiae*. Biochem Biophys Res Commun, 263: 47-51.

Pereira, S.L., Leonard, A.E. and Mukerji, P. 2003. Recent advances in the study of fatty acid desaturases from animals and lower eukaryotes. Prostaglandins, Leukotrienes and Essential Fatty Acids, 68: 97-106.

Phadtare, S. 2004. Recent developments in bacterial cold-shock response. Curr Issues Mol Biol, 6: 125-136.

Poirier, I., Marechal, P.A., Richard, S. and Gervais, P. 1999. Saccharomyces cerevisiae viability is strongly dependant on rehydration kinetics and the temperature of dried cells. Journal of Appl Microbiol, 86: 87-92.

Prasitchoke, P., Kaneko, Y., Bamba, T., Fukusaki, E., Kobayashi, A. and Harashima, S. 2007a. Identification and characterization of a very long-chain fatty acid elongase gene in the methylotrophic yeast, *Hansenula polymorpha*. Gene, 391: 16-25.

Prasitchoke, P., Kaneko, Y., Sugiyama, M., Bamba, T., Fukusaki, E., Kobayashi, A. and Harashima, S. 2007b. Functional analysis of very long-chain fatty acid elongase gene, *HpELO2*, in the methylotrophic yeast *Hansenula polymorpha*. Appl Microbiol Biotechnol, 76: 417-427.

Quinn, P.J. 1981. The fluidity of cell membranes and its regulation. Prog Biophys Mol Biol, 38: 1-104.

Rattray, J.B., Schibeci, A. and Kidby, D.K. 1975. Lipids of yeasts. Bacteriol Rev, 39: 197-231.

Rep, M., Krantz, M., Thevelein, J.M. and Hohmann, S. 2000. The transcriptional response of *Saccharomyces cerevisiae* to osmotic shock. Hot1p and Msn2p/Msn4p are required for the induction of subsets of high osmolarity glycerol pathway-dependent genes. Journal of Biol Chem, 275: 8290-8300.

Reyes Mateo, C., Tauc, P. and Brochon, J.C. 1993. Pressure effects on the physical properties of lipid bilayers detected by trans-parinaric acid fluorescence decay. Biophys Journal, 65: 2248-2260.

Rilfors, L. and Lindblom, G. 2002. Regulation of lipid composition in biological membranes-biophysical studies of lipids and lipid synthesizing enzymes. Colloids and Surfaces B: Biointerfaces, 26: 112-124.

Rodriguez-Vargas, S., Sanchez-Garcia, A., Martinez-Rivas, J.M., Prieto, J.A. and Randez-Gil, F. 2007. Fluidization of membrane lipids enhances the tolerance of *Saccharomyces cerevisiae* to freezing and salt stress. Appl Environ Microbiol, 73: 110-116.

Rodriguez, R.J., Low, C., Bottema, C.D. and Parks, L.W. 1985. Multiple functions for sterols in *Saccharomyces cerevisiae*. Biochim Biophys Acta, 837: 336-343.

Rossler, H., Rieck, C., Delong, T., Hoja, U. and Schweizer, E. 2003. Functional differentiation and selective inactivation of multiple *Saccharomyces cerevisiae* genes involved in very-long-chain fatty acid synthesis. Mol Genet Genomics, 269: 290-298.

Russell, N.J. 1989a. Adaptive modifications in membranes of halotolerant and halophilic microorganisms. Journal of Bioenerg Biomembr, 21: 93-113.

Russell, N.J. 1989b. Structural and functional role of lipids, pp. 279-349. In: Microbial lipids. Vol. 2. C. Ratledge and S.G. Wilkinson (eds.). Academic Press, New York, USA.

Russell N.J., Evans R.I., ter Steeg P.F., Hellemons J., Verheul A. and Abee T. (1995). Membranes as a target for stress adaptation. *Int J Food Microbiol,* 28: 255-261.

Sakai, H. and Kajiwara, S. 2003. A stearoyl-CoA-specific Delta 9 fatty acid desaturase from the basidiomycete *Lentinula edodes*. On the Bioscience, Biotechnology, and Biochemistry, 67: 2431-2437.

Sakai, H. and Kajiwara, S. 2005. Cloning and functional characterization of a Delta12 fatty acid desaturase gene from the basidiomycete *Lentinula edodes*. Mol Genet Genomics, 273: 336-341.

Sakuradani, E., Kobayashi, M., Ashikari, T. and Shimizu, S. 1999a. Identification of Delta12-fatty acid desaturase from arachidonic acid-producing mortierella fungus by heterologous expression in the yeast *Saccharomyces cerevisiae* and the fungus *Aspergillus oryzae*. Eur J Biochem, 261: 812-820.

Sakuradani, E., Kobayashi, M. and Shimizu, S. 1999b. Delta 9-fatty acid desaturase from arachidonic acid-producing fungus. Unique gene sequence and its heterologous expression in a fungus, *Aspergillus*. Eur J Biochem, 260: 208-216.

Sakuradani, E., Kobayashi, M. and Shimizu, S. 1999c. Delta6-fatty acid desaturase from an arachidonic acid-producing *Mortierella fungus*. Gene cloning and its heterologous expression in a fungus, *Aspergillus*. Gene, 238: 445-453.

Sakuradani, E., Abe, T., Iguchi, K. and Shimizu, S. 2005. A novel fungal omega3-desaturase with wide substrate specificity from arachidonic acid-producing

Mortierella alpina 1S-4. Appl Microbiol Biotechnol, 66: 648-654.

Schade, B., Jansen, G., Whiteway, M., Entian, K.D. and Thomas, D.Y. 2004. Cold adaptation in budding yeast. Mol Biol Cell, 15: 5492-5502.

Schneiter, R. and Kohlwein, S.D. 1997. Organelle structure, function, and inheritance in yeast: a role for fatty acid synthesis? Cell, 88: 431-434.

Schneiter, R., Brugger, B., Sandhoff, R. et al. 1999. Electrospray ionization tandem mass spectrometry (ESI-MS/MS) analysis of the lipid molecular species composition of yeast subcellular membranes reveals acyl chain-based sorting/remodeling of distinct molecular species en route to the plasma membrane. Journal of Cell Biol, 146: 741-754.

Schneiter, R., Tatzer, V., Gogg, G., Leitner, E. and Kohlwein, S.D. 2000. Elo1p-dependent carboxy-terminal elongation of C14:1Delta(9) to C16:1Delta(11) fatty acids in *Saccharomyces cerevisiae*. Journal of Bacteriol, 182: 3655-3660.

Schuller, H.J., Schutz, A., Knab, S., Hoffmann, B. and Schweizer, E. 1994. Importance of general regulatory factors Rap1p, Abf1p and Reb1p for the activation of yeast fatty acid synthase genes *FAS1* and *FAS2*. Eur J Biochem, 225: 213-222.

Schweizer, E. and Hofmann, J. 2004. Microbial type I fatty acid synthases (FAS): major players in a network of cellular FAS systems. Microbiol Mol Biol Rev, 68: 501-517.

Semenkovich, C.F. 1997. Regulation of fatty acid synthase (FAS). Prog Lipid Res, 36: 43-53.

Shanklin, J., Whittle, E. and Fox, B.G. 1994. Eight histidine residues are catalytically essential in a membrane-associated iron enzyme, stearoyl-CoA desaturase, and are conserved in alkane hydroxylase and xylene monooxygenase. Biochemistry (Mosc), 33: 12787-12794.

Sharma, S.C., Raj, D., Forouzandeh, M. and Bansal, M.P. 1996. Salt-induced changes in lipid composition and ethanol tolerance in *Saccharomyces cerevisiae*. Appl Biochem Biotechnol, 56: 189-195.

Shirra, M.K., Patton-Vogt, J., Ulrich, A., Liuta-Tehlivets, O., Kohlwein, S.D., Henry, S.A. and Arndt, K.M. 2001. Inhibition of acetyl coenzyme A carboxylase activity restores expression of the INO1 gene in a snf1 mutant strain of *Saccharomyces cerevisiae*. Mol Cell Biol, 21: 5710-5722.

Skanes, I.D., Stewart, J., Keough, K.M. and Morrow, M.R. 2006. Effect of chain unsaturation on bilayer response to pressure. Phys Rev E Stat Nonlin Soft Matter Phys, 74: 051913.

Stahl, P.D. and Klug, M.J. 1996. Characterization and Differentiation of Filamentous Fungi Based on Fatty Acid Composition. Appl Environ Microbiol, 62: 4136-4146.

Steels, E.L., Learmonth, R.P. and Watson, K. 1994. Stress tolerance and membrane lipid unsaturation in *Saccharomyces cerevisiae* grown aerobically or anaerobically. Microbiol, 140 (Pt 3): 569-576.

Storz, G., Tartaglia, L.A. and Ames, B.N. 1990. Transcriptional regulator of oxidative stress-inducible genes: direct activation by oxidation. Science, 248: 189-194.

Stukey, J.E., McDonough, V.M. and Martin, C.E. 1989. Isolation and characterization of *OLE1*, a gene affecting fatty acid desaturation from *Saccharomyces cerevisiae*. Journal of Biol Chem, 264: 16537-16544.

Stukey, J.E., McDonough, V.M. and Martin, C.E. 1990. The *OLE1* gene of *Saccharomyces cerevisiae* encodes the delta 9 fatty acid desaturase and can be functionally replaced by the rat stearoyl-CoA desaturase gene. Journal of Biol Chem, 265: 20144-20149.

Suutari, M. 1995. Effect of growth temperature on lipid fatty acids of four fungi (Aspergillus niger, Neurospora crassa, Penicillium chrysogenum, and Trichoderma reesei). Archives in Microbiology, 164: 212-216.

Suutari, M. and Laakso, S. 1994. Microbial fatty acids and thermal adaptation. Crit Rev Microbiol, 20: 285-328.

Suutari, M., Liukkonen, K. and Laakso, S. 1990. Temperature adaptation in yeasts: the role of fatty acids. Journal of Gen Microbiol, 136: 1469-1474.

Swan, T.M. and Watson, K. 1997. Membrane fatty acid composition and membrane fluidity as parameters of stress tolerance in yeast. Can J Microbiol, 43: 70-77.

Swan, T.M. and Watson, K. 1999. Stress tolerance in a yeast lipid mutant: membrane lipids influence tolerance to heat and ethanol independently of heat shock proteins and trehalose. Can J Microbiol, 45: 472-479.

Šajbidor, J. and Grego, J. 1992. Fatty acid alterations in Saccharomyces cerevisiae exposed to ethanol stress. FEMS Microbiol Lett, 72: 13-16.

Tehlivets, O., Scheuringer, K. and Kohlwein, S.D. 2007. Fatty acid synthesis and elongation in yeast. Biochim Biophys Acta, 1771: 255-270.

Toke, D.A. and Martin, C.E. 1996. Isolation and characterization of a gene affecting fatty acid elongation in *Saccharomyces cerevisiae*. Journal of Biol Chem, 271: 18413-18422.

Trotter, P.J. 2001. The genetics of fatty acid metabolism in *Saccharomyces cerevisiae*. Annu Rev Nutr, 21: 97-119.

Tuller, G., Nemec, T., Hrastnik, C. and Daum, G. 1999. Lipid composition of subcellular membranes of an FY1679-derived haploid yeast wild-type strain grown on different carbon sources. Yeast, 15: 1555-1564.

Tunblad-Johansson, I. and Adler, L. 1987. Effect of sodium chloride concentration on phospholipid fatty acid composition of yeasts differing. FEMS Microbiol Lett, 43: 275-278.

Tunblad-Johansson, I., Andre, L. and Adler, L. 1987. The sterol and phospholipid composition of the salt-tolerant yeast *Debaryomyces hansenii* grown at various concentrations of NaCl. Biochim Biophys Acta, 921: 116-123.

Turk, M., Mejanelle, L., Šentjurc, M., Grimalt, J.O., Gunde-Cimerman, N. and Plemenitaš, A. 2004. Salt-induced changes in lipid composition and membrane fluidity of halophilic yeast-like melanized fungi. Extremophiles, 8: 53-61.

Turk, M., Abramovic, Z., Plemenitas, A. and Gunde-Cimerman, N. 2007a. Salt stress and plasma-membrane fluidity in selected extremophilic yeasts and yeast-like fungi. FEMS Yeast Res, 7: 550-557.

Turk, M., Montiel, V., Zigon, D., Plemenitas, A. and Ramos, J. 2007b. Plasma membrane composition of *Debaryomyces hansenii* adapts to changes in pH and external salinity. Microbiol, 153: 3586-3592.

van der Rest, M.E., Kamminga, A.H., Nakano, A., Anraku, Y., Poolman, B. and

Konings, W.N. 1995. The plasma membrane of *Saccharomyces cerevisiae*: structure, function, and biogenesis. Microbiol Rev, 59: 304-322.

van Roermund, C.W., Waterham, H.R., Ijlst, L. and Wanders, R.J. 2003. Fatty acid metabolism in *Saccharomyces cerevisiae*. Cell Mol Life Sci, 60: 1838-1851.

Varela, J.C. and Mager, W.H. 1996. Response of *Saccharomyces cerevisiae* to changes in external osmolarity. Microbiol, 142: 721-731.

Vemula, M., Kandasamy, P., Oh, C.S., Chellappa, R., Gonzalez, C.I., Martin, C.E. 2003. Maintenance and regulation of mRNA stability of the Saccharomyces cerevisiae OLE1 gene requires multiple elements within the transcript that act through translation-independent mechanisms. Journal of Biol Chem, 278: 45269-45279.

Viegas, C.A., Cabral, M.G., Teixeira, M.C., Neumann, G., Heipieper, H.J. and Sa-Correia, I. 2005. Yeast adaptation to 2,4-dichlorophenoxyacetic acid involves increased membrane fatty acid saturation degree and decreased OLE1 transcription. Biochem Biophys Res Commun, 330: 271-278.

Watanabe, Y. and Takakuwa, M. 1987. Change of lipid composition of *Zygosaccharomyces rouxii* after transfer to high sodium chloride culture medium. Journal of Ferment Technol, 365-369.

Watanabe, K., Oura, T., Sakai, H. and Kajiwara, S. 2004. Yeast Delta 12 fatty acid desaturase: gene cloning, expression, and function. Biosci Biotechnol Biochem, 68: 721-727.

Weete, J.D. 1989. Structure and function of sterols in fungi. Adv Lipid Res, 23: 115-167.

Weete, J.D. and Gandhi, S.R. 1996. Biochemistry and molecular biology of fungal sterols, pp. 421-438. In: Biochem mol Biol. R. Brambl and G.A. Marzluf (eds.), Springer-Verlag, Berlin Heidelberg.

Wei, D., Li, M., Zhang, X., Ren, Y. and Xing, L. 2004. Identification and characterization of a novel delta12-fatty acid desaturase gene from *Rhizopus arrhizus*. FEBS Lett, 573: 45-50.

Wei, D., Li, M., Zhang, X.X., Zhou, H. and Xing, L.J. 2006a. A novel Delta12-fatty acid desaturase gene from methylotrophic yeast *Pichia pastoris* GS115. Acta Biochim Pol, 53: 753-759.

Wei, H., Dhanaraj, A.L., Arora, R., Rowland, L.J., Fu, Y. and Sun, L. 2006b. Identification of cold acclimation-responsive Rhododendron genes for lipid metabolism, membrane transport and lignin biosynthesis: importance of moderately abundant ESTs in genomic studies. Plant Cell Environ, 29: 558-570.

Wei, D., Zhou, H., Yang, Z., Zhang, X., Xing, L. and Li, M. 2007. Identification of a novel delta9-fatty acid desaturase gene and its promoter from oil-producing fungus *Rhizopus arrhizus*. Mol Biol Rep.

Wenz, P., Schwank, S., Hoja, U. and Schuller, H.J. 2001. A downstream regulatory element located within the coding sequence mediates autoregulated expression of the yeast fatty acid synthase gene FAS2 by the FAS1 gene product. Nucleic Acids Res, 29: 4625-4632.

Wilson, R.A., Calvo, A.M., Chang, P.K. and Keller, N.P. 2004. Characterization of the *Aspergillus parasiticus* delta12-desaturase gene: a role for lipid metabolism in the *Aspergillus*-seed interaction. Microbiol, 150: 2881-2888.

Wiseman, H. Halliwell, B. 1996. Damage to DNA by reactive oxygen and nitrogen species: role in inflammatory disease and progression to cancer. Biochem Journal, 313 (Pt 1): 17-29.

Wongwathanarat, P., Michaelson, L.V., Carter, A.T., Lazarus, C.M., Griffiths, G., Stobart, A.K., Archer, D.B. and MacKenzie, D.A. 1999. Two fatty acid delta9-desaturase genes, ole1 and ole2, from *Mortierella alpina* complement the yeast ole1 mutation. Microbiol, 145 (Pt 10): 2939-2946.

Yale, J. and Bohnert, H.J. 2001. Transcript expression in *Saccharomyces cerevisiae* at high salinity. Journal of Biol Chem, 276: 15996-16007.

Yamada, T., Shimoi, H. and Ito, K. 2005. High expression of unsaturated fatty acid synthesis gene *OLE1* in sake yeasts. Journal of Biosci Bioeng, 99: 512-516.

Ykema, A., Verbree, E.C., Verwoert, IIG.S., Linden, K.H., Nijkamp, H.J.J. and Smit, H. 1990. Lipid production of revertants of Ufa mutants from the oleaginous yeast *Apiotrichum curvatum*. Appl Microbiol Biotechnol, 33: 176.

Yokoyama, K., Saitoh, S., Ishida, M. et al. 2001. Very-long-chain fatty acid-containing phospholipids accumulate in fatty acid synthase temperature-sensitive mutant strains of the fission yeast Schizosaccharomyces pombe fas2/lsd1. Biochim Biophys Acta, 1532: 223-233.

Yoshikawa, S., Mitsui, N., Chikara, K.-I., Hashimoto, H., Shimosaka, M. and Okazaki, M. 1995. Effect of salt stress on plasma membrane permeability and lipid saturation in the salt-tolerant yeast Zygosaccharomyces rouxii. J Ferment Bioeng, 80: 131-135.

You, K.M., Rosenfield, C.L. and Knipple, D.C. 2003. Ethanol tolerance in the yeast *Saccharomyces cerevisiae* is dependent on cellular oleic acid content. Appl Environ Microbiol, 69: 1499-1503.

Zhang, Q., Li, M., Ma, H., Sun, Y. and Xing, L. 2004. Identification and characterization of a novel Delta6-fatty acid desaturase gene from *Rhizopus arrhizus*. FEBS Lett, 556: 81-85.

Zhang, S., Skalsky, Y. and Garfinkel, D.J. 1999. *MGA2* or *SPT23* is required for transcription of the delta9 fatty acid desaturase gene, *OLE1*, and nuclear membrane integrity in *Saccharomyces cerevisiae*. Genetics, 151: 473-483.

Zhang, S., Sakuradani, E., Ito, K. and Shimizu, S. 2007. Identification of a novel bifunctional delta12/delta15 fatty acid desaturase from a basidiomycete, *Coprinus cinereus* TD#822-2. FEBS Lett, 581: 315-319.

Zinser, E., Sperkagottlieb, C.D.M., Fasch, E.V., Kohlwein, S.D., Paltauf, F. and Daum, G. 1991. Phospholipid-Synthesis and Lipid-Composition of Subcellular Membranes in the Unicellular Eukaryote *Saccharomyces cerevisiae*. Journal of Bacteriol, 173: 2026-2034.

Zlatanov, M., Pavlova, K. and Grigorova, D. 2001. Lipid composition of some yeast strains from Livingston Island, Antarctica. Folia Microbiol (Praha), 46: 402-406.

13

Microsporum canis—A Pathogen of Cats and Its Control Through Environmental Management: A Review

R. Papini

Dipartimento di Patologia Animale,
Profilassi e Igiene degli Alimenti, Facoltà di Medicina Veterinaria,
Viale delle Piagge 2, 56124 Pisa, Italy
E-mail: rpapini@vet.unipi.it

Abstract

Decontamination of the environment is a very important part of the control of *Microsporum canis* feline dermatophytosis. However, there is a very limited number of chemical agents that can reach a complete effectiveness against *M. canis* environmental contaminants. In addition, results of research studies often show discrepancies and differences, and the efficacy of many products appears to be controversial, since it was not always confirmed in the different experimental trials. The purpose of this chapter is to provide an overview of compounds that show many similarities among the different studies, and that can be currently proposed for the environmental control of *M. canis* infections in cats. Mainly, these included chlorine-based compounds, enilconazole, and lime sulphur. The practical applications that have proved successful are also reviewed.

INTRODUCTION

Over 20 species of dermatophytes have been reported to cause clinical disease in cats and dogs. However, *Microsporum canis* is the most commonly isolated dermatophyte in cats with dermatophytosis (or dermatomycosis or ringworm), whereas pathogens less commonly isolated include both

Microsporum gypseum and *Trichophyton mentagrophytes*. It is a matter of fact that *M. canis* is the most prevalent dermatophytic species in many countries, even in apparently healthy cats: 2.2% in Great Britain (Sparkes et al., 1994); 8% in warm humid areas of the United States (Moriello et al., 1994); 26% in Iran (Khosravi, 1996); 37.3% in Brazil (Larsson et al., 1994); 40.4-47.4% to 83% in Italy (Filipello Marchisio et al., 1995; Romano et al., 1997; Mancianti et al., 2002). This dermatophyte rapidly spreads by direct contact between animals. It also indirectly spreads by contact with surfaces contaminated with fungal hyphae and spores (or conidia or arthroconidia) shed from coats of infected animals.

The first step in the control of feline dermatophytosis is the identification and treatment of infected cats. A variety of antifungal systemic treatments has been reported, ranging from the old oral treatment with griseofulvin to the newer oral treatments with itraconazole and terbinafine (Papini, 2004).

A second step is the antifungal disinfection of the environment where animals are living. Infective fungal spores may be shed from an infected cat even under systemic therapy, and up to 1,000 *M. canis* spores per m^3 can be found in a house with an infected cat (Mignon and Losson, 1997; Symoens et al., 1989). In one study, 85% of homes with infected cats had hairborne *M. canis* organisms (Mancianti et al., 2003). It is known that fungal spores can remain viable and contaminate the environment for over one year (Sparkes et al., 1994). Therefore, the contaminated animal environment can serve as an important permanent source of infection and re-infection, and environmental decontamination is an important element in the control of cat ringworm. The purpose of this chapter thus is to critically summarize current knowledge on this topic. It is important to note that this chapter mainly discusses environmental disinfectants, although some of these (i.e., enilconazole and lime sulphur) can be used as both topical compounds on animals and environmental treatments. In these cases some additional information is given, since topical therapy in turn can play a role in reducing environmental contamination (Paterson, 1999).

Evaluation studies

It is important that disinfectants are chosen on the basis of their activity against *M. canis*. Unfortunately, the sophistication with which disinfectants are chosen often leaves much to be desired. Although new antifungal drugs are assessed by measurement of minimum inhibitory concentrations and relating these to serum and tissue levels, disinfectants are chosen on the basis of manufacturers claims of activity against target pathogens. The conditions of contact between the disinfectant and the organism have a significant impact on the activity of the disinfectant. There are two main conditions of test design in vitro: the first, tests the antimicrobial activity of the product by a mixture of disinfectant and organisms in a liquid medium (suspension test); the second, tests the application of disinfectant to a surface contaminated with organisms (carrier test). In particular, carrier tests can be performed as practical tests on carrier models (e.g., leather, clothing) or on surfaces (e.g., floors) in a simulated real life situation.

In the suspension test, the disinfectant may be diluted in autoclaved phosphate buffered saline (PBS) or distilled water, and the *M. canis* suspension may be prepared in sterile PBS or solution of neutral detergent (Rycroft and McLay, 1991; Gupta et al., 2001). The two solutions are then mixed and, following neutralization of the active agent, the suspension may be sub-cultured on Sabouraud's agar medium to establish fungicidal activity.

In the carrier test, the fungal suspension is applied to the carrier surface and dried. The disinfectant is then applied to the carrier and activity is assessed by neutralization and subculture (Moriello and DeBoer, 1998).

It is well known that the presence of organic matter can have an effect on the activity of a disinfectant. Disinfectants vary in their susceptibility to the neutralizing effect of organic matter and this is therefore an important part of the assessment of a given disinfectant. The simplest and most frequently used method is to perform suspension and carrier tests under dirty and clean conditions. Urea at a final concentration of 10 mM may be used as an organic load for dirty conditions (Rycroft and McLay, 1991).

These methods may be not truly indicative of how a disinfectant will perform in practice. Many studies have determined fungicidal activity of disinfectants against *M. canis* by testing the compounds against the mycelial form of the dermatophyte and not against the naturally infective state, that is infected hairs and spores. Moreover, there are many variables that may affect the test outcome, including *M. canis* source (dog or cat) and strain, sampling technique, method of preparation of the organism, treatment protocols, treatment length, amount of infective material, subculture method used and so on. Test organisms may undergo spontaneous inactivation on certain carriers, making the disinfectant more active than it is. Formulation of disinfectants may contain one or more active compounds together with inactive ingredients such as perfumes and colourants. The product formulation can affect the properties of the chemical; thus, it is not adequate to assume that because a specific amount of chemical has been included, a product will be effective in use. For instance, two different iodine based compounds required a different number of treatments to inactivate *M. canis* infective hairs in the same experimental study (Moriello and DeBoer, 1998).

There are many published studies on the efficacy of various disinfectants against dermatophytes. Sporulating cultures of two strains of *Trichophyton rubrum*, two strains of *T. mentagrophytes* and one strain of *M. canis* were inactivated after a 5-minute treatment, by a simple dipping technique, with 1% and 2% formaldehyde solution, a 20% iodine based compound solution, and a benzalkonium chloride solution with 10% active substance. Antifungal effects on all cultures of dermatophytes were observed already after a 1-minute treatment with 0.5%-4% peracetic acid, 4% formaldehyde solution, and concentrated solution of a product based on tributyltin oxide, which is a very toxic tin compound chiefly used as a fungicide and molluscicide, especially as a wood preservative (Rybnikar et al., 1985).

In one study, spores and hyphal cells in fungal cultures were scraped off and exposed to various disinfectants using the suspension method (Terleckyj and Axler, 1987). In that study, phenolic mixture, iodophor mixture, chlorine dioxide, 2% glutaraldehyde and alcohol mixture were effective in their fungicidal activity after 15 to 30 minute interaction times against one laboratory strain of *M. canis*. A quaternary ammonium mixture was unable to kill *M. canis* even after a 60-minute exposure to this agent.

Two chlorine derivatives (sodium hypochlorite and electrolytic chloroxidant) were equally active against cell suspensions of *M. canis* and other potentially human pathogenic fungi (Bianchi et al., 1989).

The antifungal activity of glutaraldehyde was determined, by microscopic observation, against *M. canis, M. gypseum, T. mentagrophytes* and other dermatophytes isolated from animals (Kosuge et al., 1996).

In another study, spores were scraped off and exposed to a quaternary ammonium salt using the suspension method (Ohta et al., 1996). In that study, it killed one strain of *T. rubrum* in 30 minutes at 0.4%.

Rycroft and McLay (1991) evaluated 12 compounds for their ability to kill *M. canis* spores collected from infected cat hair. The disinfectants were diluted to the concentration recommended for the disinfection of clean surfaces and the potency of each substance was determined by the degree to which it could be further diluted before losing its fungicidal action. Results of this study showed that hypochlorite, benzalkonium chloride (the most commonly used quaternary ammonium compound), and glutaraldehyde based compounds were the most effective agents, whereas phenolics, alcohol and anionic detergents were inadequate. Chlorexidine gluconate was weakly fungicidal.

Phenol-glutaraldehyde, chlorine dioxide and denaturated ethyl alcohol formulations were affective in 15-30 minutes against cell suspensions of a wide range of pathogenic fungi isolated from patients with tinea pedis, onychomicosis or tinea capitis. Whereas, quaternary ammonium compounds, iodophors and phenolics were not fungicidal against all the tested fungi even after 60 minutes of exposure (Terleckyj and Axler, 1993).

White-Weithers and Medleau (1995) evaluated seven commonly used topical antifungal products for their antifungal activity on *M. canis* infected hairs from dogs and cats. Hairs were soaked or shampooed in each product for five minutes twice a week for four weeks. Lime sulphur and enilconazole solutions were superior in inhibiting fungal growth: no growth occurred on fungal cultures after two treatments with both products. Chlorexidine and povidone iodine solutions were effective after four treatments. Sodium hypochlorite solution and ketokonazole shampoo inhibited fungal growth after eight treatments. Captan, a horticulture fungicide, did not inhibit fungal growth during the test period.

Moriello and DeBoer (1998) evaluated the fungicidal activity of various disinfectants against *M. canis* infected hairs and spores in two experiments. In the first experiment, nine rooms heavily contaminated with infected hairs and spores were used. Ten disinfectants, used at the manufacturers' recommended dilution and wetting time, were applied as a single wipe and allowed to dry. Tap water was used as a control. Only 3 compounds showed 100% fungicidal activity at one or more points in the study: undiluted household bleach (at 2, 8, and 24 h post-treatment), 1% formalin (2 h) and enilconazole (8 h). Bleach 0.525% also showed significant differences in the number of culture positive rooms when compared to water. In the second experiment, the mean number of topical applications required to inactivate *M. canis* infected hairs and spores was determined for 14 disinfectants. None of the disinfectants was fungicidal after a single treatment when used at the recommended dilution and wetting time. The stabilized chlorine dioxide, glutaraldehyde, potassium monoperoxysulfate, and bleach 0.525% were superior to the other disinfectants and required between 2 and 3.3 applications to inactivate *M. canis* infected hairs.

Some organic compounds (a phenolic compound, quaternary ammonium compounds, a glutaraldehyde compound and formalin) were more effective fungicides than were inorganic ones (iodine and calcium hydroxide) against *M. canis* and other animal dermatophytes. Some qualitative differences were observed in the resistance of *Trichophyton* and *Microsporum* species to the disinfectants. Moreover, some differences were noted, within the same fungal species, in the susceptibility to the same disinfectant (Sotohy and Mohamed, 1999).

Chlorexidine digluconate in water-ethanol was totally effective against vegetative forms or spores of *M. canis* and other microorganisms (Odore et al., 2000).

Chlorine (1%) and terbinafine (0.01%) were found to be high level disinfectants bringing about a rapid inactivation of conidia from five fungal strains belonging to three *Trichophyton* species. Phenol and quaternary ammonium compounds were fungicidal or fungistatic, while sodium dodecyl sulphate (0.5%) was largely ineffective (Gupta et al., 2001).

In a further report, *M. canis* infected hairs and crusts were soaked and processed in order to remove and isolate ectothrix spores (Moriello et al., 2002). Eight dilutions of five disinfectants were tested. Sterile distilled water was used as a control. An equal volume of spore suspension and disinfectant solution were incubated for 5 minutes and then plated onto fungal culture plates. Fungal cultures were counted at days 7 to 10 and data were evaluated using an end-point dilution at which there was 100% fungicidal activity, i.e. no growth on the plates. The samples showed identical results. Chlorexidine and potassium monoperoxysulphate were ineffective even when used at four times the manufacturer's recommended dilution. Lime sulphur (1:33), enilconazole (20 $\mu l/ml^{-1}$), and bleach (1:10) were consistently effective when used at the recommended dilution. In addition, lime sulphur and enilconazole were 100% fungicidal even when the recommended concentration was diluted 1:4. In this study, the authors were unable to detect strain variability to the tested disinfectants.

A commercially available product based on enilconazole in fumigant form was equally effective both against spores and mycelia specimens of *M. canis* in an in vitro study. The possible association of enilconazole with other disinfectants for household disinfections was suggested in this study, though some inconvenience can occur (Mancianti and Nardoni, 2004).

More recently, chlorexidine in association with miconazole showed a synergistic and potent antimycotic effect, by an agar dilution technique, against the growth of *M. canis, T. mentagrophytes, Trichophyton erinacei* and *Microsporum persicolor* in vitro (Perrins and Bond, 2003; Perrins et al., 2005).

Finally, two commercially available benzalkonium chloride and potassium peroxymonosulphate based compounds were successfully used at the manufacturers' recommended dilutions (2% and 1%, respectively), for a 10-minute disinfection of brushes previously used for diagnosis of *M. canis* infection in 70 cats. Benzalkonium chloride, a quaternary ammonium compound, appeared to be significantly more effective than potassium peroxymonosulphate (Marchetti et al., 2006).

Effective agents

Environmental decontamination against *M. canis* is achieved by a combination of two approaches: physically removing the infected hairs from the environment and the use of chemical agents in the environment to kill the spores. In general, there are different categories of physical and chemical means for decontamination: heat, liquid disinfectants, gases and radiation. There are also different classes of disinfectants based on the type of agents they contain. These can be classified as halogens, acids, alkalis, heavy metal salts, quaternary ammonium compounds, phenolic compounds, aldehydes, ketones, alcohols and amines. Liquid disinfectants are available under a wide variety of trade names. Choosing which disinfectant is to be used is a decision that should not be made lightly. An ideal disinfecting agent acts rapidly, has a broad spectrum of activity, and penetrates adequately (as contrasted with simply superficial activity). In addition, such an agent should be non-irritating, non-toxic to animals and humans, non-staining, non-corrosive upon materials, chemically stable, inexpensive, and not readily inactivated after application. However, it is unfortunate that no disinfectant totally fulfills these criteria and is equally useful or effective in all situations. In general, the more active a compound is, the more likely it is to have undesirable characteristics such as corrosiveness, human health hazard or potential toxicity. The main requirements of a product necessary for environmental applications in feline dermatophytosis are its specific fungicidal efficacy and easy application by man. Since the complete removal of infective material is required in practice, the most effective environmental cleansers must be able to kill 100% of *M. canis* spores. However, although environmental control is an important part of treating feline dermatophytosis, it remains the most frustrating because of the limited number of effective products.

Many chemical disinfectants claim to have good activity against fungi but there are very few products that will be actually effective against *M. canis*. In fact, the efficacy of a great number of commercially available disinfectants against cat ringworm has been experimentally tested, but it has been shown that a very limited number of agents can reach complete effectiveness. Furthermore, though there are many published studies on the efficacy of various disinfectants against *M. canis*, the different testing methodologies make it difficult to compare data and may account for discrepancies between results. In any case, when data from different studies are compared, some similarities and differences can be found. Some studies reported fungicidal effectiveness with products such as phenolic compounds, various quaternary ammonium compounds, potassium peroxymonosulphate, iodine-based compounds and chlorexidine (Rycroft and McLay, 1991; Terleckyj and Axler, 1993; White-Weithers and Medleau, 1995; Moriello and DeBoer, 1998; Sotohy and Mohamed, 1999; Odore et al., 2000; Perrins and Bond, 2003; Perrins et al., 2005; Marchetti et al., 2006). Together with other therapeutical control measures, chlorexidine (Ando et al., 1997), ammonium compounds (Woloszyn et al., 1996) and iodophors (Englund et al., 1990) were successfully used for environmental disinfection of *M. canis* or *T. mentagrophytes* infected premises. Didecyldimethyl-ammonium chloride (Osame et al., 1991), which is a quaternary ammonium compound, and potassium peroxymonosulphate (Cafarchia et al., 2000) were used even for topical treatment on infected calves and horses, respectively. Despite this, the efficacy of all these products appears to be controversial, since it was not always confirmed in the different experimental trials. For instance, in one study, chlorexidine and potassium peroxymonosulphate were ineffective even when used at four times the manufacturer's recommended dilution. Therefore, the authors concluded that they can no longer be recommended for use as fungicidal disinfectants (Moriello et al., 2002). By contrast, experimental comparison works have demonstrated that the data regarding some chemical products, effective against *M. canis*, showed many similarities among the different studies. Based on a review of the literature, these compounds fall into the following group: chlorine-based compounds (sodium hypochlorite and chlorine dioxide), aldehydes (formaldehyde and glutaraldehyde), enilconazole, and lime sulphur (Bianchi et al., 1989; Rybnikar et al., 1985; Rycroft and McLay, 1991; Terleckyj and Axler, 1993; White-Weithers and Medleau, 1995; Kosuge et al., 1996; Moriello and DeBoer, 1998; Sotohy and Mohamed, 1999; Gupta et al., 2001; Moriello et al., 2002; Mancianti and Nardoni, 2004). Understanding the general characteristics of the primary classes of effective products is essential.

Chlorine releasing agents

Commonly used chlorine releasing agents include sodium hypochlorite, chlorine dioxide and sodium dichloroisocyanourate. Chlorine disinfectants, as well as iodine disinfectants, belong to the halogen group. Chlorine- and iodine-based compounds are the most significant halogens used in the clinic and have been traditionally used for both antiseptic and disinfectant purposes (McDonnell and Russel, 1999).

Chlorine rapidly eliminates both enveloped and non-enveloped viruses, and also it is effective against fungi, bacteria, and algae. Chlorine releasing agents are active oxidizing agents. The bactericidal activity seems to be due to oxidization of key enzymes within the cell membrane or cytoplasm. Their mechanism of action is mediated by release of hypochlorous acid in aqueous solutions, which interacts with key metabolic processes via protein denaturation. Active concentrations are expressed in terms of available free chlorine. Effective concentrations range from 0.05 to 5 ppm of available chlorine, which will kill vegetative bacteria within 15 seconds to 5 minutes, to 200 ppm, which will kill a substantial number of bacterial spores within 5 minutes (Love and Hirsh, 1989).

Chlorine in high concentration deteriorates fabrics, is harmful to clothing or rubber goods, and very corrosive, and thus cannot be used on many metallic objects. The chlorine fumes can be damaging to the animal respiratory system. They do have an offensive odour, which is a disadvantage in confined or poorly ventilated areas. Chlorine in high concentration can be somewhat irritating to mucous membranes, eyes and skin, and should never be used around animals.

Chlorine releasing agents demonstrate poor activity under dirty conditions because they are relatively easily inactivated by organic matter, such as faeces, and thus they may not deliver the desired effectiveness without proper environmental management. It is important, therefore, that they are used on very clean surfaces (Fraise, 1999). In order to obtain maximum results with chlorine disinfectants they must remain in contact with surfaces for several minutes. The pH of the water used for dilution should be between 6 and 8 to be effective.

Undiluted household bleach (5.25% NaClO) is a commonly used, effective, cheap and readily available disinfectant. It is typically diluted with water. Sodium hypochlorite solutions are widely used for hard surface disinfection. Sodium dichloroisocyanourate can also be used for this purpose and has the advantage of providing a higher concentration of available chlorine and being less susceptible to inactivation by organic matter. In water, sodium hypochlorite ionizes to produce Na^+ and the hypochlorite ion, ClO^-, which establishes an equilibrium with hypochlorous acid, HClO (McDonnell and Russel, 1999). Sodium hypochlorite solutions are relatively unstable and decompose easily, and thus need to be frequently replaced. Factors that decrease stability include an acidic pH, exposure to light, heavy metal ions, and high temperatures. Formulations are usually alkaline, and are provided in lightproof containers to increase shelf life. Solutions of sodium hypochlorite do not wet surfaces well and as a result have been formulated with detergents, which do not affect their microbicidal activity. The brand names are Clorox® and Purex®. Sodium hypochlorite is generally compatible with soaps but should never be mixed with acids, as this results in the release of toxic chlorine gas. The instability of sodium hypochlorite at acidic pH results in a more effective microbicidal activity in these conditions (pH below 7). This is due to the potenziation of oxidation and easier penetration through the cell membrane. However, the sporicidal activity is enhanced in the presence of 1.5 to 4% of NaOH. Factors that decrease the activity of hypochlorite solutions are cationic detergents and the presence of organic matter, which is readily oxidized, thus reducing the oxidizing power against microorganisms. The calcium and magnesium ions in hard water do not inactivate chlorine disinfectants, but ferrous or manganous cations, and nitrate and sulphide anions do. Sodium hypochlorite solutions are cytotoxic in vitro (Love and Hirsh, 1989).

Sodium hypochlorite is the treatment of first choice for feline dermatophytosis in the environment. Sodium hypochlorite in water solution with

NaCl was active against *M. canis* and other microorganisms by cell suspension test (Bianchi et al., 1989). However, attempts to prevent spread of *T. mentagrophytes* infections by disinfection of rabbit hutches with hypochlorite and copper sulphate, together with treatment of the infected animals with local iodine and griseofulvin, was unsuccessful (Simon et al., 1996). Nevertheless, multiple applications of a 0.525% solution with prolonged contact time are recommended to achieve levels capable of killing *M. canis* spores. This corresponds to 1:10 dilution of household bleach (Moriello and DeBoer, 1998). This concentration is non toxic and has little environment impact. Undiluted household bleach was reported as the most fungicidal solution tested against *M. canis* and appeared to have a residual activity of at least 24 h (Moriello and DeBoer, 1998). However, it is too harsh to be routinely used in the home and cannot be safely used in catteries in quantities sufficient to be useful. All objects treated with bleach must be rinsed thoroughly afterwards and allowed to dry before cats are allowed to contact them.

Stabilized chlorine dioxide is a chlorine derivative, which is a powerful oxidizing agent. It can destroy many pathogens, including bacteria, viruses, fungi and protozoa. Many studies have suggested that stabilized chlorine dioxide is a superior disinfecting agent to bleach. It has been shown that stabilized chlorine dioxide is able to inactivate *M. canis* infective hairs and spores after 2 treatments with a contact time of 2 minutes (Moriello and DeBoer, 1998). It was effective against cell suspensions of a wide range of dermatophytic and non-dermatophytic fungi in 15-30 minutes (Terleckyj and Axler, 1993). Its mode of action has not been thoroughly investigated but it is likely that it acts by production of free chlorine. It is commonly used both in Europe and in the United States to treat drinking water. There are very few products that are effective and can be safely used around animals, and stabilized chlorine dioxide is considered to be the best choice by avian breeders. Oxyfresh Cleansing Gele® is a cleaner containing stabilized chlorine dioxide. It is considered to be excellent for routine cleaning as it will clean and provide disinfectant protection and is not harmful. It is also an excellent washing/soaking solution for syringes, food dishes, and water containers. For hard surfaces, the solution is sprayed on and then wiped off after a 5-minute exposure. Rinsing is not necessary. Oxyfresh Denta-A-Gene® has originally been created as a disinfectant for dentists and its effectiveness seems to be well known. It is a very strong stabilized chlorine dioxide disinfectant and is a two-part product. When the two parts are mixed, it produces toxic fumes, but once stabilized it is safe for use. It is however rapidly deactivated by organic debris and exposure to sunlight, so all surfaces to be treated should be cleaned first. Once mixed, a solution can be used for 7 days if sealed tightly and kept out of light. A spray bottle of Oxyfresh Denta-A-Gene® solution can be kept near the animals' room and used on hands and shoes as fomite controller. It does not irritate the skin and is also an excellent deodorizer.

Aldehydes

The most common aldehyde-based disinfectant agents are glutaraldehyde and formaldehyde. They are predominantly used in healthcare settings. Aldehydes are irritating and cytotoxic, which restricts their use to inanimate objects as much as possible. They have a wide germicidal spectrum and are effective against bacteria, spores, viruses, and fungi. However, their effectiveness may depend on exposure times (Lemariè and Hosgood, 1995). They have a moderate residual activity and are effective in the presence of moderate amounts of organic material. Their activity is due to alkylation of sulphydryl and other groups, which causes alterations to proteins and nucleic acids (Fraise, 1999).

Glutaraldehyde is an important dialdehyde that has found usage as a disinfectant and sterilizant. In particular, this is one of the agents of choice for low-temperature disinfection and sterilization of surgical and medical equipment, such as endoscopes, and as a fixative in electron microscopy. Glutaraldehyde has a broad spectrum of activity against bacteria and their spores, fungi and viruses (McDonnell and Russel, 1999). The in vitro fungicidal activity of glutaraldehyde based compounds against animal dermatophytes, including *M. canis*, was shown by many authors (Rycroft and McLay, 1991; Terleckyj and Axler, 1993; Kosuge et al., 1996; Moriello and DeBoer, 1998; Sotohy and Mohamed, 1999). Its activity is non-specific, and is just as effective on mammalian cells as on microorganisms. However, glutaraldehyde solutions can be used to treat dermatophytes in human patients, with no adverse effects noted. Prolonged exposure, time of several hours may be required (Love and Hirsh, 1989). For cold sterilization of medical equipment, a 2 to 3% solution is used and up to 3 hours of contact times are recommended for complete destruction of bacterial spores. Glutaraldehyde causes toxicity, the most severe forms of which are respiratory and ocular irritation from vapours, if not used in extremely well ventilated areas, one contact dermatitis. Due to the cytotoxicity of glutaraldehyde, instruments should be rinsed thoroughly with sterile (not tap) water to remove all residual glutaraldehyde before being used. The activity of glutaraldehyde is strengthened by cations, such as calcium and magnesium, and by ultrasonics (Love and Hirsh, 1989). The effectiveness of glutaraldehyde is enhanced by alkaline solutions; however, polymerization occurs more rapidly at a high pH, which results in a shorter shelf life. For this reason, glutaraldehyde is generally supplied as an acid solution that must be activated by conversion to an alkaline solution at the time of use (Lemarié and Hosgood, 1995; Love and Hirsh, 1989). Organic soiling has little effect on the activity of glutaraldehyde. Glutaraldehyde disinfectants include Lysofume®, Wavicide®, Cidex®, Sporcide®, Banacide®, Sterol®, and are available in many forms including sprays, concentrates and bulk volumes. Solutions are effective in water of any temperature or hardness. Aldehydes have also been formulated with quaternary ammonium compounds and phenols to achieve a synergistic effect. Glutaraldehyde-based disinfectants are more expensive to purchase initially, compared to other disinfectants, but last a long time when mixed in solution, making the cost per use fairly low. Some glutaraldehyde formulas are caustic or corrosive to metals, others are not. It is advisable to read the label of a particular product to find the corrosive or caustic properties, and never to mix glutaraldehydes with any other cleaning or disinfectant product.

Formaldehyde is a mono-aldehyde that exists as a freely water-soluble gas. Formalin is a formaldehyde aqueous solution containing 34 to 38% (wt/wt) CH_2O with methanol to delay polymerization. Its clinical use is generally as a disinfectant and sterilizant in liquid or in combination with low-temperature steam. Formaldehyde is a very potent disinfectant and has an excellent microbicidal activity including activity against vegetative bacteria, mycobacteria, viruses, fungi and bacterial spores, but it works more slowly than glutaraldehyde (McDonnell and Russel, 1999). The fungicidal effect of formaldehyde solutions against *M. canis* and other dermatophytes was demonstrated (Rybnikar et al., 1985; Moriello and DeBoer, 1998; Sotohy and Mohamed, 1999). Due to the concerns that this product can be a highly toxic and a carcinogen agent to people and animals, there are limits for acceptable environmental levels of formaldehyde. Consequently, it cannot be recommended as a routine wipe/spray disinfectant and is no longer used except for fumigation of high-risk areas (Fraise, 1999). Gaseous formaldehyde canisters are used as disinfectants in swine units. It is recommended to use them only as a last resort and then under trained supervision in a well ventilated setting. Target areas are cleaned thoroughly and then the confined area is evacuated for up to 72 hours after activating the canisters. A spray containing 2.0% formaldehyde and 1.0% caustic soda was successfully used for environmental control measures against ringworm in donkeys infected by *M. canis*, *Trichophyton equinum* and *Trichophyton verrucosum* (Abou-Zaid, 2001). This may be a reasonable alternative in small animal facilities that can be cleaned and vacated for an appropriate length of time, but gaseous formaldehyde is not suitable for use by most pet owners (Moriello and DeBoer, 1998).

Enilconazole

Imidazoles bind to the cytochrome P-450 enzymes and block the hydroxylation of 14 alpha-methyl groups. This blocks the synthesis of ergosterol from lanosterol, thereby altering permeability of fungal cell membranes causing leakage and cell death. They can be fungistatic or fungicidal, depending on the concentration.

Enilconazole is a beta-substituted imidazole, which is reported to have potent fungicidal activity against a wide range of fungi, including the major dermatophytes, yeasts (e.g., *Malassezia*), *Penicillium* spp., and mostly *Aspergillus* spp., where the in vitro destruction rate is close to 100% (van Gestel et al., 1981). For instance, *Aspergillus* spp. growth is completely inhibited at concentrations of 0.1-1 µg/ml (Desplender, 1989). The effectiveness of enilconazole against spores and mycelia samples of *M.canis* has been demonstrated in an in vitro study, and the possible association of enilconazole with other disinfectants has been proposed, even though with great caution (Mancianti and Nardoni, 2004)

The biochemical and morphological effects of enilconazole demonstrate irreversible degenerative changes specific to the fungal cell membrane in all stages of the fungal development. This is achieved not only by direct contact but also during the vapour phase activity of enilconazole.

Its safety has been demonstrated in reproduction studies on acute, subacute and chronic mutagenicity. Ocular and dermal applications of enilconazole and its various pharmaceutical formulations are devoid of side effects.

A specific formulation has been developed for the application of enilconazole as a topical treatment (Imaverol®). It is approved for use in dogs and horses in Europe and Canada. Although not licensed for use in cats (except in France), enilconazole is being often used extra-label in cats too for topical treatment of dermatophytosis. Imaverol® (enilconazole 10%) is applied as a 0.2% (1:50 dilution in tap water) bath or spray, not only to the affected lesions, but also to the entire hair coat. Treatment schedule in dogs and cats consists of four whole-body applications with 3-4 day intervals. Enilconazole should not be rinsed from the hair coat. The product is highly active against dermatophytes and also against *Malassezia* (Rochette et al., 2003). In general, the treatment appears to be well tolerated, except in some cats that dislike this type of treatment. Anecdotal reports have stated possible hepatotoxicity or idiosyncratic reactions following topical utilization of diluted enilconazole emulsion in Persian cats. Nevertheless, successful treatment in cats has been reported on several occasions. A study, for instance, showed that 0.2% enilconazole used twice weekly for 8 weeks was effective against *M. canis* and safe in 14 Persian cats

(Hnilica and Medleau, 2002). Enilconazole is also used for sinus irrigation in dogs with nasal aspergillosis (Rochette et al., 2003) and has been evaluated in association with lufenuron for the management of dermatophytosis in catteries (Guillot et al., 2002).

As part of a strategic control programme, Enilconazole is also available in special formulation as an antifungal disinfectant (Clinafarm Spray® and Clinafarm Smoke®) for decontamination of the environment. Janssen Pharmaceutica have developed new fogger formulations of enilconazole with tensio-active properties that increase the contact between spores and the product. Since the product is active in the vapour phase as fungicidal, fungistatic and sporicidal, these formulations are highly effective and can be used to treat air sac aspergillosis in poultry houses or against spores of dermatophytes in homes (Hay, 1991).

For antimycotic disinfection of rooms, kennels, catteries, and equipment, Clinafarm Spray® (emulsifiable concentrated 15% enilconazole) may be diluted 1 to 100 in tap water and vaporized in rooms or buildings, or sprayed onto surfaces. The application volume is 100 ml of Clinafarm in 10 litres of water for a room of 3,000 m^3 (vaporization) or a surface of 750 m^2 (spray). Clinafarm Spray® is non-irritating and non-corrosive to materials or linen and can be combined with any other antibacterial or antiviral disinfectant. The tensio-active components of the formulation guarantee optimal contact and moistening of the hydrophobic fungal spores.

In rooms that can be tightly closed or where spray is contraindicated for practical reasons, a Clinafarm Smoke® (enilconazole 5 grams) generator can be ignited for a volume of 50 m^3. The active enilconazole is distributed in the smoke phase throughout the entire room. Inhalation of the smoke is not toxic, but should be avoided. The frequency of disinfection by spray or smoke depends upon the degree and risk of contaminated environment.

The fumigation with enilconazole reduced 30% of *M. canis* spores in the air, and 80 to 100% on surfaces in rooms where an infected cat was living (Symoens et al., 1989). A further study was conducted in 3 catteries with 18, 10 or 8 infected cats, respectively. After scrubbing and disinfection with enilconazole in liquid and fumigant form, all samples from the environment were negative (Palsson, 1991).

In rabbit farms, six clinical trials carried out in several European countries showed that disinfection by spraying batteries (walls, ceiling, cages, etc.) with enilconazole is an effective way of controlling *M. canis* dermatophytosis. In this study, a high dosage of 50 mg/m^2 enilconazole was applied because caging, nest materials and other litter were present, but no side effects were observed in animals (Rochette and van Meirhaeghe, 1997).

In chicken houses, enilconazole has been employed in the presence of birds as treatment and no toxicity has been observed (Desplender, 1988).

In human environments, it has been employed in the fumigation of swimming pools to control transmission of athlete's foot (van Cutsem et al., 1988).

Lime sulphur

Lime sulphur, a fungicide composed by a combination of inorganic sulphur and hydrate lime (CaO), is a liquid commonly used today as a pesticide/ fungicide in horticulture, and sprayed on plants to control a variety of diseases. Lime-sulphur was originally developed in 1851 by Grison who was the head gardener at the vegetable houses in Versailles, France. Grison boiled 'flowers of sulphur', freshly slaked lime, and water for 10 minutes, drew off the clear liquid and mixed it with water. He then used this solution to protect plants against mildews. The solution was originally known as the 'Grison Liquid' or 'Eau Grison'. It is usually yellow in colour and has a strong sulphur odour. It is commonly available at most garden stores. Sulphur is the only ingredient in the mixture that is toxic to pathogens. It is able to kill pathogens through direct contact or fumigation (sulphur vapours). The vapour action of sulphur allows the fungicide to be effective from a distance and it is important in killing spores. Once taken up by the fungus, sulphur disrupts the transfer of electrons causing the reduction of sulphur to hydrogen sulphide (H_2S), which is toxic to most cellular proteins.

In studies in which various environmental disinfectants were evaluated for efficacy against *M. canis*, lime sulphur was found to be one of the most effective treatment (White-Weithers and Medleau, 1995; Moriello et al., 2002). Therefore, a specific formulation for environmental use in veterinary premises is advisable.

Lime sulphur is licensed in the United States for topical treatment on cats and dogs (Lym Dyp®). The product is a scented sulphur concentrate, which provides antimicrobial and antiparasitic activity. It works at the surface of the skin and is used in the treatment of dermatoses responsive to the active ingredient (sulphurated lime 97.8% solution), such as sarcoptic and demodectic mange (McDonald and Lavoipierre, 1980; Ackerman, 1984; Chalmers et al., 1989). Lym Dyp® is extremely effective in the treatment of dermatophytosis too, mostly when used in conjunction with other treatments such as appropriate systemic antifungal therapy and/or diligent environmental decontamination. Lime sulphur has been used as the topical therapy of choice, and in some cases as the sole therapy against *M. canis* feline dermatophytosis (Moriello, 2004). In addition to killing dermatophytes and mange mites, it repels fleas and ticks for a while. It is antiseptic, keratinolytic and offers a high level of safety and efficacy. It can be used in adult dogs and cats, as well as in puppies and kittens. It is recommended to shake it well and dilute 4 ounces (approximately 113.5 grams) to 1 gallon (approximately 4 litres). It is applied as a rinse or dip at 5 to 7 day intervals. Lime sulphur should not be rinsed from the hair coat. For chronic and more resistant cases the use of a more concentrated solution, 8 ounces (approximately 227 grams) per gallon, is suggested together with a twice-weekly treatment (Moriello, 2004).

The product may stain the coat of light coloured dogs and cats yellow-green in colour, which will also smell like sulphur for a while, but that's the main downside and this colour commonly rapidly disappears once animals are dry. Some people prefer to wear gloves as the mixture may stain or dry out the hands. Lime sulphur may stain porous surfaces (cement) and will change the colour of jewellery. It may be irritating to mucous membranes and may cause skin irritation as a side effect, and thus cats should not be allowed to lick themselves until the coat is dry. If irritation develops, it is mandatory to decrease the frequency of use or discontinue use. Contact with eyes must be avoided. If contact occurs, eyes must be rinsed thoroughly with water. If a precipitate forms, it is suggested to immerse the sealed container in warm water for 15 minutes, to shake well, and then to use according to the package instructions. It is recommended to store the container at room temperature, and to keep out of the reach of children.

General procedures

The necessity of disinfection depends on the nature of the infection: in a single animal household in which no new pets or humans are to be introduced, minimal efforts are necessary. By contrast, disinfection is crucial in a cattery or households in which new animals and humans come and go (DeBoer and Moriello, 1995).

Although no controlled study exists that proves that clipping of the hair coat shortens the duration of infection, clinical studies strongly support the recommendation that cats with long hair and/or generalized dermatophytosis should be clipped. Although not necessary in all cases of dermatophytosis, clipping of the hair coat is optimum (DeBoer and Moriello, 1995; Moriello, 2003, 2004). It is required in cats who live with children and elderly or immune-suppressed people, as well as for short-haired cats with generalized lesions and longhaired cats regardless of lesion severity. Clipping may temporarily worsen lesions by causing microtrauma to the skin. However, clipping, discarding infected hairs and removing them from the environment will decrease the potential for the spread of the infection. Therefore, this would generally be indicated in a shelter or rescue situation. Clipping should be performed gently with a No. 10 electric clipper to a length less than 2 cm, and hair should be carefully disposed off afterward. Clipping with a closer blade causes excessive trauma and increases the chance of worsening of lesions (Moriello, 2003). It should be performed in a room that can easily be cleaned since it causes heavy environmental contamination. Instruments used should be carefully cleaned, dedicated only to that purpose, and never used on healthy animals (i.e., surgery clippers should not be used for *M. canis* infected cats). It is important to wear protective clothing whenever handling a ringworm infected cat in a shelter, especially when clipping.

If possible, to reduce environmental contamination, infected cats should be restricted to a easy cleanable room without carpets, until they become negative. This makes decontamination much easier and will reduce exposure of humans to the cats and sources of infection. All areas of the house to which infected animals have had access will require decontamination, but the majority of effort can then be concentrated on the room in which the cats are confined. A massive cleanup is mandatory if a breeder wishes to remove *M. canis* from the premises. Traditional recommendations for environmental control are based on thorough cleaning to remove infected material mechanically, followed by saturation of every possible environmental surface with fungicide (DeBoer and Moriello, 1995).

All contaminated equipment, toys, feed containers, scratching posts, grooming supplies, bedding, brushes, combs, rugs, cages, carriers and so forth should be removed from the environment, vacuumed, scrubbed with a detergent, and disinfected (DeBoer and Moriello, 1995). Ideally, any item that cannot be repeatedly vacuumed, scrubbed and thoroughly disinfected should be discarded. Washable items may be placed in a washing machine, bathtub, or laundry wash sink as appropriate to the item. The items should then be washed with hot water and soap, rinsed, and then soaked in 1:10 dilution of sodium hypochlorite for at least 10 minutes. This should be possibly repeated a minimum of three times. Soaps and detergents do not disinfect, but they help remove surface organic debris so it does not interfere with the function of disinfectants. It is recommended to always rinse soap or detergent off completely before disinfecting, and never mix with disinfectant unless the disinfectant instructions specifically state that is safe. Accidental oral ingestion of soaps or detergents must be avoided, as they can cause intestinal upset and irritate mucous membranes.

All possible hard surfaces should be thoroughly vacuumed, scrubbed with a detergent that is safe to use around cats, rinsed, and disinfected with 1:10 dilution of sodium hypochlorite, including floors, walls, storm doors, windowsills and so forth. Ideally, after rinsing, a wet vacuum may be used to remove dirty water. In a private house, lamps, bric-a-brac, bed linens, and furniture should also be vacuumed and wiped at least once a week with an antifungal liquid such as 1:10 dilution of sodium hypochlorite for at least 10 minutes to reach maximal fungicidal action. The label should be consulted for contact time when trying to use other products. Disinfectants can be used daily but they are often harsh and irritating to people and cats. One room at a time is decontaminated and allowed to dry thoroughly before persons or cats re-enter: sodium hypochlorite is very irritating to the eyes, skin, and mucous membranes. In addition, this disinfectant can damage clothing, shoes and rubber goods, and may be corrosive to steel surfaces. When permitted, the use of enilconazole (Clinafarm®) is recommended.

In addition, all surface and ledges should be dusted with a disposable electrostatic cloth. These disposable cloths can be used regularly to trap spores and dust that may be missed by the vacuuming process. In most cases, thorough and repeated vacuuming and use of disposable electrostatic dust-trapping cloths to remove dirt and spores on a daily basis should guarantee effective control of contamination of the home. Depending on the number of cats in a room, it is also suggested to scrub floors daily and any surfaces contacted by cats. Routine cleaning and disinfection of the home should be continued until cats are mycologically cured. Multicat households and catteries should be vacuumed preferably twice daily. Vacuum bags should be saturated with 1:10 dilution of sodium hypochlorite and destroyed, possibly by burning, each time. A vacuum cleaner with hose attachment that can be thoroughly cleaned is recommended. The vacuum should be sprayed down with 1:10 bleach after each use and will be ultimately discarded at the end of the treatment (Moriello, 2003).

Vigorous and frequent vacuuming mechanically removes many but not all spores (DeBoer and Moriello, 1995). Steam cleaning, that is hot water and high pressure, has been recommended as a method of decontamination for walls and tiled or cemented floors because of the mechanical action and destruction of spores (Guillot and Chermette, 1997). It can be costly due to equipment purchase charges and it is best to thoroughly wash all equipment prior to steaming it. Pressurized steam directed into cracks and corners is an excellent sterilizant. However, aggressive steam cleaning of carpets is of limited use because the temperature of the water delivered at the carpet level is insufficient to kill spores. To kill fungal spores the temperature of the water being forced into the carpets must be at least 43.3°C (110°F). To reach a temperature of 43.3°C (110°F) at the carpet level, the water chamber has to be filled with water exceeding 76.6°C (170°F). In a household situation, achieving this requires boiling water to reach this temperature. Unfortunately, the temperature of the water rapidly cools in the clean water reservoir and, within 15 minutes of filling the chamber, the temperature of the water being forced out the nozzle is less than 37.7°C (110°F). Therefore, steam cleaning of carpets may not be a reliable method of killing *M. canis* spores unless an antifungal disinfectant is added to water (Moriello, 1990). Always remember to test the carpet for colourfastness before applying any treatment. Since carpetted areas are problematic, in the lack of an effective disinfectant that preserves carpetting it is best to discard the carpet (DeBoer and Moriello, 1995). If one does not wish to rip the carpet out, then thorough vacuuming daily is the only practical method of reducing contamination. As already mentioned, the vacuum bag should be destroyed each time, possibly by burning. Alternatively, while searching web sites, some suggestions can be found. DC3 (Triple Jet) Dyna-Fog Hurricane Fogger® may be used to fog a carpetted room with a disinfectant that is colour-safe and non-staining. It is capable of dispensing both oil and water-based products, such as disinfectants, deodorizers, germicides and insecticides, both in large and small areas, wherever aerosol particle chemical treatment is needed. It may also be used to control fleas, ticks and other insects in the environment. In such a case, however, always remember that permethrin and pyretroids are very toxic to cats! In addition, it is suggested that Health Guard Laundry Additive (ViBAX-5) can be mixed as a spray and sprayed on the carpet. It may also be used to clean floors, walls, cages, kennels, and surfaces. However, it is very important to note that the efficacy of these treatments for environmental decontamination against *M. canis* is still to be proven in experimentally controlled trials.

Draperies should be dry-cleaned and not replaced until the infection is eradicated (DeBoer and Moriello, 1995). Inadequately decontaminated clothing may be a source of re-infection too, following therapy of feline dermatophytosis. Domestic laundering appears to be suitable for cleansing mycologically contaminated garments. A study determined that, regardless of the textiles and detergents used, reliable decontamination was achieved by laundering at 60°C (Ossowki and Duchmann, 1997). While searching web sites, it is possible to find that some laundry additives (NanoBio®, Health Guard Laundry Additive ViBAX-5®, Canesten Hygiene Laundry Rinse®) are claimed to be effective against dermatophytes in infected clothing. However, their efficacy on *M. canis* contaminated textiles still remains to be demonstrated in experimentally controlled trials.

Appropriate ventilation is always to be used. The use of a portable dehumidifier is advisable in cat rooms to keep humidity low, because humid environments favour spores viability (Moriello, 2003).

If possible, to decrease the number of spores that are re-circulated in the air, a fan should be put in the window to draw air out of the room. In addition, especially in multicat households and catteries, all heating or cooling vents and furnaces should be initially vacuumed and cleaned by a commercial company with high powder suction equipment. Furnace and air conditioner filters will need to be changed and discarded. Disposable house dust filters should be installed before replacing them. These can be purchased at home improvement stores and will help keep spores out of the heating vents. Air conditioners should not run in the room if this blows air throughout the house (Moriello, 2003).

All vehicles used to transport cats must likewise be decontaminated (DeBoer and Moriello, 1995).

In catteries, plastic sheeting should be placed on the inside of doorways to prevent spores from escaping. Disposable trash bags should be worn on clothing when treating cats and cleaning rooms. Shoes should be changed before and after leaving cat treatment areas (Moriello, 2003).

Discussion

There a number of commercially available chemical disinfectants. Veterinary hospitals and animal housing facilities are under constant threat of contamination with pathogenic fungi. The largest threat to cat facilities include dermatophytes, like *M. canis*, which are usually more resistant to inactivation than other pathogenic fungi. *M. canis* seems particularly resistant to disinfection and poses the greatest challenge to both clinics and manufacturers of disinfectants. In order to prevent the spread of *M. canis* infections at a facility, an effective treatment programme must be put into place including a regular programme of cleaning and disinfection with products proven to kill all of the spores that may be present. Choosing which disinfectant will be used in infected cat facilities is a decision that should not be made lightly.

Some important points about disinfection should be made before choosing a disinfectant for routine use in infected homes and catteries. Most disinfectants do not work if the surface to be disinfected is not clean before applying the disinfectant. Steam and high-pressure washers can be very useful to clean porous surfaces. Organic materials such as faeces, urine, food or milk may often inactivate some disinfectant or protect pathogens from the disinfectant's active agents. In addition, even hard water can reduce or destroy the activity of some disinfectants. Likewise, some disinfectant solutions are only active for a few days after mixing or preparing. Failure to make a fresh solution of disinfectant after it has been prepared longer than a few days, or after it has become contaminated by organic material, may result in a product that does not really work. It is true that sufficient concentration and contact time can overcome some of these problems with certain classes of disinfectants, but often increasing the concentration or contact time makes use of the product impractical, costly or caustic.

Disinfectants vary considerably in their activity. Most commonly used disinfectants are not active against the naturally infective state of *M. canis*, that is infected hairs and spores. Disinfectants must have sufficient contact time with the surfaces to which they are applied in order to allow them to inactivate all the infected material. Contact time varies with the product.

In several comparison studies, one of the disinfectants that performed the best against *M. canis* was sodium hypochlorite (Bianchi et al., 1989; Rycroft and McLay, 1991; Moriello and DeBoer, 1998; Gupta et al., 2001). In particular, in one study, undiluted household bleach (5.25% sodium hypochlorite) was the most fungicidal solution tested (Moriello and DeBoer, 1998). This product demonstrated a fungicidal activity at 2, 8 and 24 hours after a single wipe. Undiluted bleach appeared to have a residual activity of at least 24 hours. Unfortunately, this disinfectant is very corrosive to some environmental surfaces and very irritating to mucous membranes, skin, eyes and the respiratory system. Therefore, it cannot be used in catteries or houses in quantities sufficient to be useful. Diluted household bleach (1:10) is consistently effective (Moriello, 2004) but needs to be frequently replaced and may be relatively quickly inactivated by hard water and organic matter. Fumes of stabilized chlorine dioxide in undiluted form may be toxic to living tissues. Organic debris and exposure to sunlight may also relatively quickly inactivate stabilized chlorine dioxide.

Aldehydes (glutaraldehyde and formaldehyde) cannot be recommended as routine disinfectants due to the concerns that they may be carcinogenetic. They may irritate the respiratory system, skin, eyes and mucous membranes if not used in extremely well ventilated areas. This is unfortunate because these compounds always showed excellent fungicidal activity against *M. canis* (Van Gestel et al., 1981; Rybnikar et al., 1985; Rycroft and McLay, 1991; Terleckyj and Axler, 1993; Kosuge et al., 1996; Moriello and DeBoer, 1998; Sotohy and Mohamed, 1999; Abou-Zaid, 2001). Aldehydes could be a reasonable alternative to decontaminate environments from which cats had been removed.

Lime sulphur (1:33) and enilconazole (20 µl/ml) are consistently effective (Moriello, 2004). In one study they were 100% fungicidal even when the recommended concentration was diluted 1:4 (Moriello et al., 2002). Lime sulphur is not licensed for environmental use other than agricultural. It may be irritating to the skin and mucous membranes and may stain environmental surfaces. Clinafarm® is licensed only for the antifungal disinfection of veterinary premises. If used extra-label to disinfect houses heavily contaminated by *M. canis*, the spray or the fogger should be used as recommended with respect to human exposure and pre-cleaning the environment. This product is corrosive and harmful or fatal if ingested.

Disinfectants are not to be applied to animals directly, unless labelled for such use. Imaverol® is not available in the United States, nor is it approved for use in cats. Clinafarm®, a different formulation of enilconazole licensed for use as an environmental disinfectant, is available in the United States. This formulation has been used off-label to treat dermatophytosis at a dilution of 55.6 ml/gallon of water as a topical antifungal (Moriello, 2004). It is important to note that using this product off-label is illegal. The label should be always consulted to be sure that there are no warnings against using a disinfectant in cat rooms. A general recommendation is to rinse disinfectants off after the appropriate amount of contact time if animals will have contact with the disinfected surfaces. Label directions should be strictly followed, and different classes of disinfectants should not be mixed.

For routine use in ringworm treatment programmes, owners and breeders should consider the major risks they are concerned about, consider the type of surface they wish to disinfect, select a disinfectant that will be active against *M. canis* under the conditions in which it will usually be used, and then select a disinfectant that best suits their needs. The conditions may include hard water, contamination with organic material, and potential for toxicity or damage to environmental surfaces, skin and clothing. Information about activity in hard water or in the presence of organic debris, contact time needed, spectrum of activity, environmental concerns, and other details are usually on the label or can be obtained from the producer. It is also important to keep solutions clean and freshly made as indicated by the manufacturer. Web sites are often good sources of information about individual products. Above all, owners and breeders should remember that disinfection is just one aspect of their *M. canis* ringworm control programme. The optimum treatment protocol for cats with dermatophytosis involves a combination of clipping of the hair coat, twice-weekly topical antifungal therapy, concurrent systemic antifungal therapy and environmental decontamination. Finally, fungal culture monitoring should be performed every 2-4 weeks until mycological cure (Moriello, 2004).

Future perspectives

Dermatophytosis is one of the most important coat infections of cats. Since the primary etiological agent, *M. canis*, can cause disease in humans, feline dermatophytosis is of public health concern as a source of human infection and thus is generally accepted as an anthropozoonosis. Infection occurs via direct transmission of infective spores to a susceptible host. Reservoirs of infection for both people and animals include contaminated environments and objects, animals with sub-clinical or clinical infections, and animals that are in-apparent carriers of the spores on their coat.

Once of the pressing concerns of cats recovering from *M. canis* dermatophytosis is the persistence of fomites in their living areas that expose them to re-infection. Similarly, owners, breeders and veterinary practitioners require adequate means of disinfection, especially in multicat households, catteries and veterinary hospitals where time constraints and large cat numbers demand a quick and reliable method of disinfecting shared environments, equipment and instruments between cats. Traditional disinfecting techniques mainly based on strong use of sodium hypochlorite appear applicable for hard surfaces like floors and so on. However, these methods cannot be recommended for soft materials such as curtains and carpet that may be fomites in many outbreaks of feline ringworm.

Terbinafine is an allylamine derivative fungicidal agent that has been found to be very effective in the treatment of human dermatophytosis. Allylamine derivatives are a new class of synthetic antifungal agents of relatively recent discovery. Their mechanism of action is fungicidal, which distinguishes them from most other cutaneous antifungal agents. Terbinafine inhibits squalene epoxidase, an early step in the pathway of ergosterol synthesis, which results in fungal cell death. The active site is not associated with the cytochrome P-450 enzyme system and terbinafine does not affect the metabolism of hormones like azole antifungal agents do. Like azole derivatives, terbinafine persists for several weeks at therapeutic levels in the stratum corneum after systemic administration (de Jaham and Paradis, 1997). Various doses of oral terbinafine, 5 to 40 mg/kg, were reportedly used to treat cats or dogs. As a result of these studies, it has been determined that terbinafine needs to be administered at a dose >20 mg/kg (i.e., 30 to 40 mg/kg) given orally once a day, to achieve a mycological cure. This dose results in significantly higher concentrations in hair when compared with lower doses. Terbinafine at this dose may be substituted for itraconazole in combined continuous/pulse therapy. Terbinafine also appears to be equivalent to griseofulvin and itraconazole in treating feline dermatophytosis. The number of treatment days to cure varied from 21 to >126 days (Moriello, 2003, 2004).

Side effects of terbinafine have been reported in humans and are widely known, the majority of them including gastrointestinal disorders and skin reactions (van Puijenbroek et al., 2001). However, very little is known about adverse drug reactions associated with terbinafine administration in cats. To the best of our knowledge, at present vomiting is the only problem that has been occasionally observed at 10-30 minutes after oral administration of the drug (Mancianti et al., 1999; Kotnik et al., 2001). Sometimes soft stool has been recorded too. Vomiting can be resolved by feeding the cat after oral administration of the drug (Kotnik et al., 2001).

Commercial chemical disinfectants and pharmaceutical antifungal agents were tested against arthroconidia from five fungal strains belonging to three *Trichophyton* species: *T. mentagrophytes, T. raubitschekii* and *T. tonsurans* (Gupta et al., 2001). The chemical disinfectants included chlorine, phenol, sodium dodecyl sulphate and several quaternary ammonium compounds, while the two pharmaceutical preparations contained 0.1% bifonazole spray (Mycospor®) and 1% terbinafine hydrochloride (Lamisil®) as active agents. Arthroconidia were exposed to the antifungal agents either in a suspension solution for a given period of time and assayed for kill rate, or on a sprayed agar plate and monitored for surviving colonies over a period of 14 days. Sodium dodecyl sulphate, a protein denaturant with high antiviral activity, was ineffective against all dermatophytic species tested. Phenol is a broad-spectrum disinfectant; however, because it is usually effective at concentrations that can potentially be toxic to humans, its use has declined rapidly in last years. Phenol was equally effective against *T. raubitschekii* and *T. tonsurans*; however, *T. mentagrophytes* cells were able to survive for up one hour in 5% phenol. Quaternary ammonium compounds were less rapid in their action against dermatophytes. An exposure period longer than 24 hours and a concentration of about 0.5% were required with these disinfectants to complete the fungicidal effect with *Trichophyton* species. When used at lower concentrations (0.1-0.3%), they produced only a fungistatic effect on *T. mentagrophytes*. Bifonazole was fungicidal against *T. raubitschekii* and *T. tonsurans* but also produced a fungistatic effect against *T. mentagrophytes*. Of the various groups of disinfectants examined in the study, chlorine and terbinafine exhibited by far the strongest fungicidal action against *Trichophyton* species. They were found to bring about a rapid inactivation of conidia in all five strains, and can be classified as high level disinfectants for dermatophytes. A 15-minute exposure to 1% chlorine was sufficient tc kill a highly dense inoculum of *Trichophyton* conidia. Chlorine is, however, highly corrosive and causes irritation in closed surroundings making it of limited use in many areas. All five strains of the three *Trichophyton* species were killed within 15 minutes of exposure to 0.01% terbinafine. The authors reported that, in other experiments, a 100% kill rate was achieved by the five strains with a terbinafine concentration as low as 0.0005% indicating a markedly high sensitivity of *Trichophyton* species against this drug. The spray formulation in Lamisil® contains 1% terbinafine and it was evidently fungicidal against all strains studied. It appears from these results that a much lower concentration of terbinafine will be sufficient to produce fungicidal spray against *Trichophyton* conidia. This antifungal efficacy of terbinafine against dermatophytic spores appears to make this drug a strong candidate for disinfection of various fomites, including contaminated environments and objects potentially involved in contagion and re-infection with *M. canis* in cats.

CONCLUSION

Disinfectants are defined as chemical agents capable of removing infectious microorganisms in addition to spores. A number of commonly used disinfectants were shown to be relatively ineffective against dermatophytes. Moreover, not all dermatophytic species are equally sensitive to a given product, and even strains of the same dermatophytic species may vary in resistance (Sotohy and Mohamed, 1999). The limited range of disinfectants currently effective does not allow a significant choice in selecting appropriate disinfectants for environmental control of *M. canis* infections. This limited range also means that care must be taken to choose the best product for this particular use. An appropriate choice is best made by evaluating the activity against *M. canis* and comparing this activity with data on toxicity, materials compatibility and cost. Adequate risk analysis and cost-benefit analysis is desirable to ensure that the safest and most cost effective agent capable of performing adequately is chosen.

The striking efficacy of terbinafine against fungal spores offers a potentially new source of non-corrosive and non-toxic disinfectant to target a broad group of environmental fungal pathogens. The feasibility of a new formulation of terbinafine for environmental use against dermatophytes, with special reference to *M. canis,* needs to be tested. Terbinafine is only moderately effective or completely ineffective against some medically important yeast species, which in turn are susceptible to some chemical compounds. Therefore, the new formulation could be a combination of both terbinafine as an antifungal drug and a chemical compound as an effective disinfectant against yeasts to achieve a broader range of activity towards a thorough antimycotic decontamination of the environment.

References

Abou-Zaid, A.A. 2001. Some studies on ringworm in donkeys. Veterinary Medical Journal Giza, 49(3): 473-48

Ackerman, L. 1984. Demodicosis in cats. Modern Veterinary Practice, 65(10): 751-752.

Ando, J., Sawada, T. and Nagata, M. 1997. Occurrence of dermatophytosis in a cattery Journal of the Japan Veterinary Medical Association, 50(6): 341-343.

Bianchi, P., Repetto, A., Bulletti, S., Mattiacci, P., Rossi, J., Pagiotti, R., Ribaldi, M. and Schaffella, F. 1989. In vitro study of the antifungal activity of two chlorine derivatives to be used in antisepsis. Annali d'Igiene 1(3-4): 827-840.

Cafarchia, C., Riccardi, L. and Puccini, V. 2000. Dermatofitosi degli equini. Tre casi di geotricosi cutanea degli equini in Puglia. Obiettivi & Documenti Veterinari, 50(1): 7-10.

Chalmers, S., Schick, R.O. and Jeffers, J. 1989. Demodicosis in two cats seropositive for feline immunodeficiency virus. Journal of the American Veterinary Medical Association, 194(2): 256-257.

DeBoer, D. and Moriello, K.A. 1995. Clinical update on feline dermatophytosis - Part II. Compendium on Continuing Education for the Practising Veterinarian, 17(12): 1471-1481.

De Jaham, C. and Paradis, M. 1997. La dermatophytie féline II: modalités thérapeutiques. Le Médecin Vétérinaire du Quebec, 27(4): 147-152.

Desplenter, L. 1988. Aspergillosis in animal health: environmental control, pp. 289-297. In: Aspergillus and Aspergillosis. H. Vanden Bossche, D.W.R. MacKenzie, and G. Cauwerbergh (eds.). Plenum Press, New York, USA.

Desplender, L. 1989. Dermatophytosis in small animals: topical treatment and environmental control with enilconazole. In: Proceedings Voorjaasdagen, Amsterdam, The Netherlands, Tijdschrift Voor Diergeneeskunde, 114 (Suppl. 1): 33S-34S.

Englund, L., Mattson, R. and Berndtson, L.T. 1990. Possible effect of vaccination against *Trichophyton mentagrophytes* infection in a Swedish fox farm. Acta Veterinaria Scandinavica, 31(1): 121-123.

Filipello Marchisio, V., Gallo, M.G., Tullio, V., Nepote, S., Piscozzi, A. and Cassinelli, C. 1995. Dermatophytes from cases of skin disease in cats and dogs in Turin, Italy. Mycoses, 38(5-6): 239-244.

Fraise, A.P. 1999. Choosing disinfectants. Journal of Hospital Infection, 43(4): 255-264.

Guillot, J. and Chermette, R. 1997. Le traitement des mycoses des carnivores domestiques. Le Point Vétérinaire, 28(185): 51-61.

Guillot, J., Malandain, E., Jankowski, F., Rojzner, K., Fournier, C., Touati, F., Chermette, R., Seewald, W. and Schenker, R. 2002. Evaluation of the efficacy of oral lufenuron combined with topical enilconazole for the management of dermatophytosis in catteries. Veterinary Record, 150(23): 714-718.

Gupta, A.K., Ahmad, I. and Summerbell, R.C. 2001. Comparative efficacies of commonly used disinfectants and antifungal pharmaceutical spray preparations against dermatophytic fungi. Medical Mycology, 39(4): 321-328.

Hay, R.J. 1991. The role of chemoprophylaxis in the prevention of airborne infection by fungal pathogens. Journal of Hospital Infection 18 (Suppl. A): 460-465.

Hnilica, K.A. and Medleau, L. 2002. Evaluation of topically applied enilconazole for the treatment of dermatophytosis in a Persian cattery. Veterinary Dermatology, 13(1): 23-28.

Khosravi, A.R. 1996. Fungal flora of the hair coat of stray cats in Iran. Mycoses, 39(5-6): 241-243.

Kosuge, J., Takatori, K. and Ichijo, S. 1996. Antifungal activities of glutaraldehyde disinfectant against animal dermatophytes. Journal of Veterinary Medicine, Japan, 49(11): 889-892.

Kotnik, T., Kožuh Erzen, N., Kužner, J. and Drobniè-Košorok, M. 2001. Terbinafine hydrochloride treatment of *Microsporum canis* experimentally-induced ringworm in cats. Veterinary Microbiology, 83(2): 161-168.

Larsson, C.E., Nahas, C.R. and Ledon, A.L.B.P. 1994. Ringworm in domestic cats in Sao Paulo, Brazil, between 1981-1990. Feline Practice, 22(3): 11-14.

Lemarié, R.J. and Hosgood, G. 1995. Antiseptics and disinfectants in small animal practice. Compendium on Continuing Education for the Practising Veterinarian, 17(11): 1339-1351.

Love, B.C. and Hirsh, D.C. 1989. Disinfectant and antiseptic use in small animal practice, pp. 258-262. In: Current Veterinary Therapy X-Small Animal Practice. R.W. Kirk and J.D. Bonagura (eds.), W.B. Saunders Company, London, UK.

Mancianti, F. and Nardoni, S. 2004. Susceptibility of *Microsporum canis* isolated from domestic animals against a commercially available enilconazole in fumigant form. Journal de Mycologie Médicale, 14(2): 73-74.

Mancianti, F., Pedonese, F., Millanta, F. and Guarnieri, L. 1999. Efficacy of oral terbinafine in feline dermatophytosis due to *Microsporum canis.* Journal of Feline Medicine and Surgery, 1(1): 37-41.

Mancianti, F., Nardoni, S., Cecchi, S., Corazza, M. and Taccini, F. 2002. Dermatophytes isolated from symptomatic dogs and cats in Tuscany, Italy during a 15-year-period. Mycopathologia, 156(1): 13-18.

Mancianti, F., Nardoni, S., Corazza, M., D'Achille, P. and Ponticelli, C. 2003. Environmental detection of *Microsporum canis* arthrospores in the households of infected cats and dogs. Journal of Feline Medicine and Surgery, 5(6): 323-328.

Marchetti V., Mancianti F., Cardini G. and Luchetti E. 2006. Evaluation of fungicidal efficacy of Benzalkonium Chloride (Steramina G u.v.) and Virkon-S against *Microsporum canis* for environmental disinfection. Veterinary Research Communications, 30(3): 255-261.

McDonald, S.E. and Lavoipierre, M.M. 1980. *Trixacarus caviae* infestation in two guinea pigs. Laboratory Animal Science, 30(1): 67-70.

McDonnell, G. and Russel, A.D. 1999. Antiseptics and disinfectants: activity, action and resistance. Clinical Microbiology Review, 12(1): 147-179.

Mignon, B.R. and Losson, B.J. 1997. Prevalence and characterisation of *Microsporum canis* carriage in cats. Journal of Medical and Veterinary Mycology, 35(4): 249-256.

Moriello, K.A. 1990. Management of dermatophyte infections in catteries and multiple-cat households. Veterinary Clinics of North America-Small Animal Practice, 20(6): 1457-1474.

Moriello, K.A. 2003. Symposium on feline dermatophytosis. Veterinary Medicine, 98(10): 844-891.

Moriello, K.A. 2004. Treatment of dermatophytosis in dogs and cats: review of published studies. Veterinary Dermatology, 15(2): 99-107.

Moriello, K.A. and DeBoer, D.J. 1998. Environmental decontamination of *Microsporum canis*: *in vitro* studies using isolated infected cat hair, pp. 309-318. In: Advances in Veterinary Dermatology, Vol. 3, Proceedings of the Third World Congress of Veterinary Dermatology, Edinburgh, Scotland, 11-

14 September, 1996. K.W. Kwochka, T. Willemse and C. Tscharner (eds.), Butterworth Heinemann, Oxford, UK.

Moriello, K.A., Kunkle, G. and DeBoer, D.J. 1994. Isolation of dermatophytes from the haircoats of stray cats from selected animal shelters in two different geographic regions in the United States. Veterinary Dermatology, 5(2): 57-62.

Moriello, K.A., DeBoer, D.J. and Volk, L. 2002. Determination of strain variability of *Microsporum canis* to disinfectants. Veterinary Dermatology, 13(4): 211-229.

Odore, R., Colombatti Valle, V. and Re, G. 2000. Efficacy of chlorexidine against some strains of cultured and clinically isolated microorganisms. Veterinary Research Communications, 24(4): 229-238.

Ohta, S., Masahiro, M., Katsuhiro, N. and Zenda, H. 1996. Comparative fungicidal activity of a new quaternary ammonium salt, N-alkyl-N-2-hydroxyethil-N,N-dimethilammonium butyl phosphate and commonly used disinfectants. Biological and Pharmaceutical Bulletin, 19(2): 308-310.

Osame, S., Ichijo, S., Takotori, K., Saitoh, R. and Kageyama, Y. 1991. Therapeutic effects of didecyldimethylammonium chloride. Journal of Veterinary Medicine, Japan, 44(11): 713-716.

Ossowki, B., and Duchmann, U. 1997. The influence of domestic laundering on mycologically contaminated textiles. Hautarzt, 48(6): 397-401.

Palsson, G. 1991. A treatment programme for ringworm in cats. Svenk Veterinartidning, 43(16): 709-712.

Papini, R. 2004. *Microsporum canis* infections in cats: approach to systemic therapeutic protocols, pp. 131-142. In: Fungi in Human and Animals Health R.K.S. Kushwaha (ed.). Scientific Publishers, Jodhpur, India.

Paterson, S. 1999. Miconazole/chlorexidine shampoo as an adjunct to systemic therapy in controlling dermatophytosis in cats. Journal of Small Animal Practice, 40(4): 163-166.

Perrins, N. and Bond, R. 2003. Synergistic inhibition of the growth in vitro of *Microsporum canis* by miconazole and chlorexidine. Veterinary Dermatology, 14(2): 99-102.

Perrins, N., Howell, S.A., Moore, M. and Bond, R. 2005. Inhibition of the growth in vitro of *Trichophyton mentagrophytes*, *Trichophyton erinacei* and *Microsporum persicolor* by miconazole and chlorhexidine. Veterinary Research Communications, 16(5): 330-333.

Rochette, F., Engelen, M. and Vanden Bossche, H. 2003. Antifungal agents of use in animal health-practical applications. Journal of Veterinary Pharmacology and Therapy, 26(1): 31-53.

Rochette, F. and Van Meirhaege, P. 1997. Enilconazole as a treatment of naturally occurring dermatophytosis in rabbit farms: a review. World Rabbit Science, 5(1): 7-11.

Romano, C., Valenti, L. and Barbara, R. 1997. Dermatophytes isolated from asymptomatic stray cats. Mycoses, 40(11-12): 471-472.

Rybnikar A., Vrzal V., Chumela J., and Harnach R. 1985. Disinfection for dermatomycoses in veterinary practice. Veterinarni Medicina (Praha), 30(2): 119-128.

Rycroft, A.N., and McLay, C. 1991. Disinfectants in the control of small animal ringworm due to *Microsporum canis*. Veterinary Record, 129(11): 239-241.

Simon, G., Torok, I. and Horvath, A. 1996. *Trichophyton mentagrophytes* infection in a rabbit farm. Revista Iberoamericana de Micologia, 13(3): 73-75.

Sotohy, S.A. and Mohamed, A.A. 1999. Fungicidal effect of the common disinfectants on the most widely spread dermatophytes with references to their differential resistance "in-vitro" study. Assiut Veterinary Medical Journal, 41(81): 113-126.

Sparkes, A.H., Werrett, G., Stokes, C.R. and Gruffydd-Jones, T.J. 1994. *Microsporum canis*: Inapparent carriage by cats and the viability of arthrospores. Journal of Small Animal Practice, 35(8): 397-401.

Symoens, F., Fauvel, E. and Nolard, N. 1989. Evolution de la contamination de l'air et des surfaces par *Microsporum canis* dans une habitation. Bulletin de la Societé Française de Mycologie Médicale, 18(2): 293-298.

Terleckyj, B. and Axler, D.A. 1987. Quantitative neutralization assay of fungicidal activity of disinfectants. Antimicrobial Agents and Chemotherapy, 31(5): 794-798.

Terleckyj, B. and Axler, D.A. 1993. Efficacy of disinfectants against fungi isolated from skin and nail infections. Journal of the American Podiatric Medical Association, 83(7): 386-393.

Van Cutsem, J.V., Van Gerven, F. and Janssen, P.A.J. 1988. In vitro activity of enilconazole against *Aspergillus* spp. and its fungicidal efficacy in a smoke generator against *Aspergillus fumigatus*. Mycoses, 31(3): 143-147.

Van Gestel, J., Van Cutsem, J. and Thienpont, D. 1981. Vapour phase activity of imazalil. Chemotherapy, 27(4): 270-276.

Van Puijenbroek, E.P., Egberts, A.C.G., Meyboom, R.H.B. and Leufkens, H.G.M. 2001. Association between terbinafine and arthralgia, fever and urticaria: symptoms or syndrome?. Pharmacoepidemiology and Drug Safety, 10(2): 135-142.

White-Weithers, N. and Medleau, L. 1995. Evaluation of topical therapies for the treatment of dermatophyte-infected hairs from dogs and cats. Journal of the American Animal Hospital Association, 31(3): 250-253.

Woloszyn, S., Andrychiewicz, J., Kostro, K., Winiarczyk, S. and Gradzki, Z. 1996. Efficacy of vaccines in the control of dermatomycoses in rabbits. Medycyna Weterynaryjna, 52(8): 818-521.

14

Thermophilic Molds in Environmental Management

Bijender Singh and T. Satyanarayana*

Department of Microbiology
University of Delhi, South Campus, Benito Juarez Road,
New Delhi 110021, India, Phone: 091-11-24112008,
Fax: 091-11-24115270, E-mail: tsnarayana@gmail.com
**Corresponding author:* Department of Microbiology
University of Delhi, South Campus, Benito Juarez Road,
New Delhi 110021, India, Phone: 091-11-24112008

Abstract

Thermophilic molds are the major microflora in heaped masses of plant materials, piles of agricultural and forestry residues and other organic materials, where warm, humid, and aerobic environment provides the basic conditions for their growth and development. The ability of these molds to degrade organic matter efficiently, produce an array of useful enzymes, antibiotics and nutritionally enriched feeds and their suitability as agents in bioconversions makes them nature-borne biotechnologists. Their enzymes are also useful in the treatment of industrial wastes and effluents that are rich in oil, heavy metals, anti-nutritional factors (e.g. phytic acid) and other polysaccharides (cellulose, hemicellulose, chitin, pectin). The phytases hydrolyze phytates, and therefore, aid in combating environmental phosphorus pollution. Thermophilic molds also play an important role in mushroom composting. Their ability to produce glucose, xylose and mannose by the hydrolysis of agro-residues and their fermentation to ethanol is a major venture in the field of biotechnology. This chapter aims at focusing attention on these activities of thermophilic molds which play an important role in the management of the environment.

INTRODUCTION

Microorganisms, their biology, economic value and pathogenic capabilities are not new to human society. They have served man from the traditional fermentation of foods and beverages to the modern production of pharmaceuticals. With the development of modern molecular and phylogenetic techniques, it is now estimated that only 1% of the extant microbial diversity on the earth is cultivable. The biological world exists under a selected set of physical conditions, and the cultivable microorganisms differ slightly from each other in these parameters. Thermophilic molds are the group of fungi that are capable of growth optimally at or above 40°C (Crisan, 1959, 1969). These are one of the major components of the microflora that develop in heaped masses of plant materials, piles of agricultural and forestry products and other organic materials, which provide a suitable environment for their growth and development. Cooney and Emerson (1964) provided a taxonomic description of 13 species known at that time and about their habitats and biology. Presently a large number of thermophilic fungal species have been described (Mouchacca, 1997), which constitute a heterogeneous physiological group of various genera in Zygomycetes, Ascomycetes, Deuteromycetes and Basidiomycetes. Thermophilic fungi grow in simple media containing carbon and nitrogen sources and mineral salts, suggesting that they do not have any special nutritional requirement for growth and are autotrophic for all vitamins (Cooney and Emerson, 1964; Emerson, 1968). On the basis of fungal growth in sealed containers with alkaline pyrogallol, Henssen (1957) demonstrated that thermophilic fungi could undergo growth as facultative anaerobes. In contrast, Deploey and Fergus (1975) recorded complete cessation of growth in the presence of 100% nitrogen, and the fungi sporulated only in the presence of oxygen, suggesting an absolute requirement of oxygen for their growth.

The biotechnological potential of these molds stems from their ability to efficiently degrade organic matter acting as biodegradants and natural scavengers, to produce an array of industrially important extracellular as well as intracellular enzymes, antibiotics, amino acids, phenolic compounds and polysaccharides; from their role in the production of single cell protein (SCP) and nutritionally enriched feeds; and from their suitability as agents of bioconversions and experimental systems for genetic manipulations. Due to these multifarious potential applications, they appear to be nature-borne biotechnologists (Satyanarayana et al., 1992). The various aspects of thermophilic fungi such as their occurrence, ecology, and production, and characteristics of enzymes have been reviewed (Satyanarayana et al., 1988, 1992; Johri et al., 1999; Maheshwari et al., 2000; Archana and Satyanarayana, 2001; Satyanarayana and Singh, 2004). This chapter aims at delineating their role in environmental management.

ECOLOGY AND DIVERSITY OF THERMOPHILIC MOLDS

Temperature is an important factor that affects growth, distribution and metabolism of the organisms. Microorganisms are known to have arrived on the earth nearly 3.5 billion years ago at a time when the temperature was likely to be in very high. This has been proved by the discovery of microorganisms from geothermal areas around the world. Prokaryotes have a wide range of temperature tolerance, while only a few species of fungi, among eukaryotes, can tolerate up to 62°C. Thermophilic molds are defined as those that are capable of growing optimally at or above 40°C (Crisan, 1959; 1969). Only a few species of fungi have the ability to survive and grow at 45-55°C.

The first two of the known thermophilic fungi were identified as chance contaminants on the organic substrates incubated at elevated temperatures. *Mucor pusillus* was isolated from bread and described over a century ago by Lindt (1886). Three years later, Tsiklinskaya (1899) discovered another thermophilic mold, *Thermomyces lanuginosus* growing on potato, which was inoculated with garden soil. The natural habitats of thermophilic fungi and the conditions which favoured their growth remained unknown until Miehe (1907) investigated the causes of self-heating and spontaneous combustion of damp haystacks. He isolated four species of thermophilic fungi, *Mucor pusillus, Thermomyces lanuginosus, Thermoidium sulfureum,* and *Thermoascus aurantiacus* from self-heating hay. Miehe (1907) explained the self-heating of hay and other plant materials. Due to the exothermic reactions of the mesophilic microflora present therein, the temperature of the material rises to ~40°C. The resultant warm environment favours germination of spores of the thermophilic microflora, and eventually the latter outgrows the mesophilic microflora; in the process, the temperature of the mass rises further to 60°C or even higher.

By the beginning of the 20th century, Miehe's (1907) work had led to the discovery of a small group of thermophilic fungi and their primary habitats. Their unique thermal adaptation attracted the attention of Noack (1920), who isolated thermophilic fungi from several natural substrates. He was intrigued by the fact that in addition to self-heating masses of hay and compost heaps of leaves, these fungi were present in places where temperatures conducive to their growth occur only infrequently, for example in soils of the temperate zone. This puzzling aspect of the ecology of thermophilic fungi provided the foundation for Noack's pioneering investigations on their physiology. Using respiration as the probe, Noack sought to determine whether thermophilic fungi had an unusually high rate of respiration whereby the released metabolic heat could warm their environment, allowing them to complete their life cycle rapidly. He confirmed that the respiration of thermophilic fungi does not confer any special advantage to them.

Cooney and Emerson (1964) provided taxonomic descriptions of 13 species known at that time, an account of their habitats, and the general biology of thermophilic molds. This monograph for the first time made mycologists generally aware of the existence of thermophilic fungi. Thermophilic molds constitute a heterogeneous physiological group of various genera included in Zygomycetes, Ascomycetes, Deuteromycetes, Basidiomycetes and Mycelia Sterilia. These molds form a small group of about 40 taxa included in 21 genera (Mouchacca, 1997).

THERMOPHILIC MOLDS IN ENVIRONMENTAL MANAGEMENT

Environmental pollution is one of the most formidable dangers that confront mankind today. Essentially a product of industrial proliferation, environmental pollution has attained diabolic proportions as a result of man's awareness and abuse of environmental and natural resources. In the past few decades, increased human activity has led to the expansion of industries at an accelerated rate which in turn has resulted in the deterioration of the environment. The large volume of sewage and industrial wastes generated due to urbanization and industrialization are discharged into the atmosphere or natural water bodies or into soils which exert a hazardous effect on the life support systems. Therefore, there is an increasing demand on the management of these pollutants. Biological cleaning procedures make use of the fact that most organic chemicals are subjected to microbial attack. This process is called bioremediation and is defined as the productive use of microorganisms to remove or detoxify pollutants, usually as contaminants of soil, water or sediments which otherwise threaten public health. Microorganisms have been extensively employed in the disposal of organic matter and toxic chemicals from domestic and industrial wastes. In natural habitats, they colonize the organic matter and degrade the major components of the organic matter such as chitin, starch, lignin, cellulose, pectin, phytic acid and hemicellulose. During this process, they secrete a large number of hydrolytic enzymes (chitinase, amylase, laccase, cellulase, phytase and hemicellulase) which degrade these polymers into monomeric and dimeric forms that are utilized by these molds for their growth and development.

Solid waste management and composting

Large amounts of organic materials are produced annually in nature, which are degraded by microbial action. The decomposition of organic matter takes place slowly on the surface of the ground at ambient temperatures. The natural process of degradation can be speeded up by gathering organic material into heaps to conserve part of the heat of fermentation so that the temperature of the mass rises and faster reaction rates are attained. This accelerated process of decomposition of organic matter by a mixed population of microorganisms in a warm, moist and aerobic environment is known as composting. A wide range of agricultural and forest residues (wood chips, bagasse, wool, hemp straw and others) are colonized and degraded by thermophilic molds. There is a great variation in the composition of these materials (Tables 1-3). Rapid growth of thermophilic molds leads to self heating of organic matter that depends on the presence of soluble organic substances in the substrate. Available substrates such as starch, cellulose, hemicellulose and lignin are utilized by these fungi (Sharma and Johri, 1992).

Several attempts have been made to produce protein-enriched upgraded feeds, single cell protein (SCP) and some enzymes by solid state fermentation (SSF) using thermophilic molds (Satyanarayana et al., 1992).

Table 1. Phytic acid contents in different plants

Plants	*Phytic acid (%)*
Bajra	0.31
Barley (Betzes)	0.56
Beans (Pinto)	0.93
Corn dent	0.72
Corn cereal (toasted)	0.07
Corn flour	0.10
Maize	0.39
Millet	0.81
Oats	0.62
Rice	0.15
Rice bran (0.55% oil)	8.70
Rye	0.57
Sorghum	0.33-0.81
Soy concentrates	10.7
Soy flakes (defatted)	1.64
Wheat (HRW)	0.77
Wheat (SWW)	0.65
Wheat	0.44
Groundnut meal	0.46
Soybean meal	0.56
Cotton seed meal	0.79
Sunflower meal	0.45

*Adapted from Reddy et al., 1982.

Employing biological treatment of cellulosic feedstuffs, digestibility and intake can be improved (Ryu, 1989). The cellulolytic thermophilic molds have certain advantages over mesophiles, such as high rates of cellulose breakdown, good sources of protein, activity over a wide range of temperatures 20-55°C, and higher growth rates (Seal and Eggins, 1976). *Chaetomium cellulolyticum* has been used in the conversion of forest and agricultural residues into protein enriched feed. The SCP products are nutritious, digestible and non-toxic in animal feed protein rations. The composition and properties of SCP produced from pulp and fiber board mill effluents by the Waterloo process using *C. cellulolyticum* is shown in Table 2.

Table 2. Composition of some lignocellulosic materials*

Materials	*Composition (%)*				
	Cellulose	*Hemicellulose*	*Lignin*	*Extractives*	*Ash*
Spruce wood	43.0	27.0	28.6	1.8	0.4
Pine wood	44.0	26.0	27.8	5.3	0.4
Birch wood	40.0	39.0	19.5	3.1	0.4
Pine craft pulp	77.0	18.0	5.0	0.2	0.4
Bagasse	33.4	30.0	18.9	6.0	2.4
Wheat straw	30.5	28.4	18.0	3.5	11.0
Rice straw	32.1	24.0	12.5	4.6	17.5
Bamboo	ND	19.6	20.1	1.2	3.3
Cotton	80-95	5-20	ND	ND	ND
Municipal refuse	76.0	ND	ND	ND	ND

*Adapted from Sharma and Johri, 1992.

Ghai et al. (1979) described the production of SCP from canning industry wastes using *C. cellulolyticum* and *Actinomucor*. The protein content increased from 0-6% when newsprint was fermented in SSF using *S. thermophile*. Leafy waste material was converted to brown coloured product with increased N, P, K contents without a foul smell by *S. pulverulentum* D-14, *Mucor* sp. C-16 and *Humicola* sp. H-37 in 90 days (Sen et al., 1979). *Thermoascus aurantiacus* and *S. thermophile* resulted in a 2-fold increase in protein content of sugar beat pulp in 48h (Grajek 1987). Thakur et al. (1993) produced fungal rennet using *Mucor miehei* in SSF on a large scale in trays. Grajek (1987) observed that *T. aurantiacus* and *H. lanuginosa* produced higher levels of xylanase in sugar beat pulp in solid state cultivation. Kalogeris et al. (1998) reported the production of endoxylanase by solid state fermentation of wheat straw. *Thermomucor indicae-seudaticae* secreted

10-fold higher glucoamylase in SSF than SmF using wheat bran as a substrate (Kumar and Satyanarayana, 2004). *S. thermophile* secreted phytase in SSF using sesame oil cake as the substrate (Singh and Satyanarayana, 2006a). In solid-state fermentation, wheat bran in combination with citrus peel supported maximum xylanolytic, pectinolytic and cellulolytic enzyme secretion by *S. thermophile* (Kaur and Satyanarayana, 2004). Cane molasses, a sugarcane industry byproduct, was utilized for glucoamylase production by a thermophilic mold *Thermomucor indicae-seudaticae* (Kumar and Satyanarayana, 2007). Among the lignocellulosic substrates tested, wheat bran supported a high xylanase secretion by *Humicola lanuginosa* in solid-state fermentation which was 23-fold higher than that in submerged fermentation (Kamra and Satyanarayana, 2004). An extracellular thermostable chitinase was purified from culture filtrate of *Thermomyces lanuginosus* (Guo et al., 2005). Chitinase of *T. emersonii* had a high optimal temperature for activity with a high thermostability (McCormack et al., 1991). Chitinases from thermophilic molds are promising for biocontrol and enzymatic conversion of chitin (Guo et al., 2005).

Table 3. Properties and composition of waterloo fungal SCP prepared by using *Chaetomium cellulolyticum* from pulp and fibre board mill effluents*

Composition		*Properties*	
Protein	45%	Colour	Gray
Carbohydrates	35%	Texture	Granular
Fat	10%	Odor	Mushroomy
Nucleic Acid	5%	Digestibility	73%
Minerals, ash	5%	Moisture	8%
Vitamins	high	Biomass content	73%
		Lignin	12%
		Ash	12%

*Adapted from Moo-Young et al., 1979.

Cellulose and hemicellulose are the chief components of lignocellulosic materials such as wheat straw, rice straw and sugarcane bagasse. These materials are utilized by thermophilic molds by secreting a large number of hydrolytic enzymes. The major applications of cellulases are in the production of sea weed jelly and in saccharification of delignified cellulosic wastes. Despite the lower cellulase activity, *S. thermophile* degraded cellulose faster than mesophilic *T. reesei* (Bhat and Maheshwari, 1987). Extracellular cellulolytic enzymes were produced by *Thermoascus aurantiacus* in solid state cultivation (Kalogeris et al., 2003).

Next to cellulose, xylan is the most abundant structural polysaccharide in nature. Its complete degradation requires the cooperative action of several hydrolytic enzymes: the endoxylanases (EC 3.2.1.8), which randomly cleave β-1,4-linked xylose; the β-xylosidases (EC 3.2.1.37) that hydrolyzes xylooligomers; and the different side chain splitting enzymes, e.g., α-glucuronidase and α-arabinosidase, acetyl xylan esterase and acetyl esterase, which liberate other sugars like glucuronic acid, arabinose, which are attached to xylose (the backbone of xylan) [Biely 1985]. *Malbranchea pulchella* var. *thermoidea* TMD-8 was shown to secrete xylanase (Banerjee et al., 1994). Wheat bran supported maximum enzyme production in *T. aurantiacus* and *H. lanuginosa* (Grajek, 1987). *Melanocarpus albomyces*, a thermophilic fungus, isolated from compost produced cellulase-free xylanase in a liquid medium containing sugarcane bagasse (Prabhu and Maheshwari, 1999). An endo-β-1,4-xylanase produced by *Sporotrichum thermophile* ATCC 34628 was purified to homogeneity (Katapodis et al., 2003). The enzyme was optimally active at pH 5 and 70°C. Xylanase was also produced in SSF using *Thermoascus aurantiacus* (dos Santos et al., 2003). Maximum production (500 U g^{-1} bagasse) was achieved on the sixth day of cultivation on sugarcane bagasse medium supplemented with 15% (v/w) rice bran extract. Xylanase production by the thermophilic fungus *Melanocarpus albomyces* IIS68 in SSF and the effects of different variables were assessed using the response surface methodology (Narang et al., 2001); this resulted in a xylanase titre of 7760 U g^{-1} dry substrate. Among the lignocellulosic substrates tested, wheat bran supported a high xylanase secretion by *Humicola lanuginosa* in solid-state fermentation which was 23-fold higher than that in submerged fermentation (Kamra and Satyanarayana, 2004). Xylanases of thermophilic fungi are also receiving considerable attention due to their application in pre-bleaching of pulp in the paper industry, where the enzymatic removal of xylan from lignin-carbohydrate complexes facilitates the leaching of lignin from the fiber cell wall, thus reducing the amount of chlorine for pulp bleaching in the brightening process.

Thermophilic molds are tolerant to a range of pH with most species growing between pH 4.0 and 8.0. The pH of the hay at baling is about 6.5 and it was mainly colonized by *Mucor miehei* and *M. pusillus* (pH optima close to hay pH) followed by *H. lanuginosa*, *M. pulchella* var. *sulfurea* and *T. thermophilus* (pH optima near to neutral) [Rosenberg, 1975]. Some thermophilic molds like *Allescheria terrestris* and *T. emersonii* had been shown to grow in acidic environment of bagasse (pH 3.4-6.0) [Rosenberg, 1975].

Mushroom composting is the biological conversion of solid organic residues into a simpler substrate for mushroom cultivation. Moreover, their high organic matter content and biological activity make composts effective in a variety of applications, including erosion control, revegetation, biofiltration and bioremediation (Straatsma et al., 1994). The active component involved in the biodegradation and conversion processes during composting is the resident microbial community, among which fungi play a very important role. The biomass ratio of fungi to prokaryotes in compost is about 2:1. In addition, fungi use many carbon sources, mainly lignocellulosic polymers and can survive in extreme conditions. They are mainly responsible for compost maturation. Fungi affect soil fertility, suppress plant diseases and promote mushroom growth (Straatsma et al., 1994). They also degrade complex polymers such as polyaromatic compounds or plastics and are being increasingly applied to bioremediate soils contaminated with a wide range of pollutants. Monitoring fungal diversity is essential to detect fungi hazardous to humans, animals and plants and to optimize compost quality.

The white button mushroom, *Agaricus bisporus* is cultivated on mushroom compost made from the mixture of wheat straw, horse manure, chicken manure, and gypsum. Compost is prepared in a sequence of processes. After mixing and moistening of ingredients, the mixture is left for a short period to start self-heating. Then phase I is implemented; this consists of uncontrolled self-heating for 6 days in windows in the open air. It is followed by phase II, controlled aerobic composting at 45°C for 6 days (Straatsma et al., 1994). Thermophilic molds are important microbes in mushroom composting. These molds have a larger area of conversion because of the way they can grow through the dense compost substrate as a fine thread of mycelium, whereas the actinobacteria are able to grow in well defined areas and their area of conversion is confined and overall much smaller. The thermophilic fungi grow and convert complex organic compounds into simpler ones (food for the mushroom). These fungi penetrate the dense and tight areas of compost substrate or into the balls of compost (Straatsma et al., 1994). Twenty-two species of thermophilic fungi were isolated from mushroom compost. *Scytalidium thermophilum* was present in the compost ingredients, fresh straw, horse droppings, and drainage from compost and dominated the thermophilic fungal biota of compost (Straatsma et al., 1994). Of the 34 species of thermophilic fungi tested, 9 strains (*Chaetomium thermophilum*, an unidentified *Chaetomium* sp., *Malbranchea sulfurea*, *Myriococcum thermophilum*, *S. thermophilum*, *Stilbella thermophila*, *Thielavia terrestris*, and two unidentified basidiomycetes) promoted the mycelial growth of *Agaricus bisporus* on sterilized compost. The presence of *S. thermophilum* in compost resulted in an enhanced hyphal extension rate of the mushroom mycelium. Carbon dioxide appeared to be responsible for the improved mycelial growth. When a mixture of wheat straw and organic nitrogen supplements was composted at 40°C in a simple small-scale, the initial inoculation with several species of thermophilic fungi resulted in a large improvement in the suitability of the compost for the cultivated mushroom (Ross and Harris, 1983). The species of *Torula thermophila* (syn. *S. thermophilum*), was particularly effective in bringing about the changes. This mold exhibited a considerable potential as a rapid composting organism.

Bioethanol from lignocellulosic materials

The most common renewable fuel today is ethanol produced from sugar or grain (starch); thisraw material base will not, however, be adequate. Consequently, large sale use of ethanol in the future will most certainly have to be based on production from lignocellulosic materials. One of the major challenges is to optimize the integration of process engineering, fermentation technology, enzyme engineering and metabolic engineering.

One of the greatest tests for society in the 21st century is to meet the growing demand for energy for transportation, heating and industrial processes, and to provide raw material for the industry in a sustainable way. An increasing concern for the security of the oil supply has been evidenced by increasing oil prices. More importantly, the future energy supply must be met with a simultaneous substantial reduction of green house gas emissions. Actions towards this aim have been initiated. The European Commission plans to substitute progressively 20% of conventional fossil fuels with alternative fuels in the transport sector by 2020, with an intermittent goal set at 5.75% in 2010. In the USA, the Energy Policy Act of 2005 requires blending of 7.5 billion gallons of alternative fuels by 2012 (Gray et al., 2006). The Petroleum Ministry in India mandated to mix petrol with 5% ethanol in nine states (Maharashtra, Andhra Pradesh, Gujarat, Goa, Karnataka, Uttar Pradesh, Punjab, Haryana, Tamil Nadu) in 2003. Furthermore, the industry experts say 20% of ethanol can be mixed in petrol without harming the vehicle. Environment will be the biggest beneficiary of such a decision as it will cut emissions by half. Liquid biofuels from renewable resources, particularly from lignocellulosic materials, will have a substantial role in meeting these goals.

Ethanol has already been introduced on a large scale in Brazil, the US and some European countries, and it is expected it to be one of the dominating renewable biofuels in the transport sector within the coming 20 years. Ethanol can be blended with petrol or used as neat alcohol in dedicated engines, taking advantage of the higher octane number and higher heat of vapourization; furthermore, it is an excellent fuel for future advanced flexi-fuel hybrid vehicles. Currently, ethanol for the fuel market is produced from sugar (Brazil) or starch (USA) at competitive prices. However, this raw material base, which also has to be used for animal feed and human needs, will not be sufficient to meet the increasing demand for fuel ethanol; and the reduction of greenhouse gases resulting from use of sugar- or starch-based ethanol is not as high as desirable (Farrell et al., 2006). Both these factors call for the exploitation of lignocellulose feed-stocks, such as agricultural and forest residues as well as dedicated crops, for the production of ethanol.

The thermophilic molds have the ability to degrade the lignocellulosic materials by secreting various hydrolytic enzymes. These convert the agro-residues into fermentable sugars such as glucose, xylose and mannose, which can be further fermented to ethanol by *Sachharomyces cerevisiae*. Thus the biofuels from lignocellulose generate low net greenhouse gas emissions, reducing environmental impacts, particularly climate change. Unlike petrol and diesel, ethanol contains oxygen which results in the improved combustion and lower emissions of unburnt hydrocarbons, CO and particulate matter (Hahn-Hagerdal et al., 2006). Therefore, the use of ethanol as biofuels would be more eco-friendly than petrol and diesel, which release a large number of greenhouse gases and other pollutants in the environment.

Phytate degradation and combating environmental phosphorus pollution

Phytic acid (*myo*-inositol hexakisphosphate) is a stored organic form of phosphorus in plant materials (cereals, legumes, nuts, pollen etc.) [Table 1]. It acts as an anti-nutritional factor by chelating metals such as Ca^{+2}, Mg^{+2}, Zn^{+2} and Fe^{+2} and making them unavailable, complexing with proteins and thus affecting their digestion, and inhibiting enzymes such as α-amylase, trypsin, acid phosphatase and tyrosinase (Fig. 1) [Maga, 1982; Harland and Morris, 1995; Wodzinski and Ullah, 1996; Pandey et al., 2001; Vohra and Satyanarayana, 2003; Vats and Banerjee, 2004; Greiner and Konietzny, 2006]. Due to the lack of adequate levels of phytases (phytate-hydrolyzing enzymes) in monogastric animals (poultry birds, pigs, fishes etc.), phytic acid is excreted in faeces, which is degraded by soil microorganisms and thus releasing phosphorus into the soil. This phosphorus reaches aquatic bodies and thus causes eutrophication. In order to overcome this problem, foods and feeds can be supplemented with phytases that help in improving the nutritional value of the food and help in combating environmental phosphorus pollution (Singh et al., 2006; Kaur et al., 2007b).

Phytases (*myo*-inositol hexakisphosphate phosphohydrolase, EC.3.1.3.8 and EC. 3.1.3.26) are the enzymes, which hydrolyze phytic acid (*myo*-inositol hexakisphosphate) to *myo*-inositol and inorganic phosphate through a series of intermediates. All phytases share a highly conserved sequence motif, RHGXRXP, which is present in the active sites of acid phosphatases (Ullah et al., 1991). Phytases are, therefore, said to form

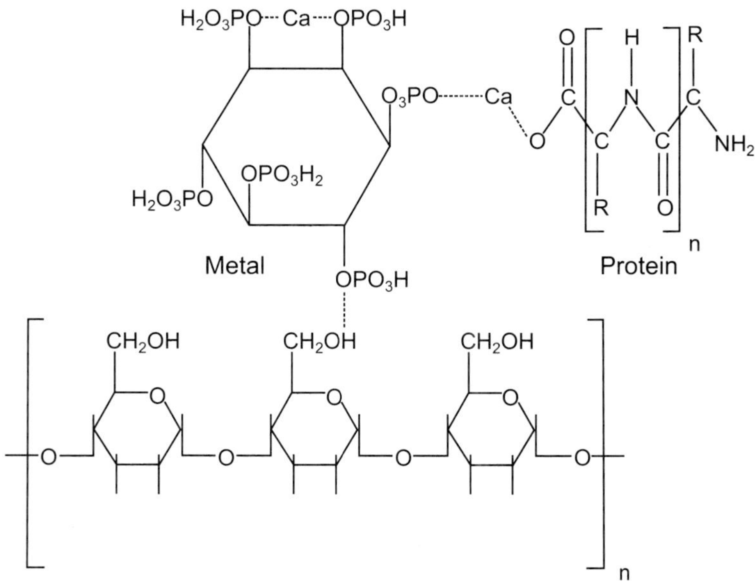

Fig.1. Structure of phytic acid showing its anti-nutritional nature

phytase subfamily of histidine acid phosphatases (Mitchell et al., 1997). Phytases have been classified into two groups depending upon their attack on phosphate group of phytic acid:

1. **3 phytase** (EC 3.1.3.8) that hydrolyzes the ester bond at the 3 position of *myo*-inositol hexakisphosphate to *myo*-inositol 1,2,4,5,6 pentakisphosphate and orthophosphate. This is typical for microorganisms except *E. coli.*

2. **6 phytase** (EC 3.1.3.26) that hydrolyzes ester bond at 6 position of *myo* inositol hexakisphosphate to *myo*-inositol 1,2,3,4,5 pentakisphosphate and orthophosphate. Subsequent ester bonds in the substrate are hydrolyzed at different rates. The 6 phytase dephosphorylates phytic acid completely. It commonly occurs in plants and *E. coli,* and was recently reported in four basidiomycetous fungi (Lassen et al., 2001).

Thermophilic molds such as *Chaetomium thermophilum* ATCC 58420, *Rhizomucor miehei* ATCC22064, *Thermomucor indicae-seudaticae* ATCC 28404, *Myceliophthora thermophila* ATCC 48102 etc. (Mitchell et al., 1997) and *Talaromyces thermophilus* (Pasamontes et al., 1997) have been shown to produce phytases. Phytase encoding genes from M*yceliopthora thermophila* (Mitchell et al., 1997) and *Talaromyces thermophilus* (Pasamontes et al., 1997) have been cloned and over-expressed in a mesophilic fungus *Aspergillus niger.* A phytase encoding gene was cloned from *Thermomyces lanuginosus* and the recombinant enzyme was expressed in *Fusarium venenatum* (Berka et al., 1998). This recombinant phytase was optimally active at 65°C and neutral pH, and needed a higher temperature (69°C) for unfolding this phytase than the mesophilic fungal phytase from *Aspergillus* spp., which requires 60°C (Berka et al., 1998). *Rhizomucor pusillus* secreted phytase in SSF using wheat bran as the substrate (Chadha et al., 2004), while *Thermoascus aurantiacus* produced phytase in semisynthetic medium containing wheat bran (Nampoothiri et al., 2004). Another thermophilic mold *Sporotrichum thermophile* secreted phytase in submerged as well as solid state fermentations. The mold secreted phytase in SSF using sesame oil cake in 120 h at 45°C, at the initial substrate to a moisture ratio of 1:2.5 and a_w of 0.95 (Singh and Satyanarayana, 2006a). This mold produced phytase in a medium containing cane molasses, a sugarcane industry byproduct. Statistical optimization of the medium components for phytase production by *S. thermophile* in cost–effective cane molasses medium resulted in a 2-fold improvement in phytase production (Singh and Satyanarayana, 2006b). The mold also secreted phytase optimally in a synthetic medium. Starch, glucose, sodium phytate and peptone were identified as most significant factors by Plackett-Burman design and these were further optimized by response surface methodology (RSM) that resulted in a 3.73-fold improvement in phytase production (Singh and Satyanarayana, 2007). Substitution of expensive sodium phytate with wheat bran resulted in a sustained enzyme titer, thus making the fermentation process more economical.

There are a few areas where phytase can be used. The first one is phytate reduction in feed and food industries, thereby mitigating its antinutritional properties, and the second is preparation of *myo*-inositol phosphates as tools for biochemical investigation. *Myo*-inositol phosphates, particularly inositol-3-phosphate, are used in signal transduction. An emphasis has also been laid on its use in aquaculture and as a soil amendment.

The phosphoserine residues in caseins can be hydrolyzed by both alkaline and acid phosphatases, and thus it can be used in reducing phosphate content of foods. The production of detectable quantities of acid phosphatases (EC 3.1.3.2) was reported in *Thermomyces lanuginosus* (Crisan, 1969). Two isoenzymes were produced and the degree of intrinsic thermostability was not greater than that commonly exhibited by acid phosphatase from mesophilic organisms. Alkaline phosphatase was purified from conidia of *Humicola grisea var. thermoidea* (Buainain et al., 1998; Cenie et al., 1998). Seven hundred and seventy five strains of thermophilic fungi were tested for the production of acid phosphatases (Bilai et al., 1985) but only isolates of 15 species were found positive. Among seven thermophilic molds tested, *Thermoascus aurantiacus* and *Chrysosporium thermophilum* produced acid phosphatase. Of the 13 thermophilic fungi tested for the production of extracellular acid and alkaline phosphatases (Satyanarayana et al., 1985), two fungi *Acremonium alabamensis* and *Rhizopus rhizopodiformis* secreted only acid phosphatase where as the others secreted both the enzymes. All the fungi tested were observed to contain intracellular acid and alkaline phosphatases.

In spent media from cultures of *Thermomyces lanuginosus*, significant amounts of acid phosphatase activity occurred and the activity was greatest at 37°C (Arnold et al., 1988). Acid phosphatase was detected on the cell-surface in the periplasmic space and cytoplasm, but the activity was mostly present intracellularly. The optimum pH for the enzyme activity was 5.0. Extracellular and intracellular alkaline phosphatases from the thermophilic fungus *Scytalidium thermophilum* were purified by DEAE-cellulose and Concanavalin A-Sepharose chromatography (Guimarães et al., 2001). These enzymes showed allosteric behaviour either in the presence or absence of $MgCl_2$, $BaCl_2$, $CuCl_2$, and $ZnCl_2$. All these ions increased the maximal velocity of both enzymes. The best substrate for the both phosphatases was *p*-nitrophenylphosphate, but glycerophosphate and other phosphorylated compounds also served as substrates. The optimum pH for phosphatases was 10.0 and 9.5, and their carbohydrate contents were about 54% and 63%, respectively. The optimum temperature for the activity was 70-75°C. The enzymes were fully stable up to 1 h at 60°C. These and other properties suggested that the alkaline phosphatases of *S. thermophilum* might be suitable for biotechnological applications. Acid and alkaline phosphatases hydrolyze the phosphoserine residues in casein and thus help in reducing the phosphate content of food. Furthermore, acid phosphatases are of potential interest for application in biogas production from plant and animal biomass (Kaur et al., 2007a).

Bioremediation of heavy metals

The existence of heavy metals and radionuclides represent a significant environmental hazard. While the major source of such pollutants is the industry, the quantities contained in waste materials from both agricultural and domestic sources can not be overlooked (Noorwez and Satyanarayana, 2002). Biomass produced by *Talaromyces emersonii* CBS814.70 was shown to have a high biosorption capacity for uranium (280mg uranium/g biomass) [Bengtsson et al., 1995]. Likewise other organisms have been shown to have biosorption capacity for different heavy metals. These organisms can be effectively used as detoxifiers in situ. Some species of *Mucor* and *Rhizopus* are useful in accumulation and removal of heavy metals and radionuclides from waste water and mining operations (Gadd, 1990).

Treatment of industrial effluents

Pollution has become a serious concern and anything which tampers with the ecosystem is a matter of great concern. Biological means of pollution control are the most favoured ones today and in this regard thermophilic molds have special applications. The waste water from pulp and paper, board and rayon mills contains suspended solids, colour, foam, and appreciable quantities of toxic, inorganic and organic constituents (Ghosh and Konar, 1980; Jones, 1973). Waste water from pulp and fiber board mills containing both water soluble mono- and oligo-saccharides and solid lignocellolosic material was fermented in an economical way with respect to protein production and water purification. The degree of purification depends on the residence time. A 15 h residence time caused a BOD (Biological Oxygen Demand) reduction of 60%. Nutritional experiments with biomass product showed digestibility equivalent to 82% crude protein and a metabolizable energy content of 72% of soyabean meal.

For the treatment of pulp and paper mill effluents, any of the standard biological treatment methods can be applied. The most common ones are storage oxidation basin, aerated lagoons, aerated storage basin, activated sludge, and trickling filters. The use of high rate trickling filter has shown a reduction in BOD of 40-60%. Activated sludge process has been found to be the most successful method for kraft pulp and paper mill waste treatment. The waste generated by pulp and paper industries contains various chemicals (chlorine and other toxic compounds) used in the processing of paper. These chemicals are hazardous to the ecosystem. The thermo- and alkali-stable xylanases and pectinases of thermophilic molds could reduce the use of these chemical in the paper and pulp industries and make the process more economical and eco-friendly (Sharma et al., 1997). The addition of thermophilic fungal phytase to these enzymes not only reduces the phytic acid content of the pulp but also improves the quality of the paper (Liu et al., 1998).

The waste water from citrus processing industry contains pectinaceous materials that are not decomposed by the microbes during activated sludge treatment. A new waste water treatment method employing alkali-tolerant thermophilic molds has been employed. Thermophilic molds are well known producers of alkaline pectinases (the enzymes which degrade pectin). Craveri et al. (1967) reported pectinase from *Penicillium duponti, Humicola stellata, Humicola lanuginosa, Mucor pusillus* and *Humicola insolens. Sporotrichum thermophile* is also known to produce pectinase in solid state as well as submerged fermentation using wheat bran and citrus peal as the substrates (Pandey, 2003; Kaur and Satyanarayana, 2004). The mold produced 330 fold higher enzyme titres in SSF than in submerged fermentation (SmF). Lactose and yeast extract (0.5%) in combination with citrus peel supported a high enzyme titres in Smf. In solid-state fermentation, wheat bran in combination with citrus peel (1:1 ratio) supported maximum xylanolytic, pectinolytic and cellulolytic enzyme secretion (Kaur and Satyanarayana, 2004). Maximum enzyme production was achieved at water activity (a_w) of 0.95. The production of xylanolytic, pectinolytic and cellulolytic enzymes was approximately 400-, 200- and 20-folds higher in SSF using wheat bran and citrus peel than SmF. The enzyme treated fruit pulps (carrot, apple and banana) yielded more juices and the juices contained a high amount of total and reducing sugars (Kaur et al., 2004). The pectinases are used in reducing the level of pectic substances in the processing of fruits and juice industries and in the reduction of pectic wastes in the effluents.

The industrial effluent from oil industries is rich in oil, and therefore, lipase-producing thermophilic molds could find application in the treatment of such effluents. Several lipase producing thermophilic mold isolates have been isolated from stored ground nuts and palm produce in Nigeria (Ogundero, 1980, 1981a, b). Thermophilic strain of *Rhizopus arrhizus* accumulates an acidic lipase in culture fluid when grown in a medium containing groundnut oil, milk powder and inorganic salts (Kumar et al., 1993). Arima et al. (1968) purified an extracellular lipase from *Humicola lanuginosa* strain Y-38, isolated from compost in Japan. The mold produced lipase in a medium containing soybean oil, starch, and corn steep liquor. *Rhizomucor miehei* and *Thermomyces lanuginosus* are well known lipase producing thermophilic molds (Rao and Divakar, 2002; Noel and Combes, 2003).

Some thermophilic molds have the ability to grow in a pigmented medium and decolourize it with the help of some secretary enzymes/ chemicals. Thermophilic *Aspergillus fumigatus* G-2-6 decolourized 75% melanoidin of molasses solution when this strain was cultivated at 45°C for 3 days in glycerol peptone medium (Ohmomo et al., 1987). *Sporotrichum pulverulentum* was shown to decolourize kraft bleach plant effluents (Sundman et al., 1981).

Soil amendment and bioremediation

Microorganisms contribute to the formation of humus either by extracellular transformation of plant and animal constituents into humic compounds or by synthesis of humic substances within the cells (Boethling and Alexander, 1979). There has been a general increase in humic substances in the final degraded material by all thermophilic fungi, although some like *A. fumigatus* and *H. insolens* formed more humic substances than other thermophilic fungi (Jain et al., 1973). This humus improves the soil structure as reflected by increased aeration, reduction in power for ploughing heavy soils, rapid germination of seeds and reduced volume weight (Banse, 1961). The use of compost as fertilizer has resulted in improvement in the yield of various crops like sorghum (Hortenstine and Rothwell, 1973), corn (Zobac and Vana, 1974), potatoes (Purvis and McKenzie, 1973; Trenel, 1961), tobacco (Duggan, 1973), tomatoes (Vlamis and Williams, 1971) beet (Schneider, 1974) and ryegrass. The incorporation of organic matter prepared by composting agricultural waste and low grade rock phosphate in the soil improved the water holding capacity and yield of the cereal crops (Singh and Yadav, 1986). Compost application to the soil had a positive effect on the microbial population and the rhizosphere microorganisms, and also contributed to the reduction of nematode population in plants (Chopra and Magu, 1985; Patrick et al., 1965; Pera et al., 1983). The practices of soil bioremediation for the degradation of pollutants involve growth of fungi on wood chips. Many species of thermophilic molds have been shown to colonize wood chips very efficiently (Tansey, 1973) as compared to their mesophilic counterparts. The immobilization of thermophilic fungal biomass for liquid waste treatment provides an attractive technology for suitable bioreactor design. Use of fungal biomass as soil conditioner is the need of the hour because several species are a good source of growth promotory substances and therefore, make the habitat ideal for plant growth that results in improved soil fertility.

FUTURE PERSPECTIVES AND CONCLUSIONS

In the modern era of biotechnology, there is an increasing demand for novel microbes with unique characteristics. Due to their potential to produce a number of hydrolytic enzymes which have immense biotechnological applications, antibiotics, and their utility in the production of SCP and as agents of bioconversions, thermophilic molds have been termed nature-borne biotechnologists. Their polysaccharide-degrading enzymes generate fermentable sugars such as glucose, xylose and mannose from agro-residues, which can be fermented to ethanol. This achieves the twin objectives of recycling of renewable agro-residues and generating wealth from the wastes. Further investigations are needed on understanding the diversity of thermophilic molds from hitherto unexplored habitats and their bioactive compounds, optimization of the production of enzymes by process optimization and genetic and molecular methods, development of appropriate methods for the pre-treatment of lignocellulosics and their bioconversion to fermentable sugars, decolourization of industrial effluents and bioremediation of heavy metals from the contaminated soils and water, in order to fully realize their utility in the environmental management.

Acknowledgements

B. Singh is grateful to the Council of Scientific and Industrial Research (CSIR) New Delhi, for providing financial support as Junior/Senior Research Fellowship during the course of writing this chapter.

References

Archana, A. and Satyanarayana, T. 2001. Biodiversity and potential applications of thermophilic moulds, pp. 70-84. In: Frontiers in Fungal Biodiversity and Plant Pathogen Relations. C. Manoharachari, G. Bagyanarayana, B. Bhadraiah, B.N. Reddy K. Satyaprasad. and A. Nagamani (eds.), Allied Publishers Limited, India.

Arima, K., Iwasaki, S. and Tamura, G. 1968. Milk clotting enzymes from microorganisms. V. Purification and crystallization of *Mucor* rennin from *Mucor pusillus* Lindt. Applied Microbiology, 16: 1727-1733.

Arnold, W.N., Garrison, R.G., Mann, L.C., and Wallace, D.P. 1988. The acid phosphatase from *Thermoascus crustaceus*, a thermophilic fungus. Microbiology, 54: 101-112.

Banerjee, S., Archana, A. and Satyanarayana, T. 1994. Xylose metabolism in a thermophilic mould *Malbranchea pulchella var. sulfurea* TMD-8. Current Science, 29: 349-353.

Banse, H.J. 1961. Beeinflussung der physikalishen bodeniegenschaften durch kompotgaben. Internationale Arbeitsgeeinschaft fur Mullforschung Informatiosblatt, 13: 30-34.

Bengtsson, L., Johansson, B., Hackett, T.J. and McHale, A.P. 1995. Studies on the biosorption of uranium by *Talaromyces emersonii* CBS 814.70 biomass. Applied Microbiology and Biotechnology, 42: 807-811.

Berka, R.M., Rey, M.W., Brown, K.M., Byun, T. and Klotz, A.V. 1998. Molecular characterization and expression of a phytase gene from the thermophilic fungus *Thermomyces lanuginosus*. Applied and Environmental Microbiology, 64: 4423-4427.

Bhat, K. M., and. Maheshwari, R. 1987. *Sporotrichum thermophile*, growth, cellulose degradation, and cellulase activity. Applied and Environmental Microbiology, 53: 2175-2182.

Biely, P. 1985. Microbial xylanolytic systems. Trends in Biotechnology, 3: 286-290.

Bilai, T.I., Chernyagina, T.B., Dorokhov, V.V., Poedinok, N.L., Zakharchenko, V.A., Ellanskaya, I.A. and Lozhkina, G.A. 1985. Phosphatase activity of different species of thermophilic and mesophilic fungi. Mikrobiologia, 47: 53-56.

Boethling, R.S. and Alexander, M. 1979. Effect of concentration of organic chemicals on their biodegradation by natural microbial communities. Applied and Environmental Microbiology, 37: 1211-1216.

Buainain, L.B., Kadowaki, M.K., Polizelli, T.M., Terenzi, H.F. and Jorge, A.A. 1998. Characterization of conidial alkaline phosphatase from the thermophilic fungus *Humicola grisea var. thermoidea*. Journal of Basic Microbiology, 38(2): 85-94.

Cenie, G., Caldini, G. and Strappini, C. 1998. Characterization of conidial alkaline phosphatase from the thermophilic fungus *Humicola grisea var. thermoidea*. Journal of Basic Microbiology, 38: 85-94.

Chadha, B.S., Gulati, H., Minhas, M., Saini, H.S. and Singh, N. 2004. Phytase production by the thermophilic fungus *Rhizomucor pusillus*. World Journal of Microbiology and Biotechnology, 20: 105-109.

Chopra, P. and Magu, S.P. 1985. Effect of selected herbicides and city compost on the rhizospheric microflora of wheat and maize. Indian Journal of Agronomy, 30: 5-9.

Cooney, D.G. and Emerson, R. 1964. Thermophilic fungi: An account of their biology, activities and classification. W.H. Freeman and Co., San Francisco, USA.

Craveri, R., Craveri, A. and Guicciardi, A. 1967. Research on the properties and activities of enzymes of eumycete thermophilic isolates of soil. Annali di Microbiologia et Enzimologia, 17: 1-30.

Crisan, E.V. 1959. The isolation and identification of thermophilic fungi. M.Sc. Thesis, Purdue University, USA.

Crisan, E.V. 1969. The proteins of thermophilic fungi, pp. 32-33. In: Current Topics in Plant Science J.E. Grunckel (eds.), Academic Press, New York and London.

Deploey, J.J. and Fergus, C.L. 1975. Growth and sporulation of thermophilic fungi and actinomycetes in O_2-N_2 atmosphere. Mycologia, 67: 780-797.

dos Santos, E., Piovan, T., Roberto, I.C. and Milagres, A.M. 2003. Kinetics of the solid state fermentation of sugarcane bagasse by *Thermoascus aurantiacus* for the production of xylanase. Biotechnology Letters, 25(1): 13-16.

Duggan, J. 1973. Utilization of municipal refuse compost. I. field scale compost demonstration. Compost Science, 14: 1.

Emerson, R. 1968. Thermophiles, pp. 105-128. In: The Fungi. Vol. 3. G.C. Ainsworth and A.S. Sussam (eds.). Academic Press, London and New York.

Farrell, A.E., Plevin, R.J., Turner, B.T., Jones, A.D., O'Hare, M. and Kammen, D.M. 2006. Ethanol can contribute to energy and environmental goals. Science, 311(5760): 506-508.

Gadd, G.M. 1990. Metal tolerance, pp. 178-210. In: Microbiology of Extreme Environments. C. Edwards (eds.), McGraw-Hill Publishing Company, New York, USA.

Ghai, S.K., Kahlon, S.S. and Chahal, D.S. 1979. Single cell protein from canning industry waste: sag waste as substrate for thermotolerant fungi. Indian Journal of Experimental Biology, 17: 789-791.

Ghosh, T.K. and Konar, S.K. 1980. Toxicity of chemicals and waste waters of paper and pulp mills to worms, plankton and mollusks. Industrial Journal of Environmental Health, 22: 278-285.

Grajek, W. 1987. Production of D-xylanases by thermophilic fungi using different methods of culture. Biotechnology Letters, 9: 353-356.

Gray, K.A., Zhao, L. and Emptage, M. 2006. Bioethanol. Current Opinion in Chemical Biology, 10(2): 141-146.

Greiner, R. and Konietzny, U. 2006. Phytase for food application. Food Technology and Biotechnology, 44(2): 125-140.

Guimarães, L.H.S, Terenzi, H.F., Jorge, J.A. and Polizeli, M.L.T.M. 2001. Thermostable conidial and mycelial alkaline phosphatases from the thermophilic fungus *Scytalidium thermophilum*. Journal of Industrial Microbiology and Biotechnology, 27(4): 265-270.

Guo, R.F., Li, D.C. and Wang, R. 2005. Purification and properties of a thermostable chitinase from thermophilic fungus *Thermomyces lanuginosus*. Wei Sheng Wu Xue Bao, 45(2): 270-274.

Hahn-Hägerdal, B., Galbe, M., Gorwa-Grauslund, M.F., Lidén, G. and Zacchi, G. 2006. Bio-ethanol—the fuel of tomorrow from the residues of today. Trends in Biotechnology, 24(12): 549-556.

Harland, B.F. and Morris, E.R. 1995. Phytate, A good or a bad food component. Nutrition Research, 15: 733-754.

Henssen, A. 1957. beitrage Zur morphologie systematic der thermophilien Actinomyceten. Archives in Mikrobiology, 26: 373-414.

Hortenstine, C.C. and Rothwell, D.F. 1973. Pelletized municipal refuse compost as a soil amendment and nutrient source for sorghum. Journal of Environmental Quality, 2: 343-345.

Jain, M.K., Kapoor, K.K. and Mishra, M.M. 1973. Cellulase activity, degradation of cellulose and lignin and humus formation by thermophilic fungi. Transaction British Mycological Society, 73: 85-89.

Johri, B.N., Satyanarayana, T. and Olsen, J. 1999. Thermophilic moulds in biotechnology. Kluwer Academic Publishers, UK.

Jones, H.R. 1973. The nature of pollution problems, pp. 11-14. In: Pollution Control and Chemical Recovery. Noyes Data Corporation, Warszawa, Poland.

Kalogeris, E., Christakopoulos, P., Kekos, D. and Macris, B.J. (1998). Studies on the solid state production of thermostable endoxylanases from *Thermoascus aurantiacus,* Characterization of two isozymes. Journal of Biotechnology, 60: 155-163.

Kalogeris, E., Christakopoulos, P., Katapodis, P., Alexiou, A., Vlachou, S., Kekos, D. and Macris, B.J. 2003. Production and characterization of cellulolytic enzymes from the thermophilic fungus *Thermoascus aurantiacus* under solid state cultivation of agricultural wastes. Process Biochemistry, 38(7): 1099-1104.

Katapodis, P., Vrsanska, M., Kekos, D., Nerinckx, W., Biely, P., Claeyssens, M., Macris, B.J. and Christakopoulos, P. 2003. Biochemical and catalytic properties of an endoxylanase purified from the culture filtrate of *Sporotrichum thermophile.* Carbohydrate Research, 338(18): 1881-90.

Kamra, P. and Satyanarayana, T. 2004. Xylanase production by the thermophilic mold *Humicola lanuginosa* in solid-state fermentation. Applied Biochemistry and Biotechnology, 119(2): 145-157.

Kaur, G. and Satyanarayana, T. 2004. Production of extracellular pectinolytic, cellulolytic and xylanolytic enzymes by a thermophilic mould Sporotrichum thermophile Apinis in solid state fermentation. Indian Journal of Biotechnology, 3: 552-557.

Kaur, G., Kumar, S. and Satyanarayana, T. 2004. Production, characterization and application of a thermostable polygalacturonase of a thermophilic mould *Sporotrichum thermophile* Apinis. Bioresource Technology, 94(3): 239-43.

Kaur, P., Lingner, A., Singh, B., Boer, E., Polajeva, J., Steinborn, G., Bode, R., Gellissen, G., Satyanarayana, T. and Kunze. G. 2007a. APHO1 from the yeast *Arxula adeninivorans* encodes an acid phosphatase of broad substrate specificity. Antonie Van Leeuwenhoek, 91(1): 45-55.

Kaur, P., Singh, B. and Satyanarayana, T. 2007b. Microbial phytases in combating environmental phosphorus pollution, pp. 150-190. In: Sustainable Resource Management. Vol. I, K.K. Singh, A. Tomar, V. Phogat and S. Phogat. (eds.), MD Publications, New Delhi, India.

Kumar, P. and Satyanarayana, T. 2007. Economical glucoamylase production by alginate-immobilized *Thermomucor indicae-seudaticae* in cane molasses medium. Letters in Applied Microbiology (In press).

Kumar, S. and Satyanarayana, T. 2004. Statistical optimization of a thermostable and neutral glucoamylase production by a thermophilic mold *Thermomucor indicae-seudaticae* in solid-state fermentation. World Journal of Microbiology and Biotechnology, 20(9): 895-902.

Kumar, K.K., Deshpande, B.S. and Ambedkar, S.S. 1993. Production of extracellular acidic lipase by *Rhizopus arrhizus* as a function of culture conditions. Hindustan Antibiotics Bulletin, 35(1-2): 33-42.

Lassen, S.F., Breinholt, J., Ostergaard, P.R., Brugger, R., Bischoff, A., Wyss, M. and Fuglsang, C.C. 2001. Expression, gene cloning, and characterization of five novel phytases from four basidiomycete fungi: *Peniophora lycii, Agrocybe pediades*, a *Ceriporia* sp. and *Trametes pubescens*. Applied and Environmental Microbiology, 67(10): 4701-4707.

Lindt, W. 1886. Mitteilungen über einige neue pathogene Shimmelpilze. Archive for Experimentelle Pathologie und Pharmakologie 21: 269-298.

Liu, B.L., Rafiq, A., Tzeng, Y.M. and Rob, A. 1998. The induction and characterization of phytase and beyond. Enzyme and Microbial Technology, 22: 415-424.

Maga, J.A. 1982. Phytate, its chemistry, occurrence, food interactions, nutritional significance, and methods of analysis. Journal of Agriculture and Food Chemistry, 30: 1-9.

Maheshwari, R., Bharadwaj, G. and Bhat, M.K. 2000. Thermophilic fungi: Their physiology and enzymes, Microbiology and Molecular Biology Reviews, 64: 461-488.

McCormack, J., Hackett, T.J., Tuohy, M.G. and Coughlan, M.P. 1991. Chitinase production by *Talaromyces emersonii*. Biotechnology Letters, 13: 677-682.

Miehe, H. 1907. Die selesterhisung des Hues Eine biologishe studie, 1-227.

Mitchell, D.B., Vogel, K., Weimann, B.J., Pasamontes, L. and van Loon, A.P. 1997. The phytase subfamily of histidine acid phosphatase; isolation of genes for two novel phytases from the *Aspergillus terreus* and *Myceliophthora thermophile*. Microbiology, 143: 245-252.

Moo-Young, M., Daugulis, A.J., Chahal, D.S. and MacDonald, D.G. 1979. The waterloo process for the process of SCP from wastes. Process Biochemistry, 14: 38-40.

Mouchacca, J. 1997. Thermophilic fungi: biodiversity and taxonomic status. Cryptogamie Mycology, 18: 19-69.

Nampoothiri, K.M., Tomes, G.J., Roopesh, K., Szakacs, G., Nagy, V., Soccol, C.R. and Pandey, A. 2004. Thermostable phytase production by *Thermoascus aurantiacus* in submerged fermentation. Applied Biochemistry and Biotechnology, 118(1-3): 205-14.

Narang, S., Sahai, V. and Bisaria, V.S. 2001. Optimization of xylanase production by *Melanocarpus albomyces* IIS68 in solid state fermentation using response surface methodology. Journal of Bioscience and Bioengineering, 91(4): 425-427.

Noack, K. 1920. Der Betriebstoffwechsel der thermophilen Pilze. Jahrb. Wissensch. Bot, 59: 593-648.

Noel, M. and Combes, D. 2003. Effects of temperature and pressure on *Rhizomucor miehei* lipase stability. Journal of Biotechnology, 102(1): 23-32.

Noorwez, S.M. and Satyanarayana, T. 2002. Microbes in extreme environments and their potential uses in industrial effluent treatment, pp. 33-42. In: Microorganisms in Bioremediation. D.K. Markandey and N.R. Markandey. (eds.), Capital Publishing Company, New Delhi, India.

Ogundero, V.W. 1980. Lipase activity of thermophilic fungi from mouldy ground nuts in Nigeria. Mycologia, 72: 118-126.

Ogundero, V.W. 1981a. Isolation of thermophilic and thermotolerant fungi from stored groundnuts in Nigeria and determination of their lipolytic activity. International Biodeterioration Bulletin, 17: 51-56.

Ogundero, V.W. 1981b. Thermophilic fungi from Nigerian palm produce. Mycologia, 73: 198-202.

Ohmomo, S., Kaneko, Y., Sirianuntapiboon, Somchai, P. and Atthasampunna, P. 1987. Decolorization of molasses waste by a thermophilic strain *Aspergillus fumigatus* G-2-6. Agriculture and Biological Chemistry, 51: 3339-3346.

Pandey, A., Szakacs, G., Soccol, C.R., Rodriguez-Leon, J.A. and Soccol, V.T. 2001. Production, purification and properties of microbial phytases. Bioresource Technology, 77(3): 203-214.

Pandey, V. 2003. Pectinolytic enzymes of a thermophilic mould *Sporotrichum thermophile* Apinis, Ph.D. Thesis. University of Delhi, South Campus, New Delhi, India.

Pasamontes, L., Haiker, M., Wyss, M., Tessier, M. and van Loon, A.P.G.M. 1997. Gene cloning, purification and characterization of a heat stable phytase from the fungus *Aspergillus fumigatus.* Applied and Environmental Microbiology, 63(5): 1696-1700.

Patrick, Z.A., Sayre, R.M. and Thorpe, H.J. 1965. Nematicidal substances selective for plant-parasitic nematodes in the extracts of decomposing rye. Phytopathology, 55: 702-704.

Pera, A., Vallini, G., Sireno, I., Bianchin, M.L. and DeBertoldi, M. 1983. Effect of organic matter on rhizosphere microorganisms and root developments of sorghum plants in two different soils. Plant and Soil, 74: 3-18.

Prabhu, K.A. and Maheshwari, R. 1999. Biochemical properties of xylanases from a thermophilic fungus, *Melanocarpus albomyces*, and their action on plant cell walls. Journal of Bioscience, 24(4): 461-470.

Purvis, D. and McKenzie, E.J. 1973. Effects of applications of municipal compost on uptake of copper, zinc and boron by garden vegetables. Plant and Soil, 39: 361-371.

Rao, P. and Divakar, S. 2002. Response surface methodological approach for *Rhizomucor miehei* lipase-mediated esterification of a-terpineol with propionic acid and acetic anhydride. World Journal of Microbiology and Biotechnology, 18(4): 345-349.

Reddy, N.R., Sathe, S.K. and Salunkhe, D.K. 1982. Phytases in legumes and cereals. Advance in Food Research, 82: 1-92.

Rosenberg, S.L. 1975. Temperature and pH optima for 21 species of thermophilic and thermotolerant fungi. Canadian Journal of Microbiology, 21: 1535-1540.

Ross, R.C. and Harris, P.J. 1983. The significance of thermophilic fungi in mushroom compost preparation. *Torula thermophila*, wheat straw and organic nitrogen supplements. Scientia Horticulturae (Amsterdam), 20(1): 61-70.

Ryu, D.D.Y. 1989. Enhancement of nutritional value of cellulosic feed resources by pretreatment and bioconversion, pp. 223-243. In: Biotechnology for Livestock Production (prepared by the Animal Production and Health Division, F.A.O.), Plenum Press, London and New York.,

Satyanarayana, T. and Singh, B. 2004. Thermophilic Moulds: Diversity and Potential Biotechnological Applications, pp. 87-110. In: Microbial Diversity: Opportunities and Challenges. S.P.Gautam, A. Sharma, S.S. Sandhu, and A.K. Pandey (eds.). Shree Publishers and Distributors, New Delhi, India.

Satyanarayana, T., Chavant, L. and Montant, C. 1985. Applicability of API ZYM for screening enzyme activity of thermophilic moulds. Transaction British Mycological Society, 85: 727-730.

Satyanarayana, T., Jain, S. and Johri, B.N. 1988. Cellulases and xylanases of thermophilic moulds. pp. 24-60. In: Perspectives in Mycology and Plant Pathology. V.P. Agnihotri, A.K. Sarbhoy, and D. Kumar (eds.), Malhotra Publishing House, New Delhi, India.

Satyanarayana, T., Johri, B.N. and Klein, J. 1992. Biotechnological potential of thermophilic fungi, pp. 729-762. In: Hand Book of Applied Mycology. Vol. 4. Fungal Biotechnology, D.K. Arora, R.P. Elander, and K.G. Mukerji (eds.), Marcel Dekker Inc, New York, U.S.A.

Schneider, F. 1974. Wirkung unter Schiedlich gerotteter Stadtkomposte auf die Pflanzenertrage bei periodischer anwendung als organischer. Dunger Arch Acker Pflanzenbau Bodenk, 18: 747-758.

Seal, K.J. and Eggins, H.O.W. 1976. The upgrading of agricultural wastes by thermophilic fungi, pp. 58-78. In: Food from wastes. G.G. Birch, K.J. Parkar, and J.T. Worgen (eds.), Applied Sciences, London, UK.

Sen, S., Abraham, T.K. and Chakrabarty, S.L. 1979. Biodeterioration and utilization of city wastes by thermophilic microorganisms. Indian Journal of Experimental Biology, 17: 1284-1285.

Sharma, A., Archana, A. and Satyanarayana, T. 1997. Xylanases in eco-friendly paper-pulp bleaching technologies. The Botanica, 47: 163-167.

Sharma, H.S.S. and Johri, B.N. 1992. The role of thermophilic fungi in agriculture, pp. 707-728. In: Hand Book of Applied Mycology. Vol. 4. Fungal Biotechnology. D.K. Arora, R.P. Elander, and K.G. Mukerji (eds.), Marcel Dekker Inc., New York, U.S.A.

Singh, B. and Satyanarayana, T. 2006a. Phytase production by a thermophilic mould Sporotrichum thermophile in solid state fermentation and its application in dephytinization of sesame oil cake. Applied Biochemistry and Biotechnology, 133(3): 239-250.

Singh, B. and Satyanarayana, T. 2006b. A marked enhancement in phytase production by a thermophilic mould Sporotrichum thermophile using statistical designs in a cost-effective cane molasses medium. Journal of Applied Microbiology, 101(2): 344-352.

Singh, B. and Satyanarayana, T. 2007. Improved phytase production by a thermophilic mould in submerged fermentation due to statistical optimization. Bioresource Technology. In press.

Singh, B., Kaur, P. and Satyanarayana, T. 2006. Fungal phytases for improving the nutritional status of foods and combating environmental phosphorus pollution, pp. 289-326. In: Microbes: Health and Environment. A.K. Chauhan, and A. Verma (eds.), IK International Publishers, New Delhi, India.

Singh, R.D. and Yadav, D.Y. 1986. Evaluation of a low grade rock phosphate composted with some agricultural wastes for use in a crop rotation. Agricultural Wastes, 18: 183-194.

Straatsma, G., Samson, R.A., Olijnsma, T.W., Camp, H.J.M.O.D., Gerrits, J.P.G., and Griensven, L.J.L.D.V. 1994. Ecology of thermophilic fungi in mushroom Compost, with emphasis on *Scytalidium thermophilum* and growth stimulation of *Agaricus bisporus* mycelium. Applied and Environmental Microbiology, 60(2): 454-458.

Sundman, G., Kirk, T.K. and Chang, H. 1981. Fungal decolorization of kraft bleach plant effluent. Tappi, 64: 145-148.

Tansey, M.R. 1973. Isolation of thermophilic fungi from alligator nesting materials. Mycologia, 65: 594-601.

Thakur, M.S., Karanth, N.G. and Krishnanand, 1993. Production of fungal rennet by *Mucor miehei* using solid state fermentation. Applied Microbiology and Biotechnology, 32: 409-413.

Trenel, M. 1961. Zur Bodenverbesserung durch torf verrateten mull und Klarschlamm. Albrecht-Thaer-Arch, 5: 199-255.

Tsiklinskaya, P. 1899. Sur les mucédinées thermophiles. Annales de I,institut pasteur (Paris) 13: 500-515.

Ullah, A.H., Cummins, B.J. and Dischinger, H.C.Jr. 1991. Cyclohexanedione modification of arginine at the active site of *Aspergillus ficuum* phytase. Biochemical and Biophysical Research Communication, 178(1): 45-53.

Vats, P. and Banerjee, U.C. 2004. Production studies and catalytic properties of phytases (*myo*-inositolhexakisphophate phosphohydrolases): an overview. Enzyme and Microbial Technology, 35(1): 3-14.

Vlamis, J. and Williams, D.E. 1971. Utilization of municipal wastes as agricultural fertilizers. Califronia Agriculture, 25: 7-9.

Vohra, A. and Satyanarayana, T. 2003. Phytases: Microbial sources, production, purification and potential biotechnological applications. Critical Reviews in Biotechnology, 23(1): 29-60.

Wodzinski, R.J. and Ullah, A.H.J. 1996. Phytase. Advances in Applied Microbiology, 42: 263-310.

Zobac, J. and Vana, J. 1974. The agronomical effectiveness of industrially produced compost in dependence of the technology of production. Rostlinna Vyroba, 20: 931-939.

Subject Index

Genus & Species Index